Vertebrate Paleobiology and Paleoanthropology Series

Edited by

Eric Delson
Vertebrate Paleontology, American Museum of Natural History
New York, NY 10024, USA
delson@amnh.org

Eric J. Sargis
Anthropology, Yale University
New Haven, CT 06520, USA
eric.sargis@yale.edu

Focal topics for volumes in the series will include systematic paleontology of all vertebrates (from agnathans to humans), phylogeny reconstruction, functional morphology, Paleolithic archaeology, taphonomy, geochronology, historical biogeography, and biostratigraphy. Other fields (e.g., paleoclimatology, paleoecology, ancient DNA, total organismal community structure) may be considered if the volume theme emphasizes paleobiology (or archaeology). Fields such as modeling of physical processes, genetic methodology, nonvertebrates or neontology are out of our scope.

Volumes in the series may either be monographic treatments (including unpublished but fully revised dissertations) or edited collections, especially those focusing on problem-oriented issues, with multidisciplinary coverage where possible.

More information about this series at http://www.springer.com/series/6978

The Postcranial Anatomy of *Australopithecus afarensis*

The Postcranial Anatomy of *Australopithecus afarensis*

New Insights from KSD-VP-1/1

Edited by

Yohannes Haile-Selassie

Cleveland Museum of Natural History, Cleveland, OH, USA

Denise F. Su

Cleveland Museum of Natural History, Cleveland, OH, USA

Springer

Editors
Yohannes Haile-Selassie
Cleveland Museum of Natural History
Cleveland, OH
USA

Denise F. Su
Cleveland Museum of Natural History
Cleveland, OH
USA

ISSN 1877-9077 ISSN 1877-9085 (electronic)
Vertebrate Paleobiology and Paleoanthropology Series
ISBN 978-94-017-7427-7 ISBN 978-94-017-7429-1 (eBook)
DOI 10.1007/978-94-017-7429-1

Library of Congress Control Number: 2015952043

Springer Dordrecht Heidelberg New York London

Cover illustration: Woranso-Mille team members excavating and sieving at the partial skeleton locality of KSD-VP-1/1, aka *Kadanuumuu* (left) and recovered elements of the partial skeleton arranged in their anatomical position (right). Photo credits: Yohannes Haile-Selassie and Elizabeth Russell.

Printed on acid-free paper

Springer Science+Business Media B.V. Dordrecht is part of Springer Science+Business Media (www.springer.com)

Foreword

I have been waiting a long time for a skeleton like this to be found. Many people have, some even longer than me. The KSD-VP-1/1 skeleton, nicknamed *Kadanuumuu*, is the first partial skeleton ever discovered of a large *Australopithecus afarensis* individual. It is the oldest partial skeleton of any *Australopithecus* individual and contains skeletal elements not previously known for *Au. afarensis*. As such, it has new stories to tell us about the biology and evolution of *Australopithecus*.

Until now, the only reasonably complete *Au. afarensis* skeleton known was the famous "Lucy" skeleton, A.L. 288-1. Lucy was discovered 40 years ago, and until the discovery of *Kadanuumuu* in 2005, she remained the earliest securely dated *Australopithecus* partial skeleton. Not only is Lucy a cultural icon within paleoanthropology and among the public worldwide, she is also significant because much of what we know about *Au. afarensis* comes from studies of her skeleton.

Lucy, along with isolated skeletal elements from the large sample of *Au. afarensis* recovered at the A.L. 333 site and other finds from the sites of Hadar, Maka, and Dikika over the years, has revealed that the small body size, nearly human-like pelvis, and bipedal locomotor adaptation seen in the more recent and less securely dated South African australopiths had appeared prior to three million years ago. These fossils have provided overwhelming and convincing evidence that the postcranial skeleton of *Au. afarensis* was adapted for habitual, committed terrestrial bipedal locomotion – long before human evolution witnessed an increase in brain size or reduction in the face and dentition characterizing the genus *Homo*.

Despite being the most well-represented, well-dated, and well-understood species of early hominin known so far, a vigorous and occasionally rancorous debate has surrounded the ways in which *Au. afarensis* differs from *Homo*. While there is no doubt that *Au. afarensis* individuals were committed terrestrial bipeds, they are not exactly like *Homo* in their postcranial anatomy. These differences have begged the question of why these differences exist; whether from active selection to retain some measure of arboreal competence, or whether these differences were adaptively neutral, and only disappeared later in the face of different selection either associated with improved terrestrial locomotor efficiency and/or positive selection for manipulatory or other behaviors with the origin of *Homo*. This debate is ongoing, and while *Kadanuumuu* does not resolve this debate, it does add key pieces of data that continue to shape it.

Kadanuumuu tells us many things that Lucy and the other fossils have not. A few in particular that stand out to me as particularly salient. Namely, *Kadanuumuu* has the first complete tibia length known for any *Australopithecus* individual, supporting the hypothesis that *Au. afarensis* did not have relatively short lower limbs like an ape, but rather long legs like humans. *Kadanuumuu* has ribs complete enough to reveal a broad upper thorax, reflecting vertebral invagination and fully upright posture in *Au. afarensis,* rather than the cone-shaped great ape-like rib cage previously inferred from less complete fossil material. *Kadanuumuu* also gives us the first fully described cervical vertebrae for any *Australopithecus*. The small upper and larger lower vertebra with long spinous processes and facet joints proportioned like those of humans appear to reflect human-like head carriage of a relatively small cranium with large prognathic face, but with larger neck musculature and likely lack of a nuchal ligament. *Kadanuumuu's* remarkably complete scapula and clavicle, the first known for an adult of this species, not only support the interpretation that the shoulder joint of *Au. afarensis* was indeed oriented more cranially than that of humans, but also show that the infraspinous fossa was large and the musculature of the clavicle and scapula were human-like, perhaps suggesting a shift from arboreal competence towards use of the upper limb in manipulatory function.

Significantly, unlike Lucy, who was much smaller than modern humans, *Kadanuumuu* is as large as a small human. As a consequence of Lucy's small size, there has been an inherent, lingering uncertainty as to whether her morphology differed from that of humans because she had different adaptations, or whether she was just small. *Kadanuumuu* nicely answers this question, confirming that *Au. afarensis* did indeed differ from those of humans, regardless of size.

This volume presents this information and much more in a series of chapters describing and analyzing the various bones of the *Kadanuumuu* skeleton, along with its geological and paleoecological context. The descriptive chapters provide thorough and detailed images, descriptions, and functional analysis. The authors then all take a further step and provide their interpretations of the implications of their findings, which are integrated and summarized in the last chapter to provide a cohesive interpretation of the biology of *Australopithecus afarensis*. From this interpretation, as in the individual analytic chapters, the authors put *Au. afarensis* into the context of other known hominin fossils and hypothesize about what this all means for the origin of hominins and the subsequent appearance of the genus *Homo*. The scenario presented here builds on analyses published recently by contributors to this volume and is consistent with much of the new fossil evidence that has been recovered over the past few years. It presents an exciting new perspective on hominin origins, and many testable predictions can be – and I am sure will be – drawn from it by these and other scholars for years to come.

So, while *Kadanuumuu* answers key questions about *Australopithecus afarensis*, like all good fossils, it raises at least as many questions as it answers. This skeleton, and this careful and thorough volume, provides a solid stepping-stone for future research on the early part of human evolution. This volume is sure to become a staple in every paleoanthropologist's library and an important reference work for generations to come.

Carol V. Ward
University of Missouri

Preface

The geological sequence in the Afar region of Ethiopia includes more than a kilometer-thick, stratified, fossiliferous sediments sampling the last six million years of vertebrate evolutionary history and provides one of the most important windows into our evolutionary past. The continuous fossil-bearing sedimentary sequence in this region is unparalleled by any other place in the world and has been a research target for a number of paleontologists for the last five decades. Since the discovery of Hadar in the late 1960s by the French Geologist Maurice Taieb and the subsequent initiation of paleontological field research at the site by the International Afar Research Expedition (IARE), a number of other projects have followed suit to extensively explore the region. Currently, there are more than eight paleontological and archeological research projects working in the Afar region; combined, they have yielded an uninterrupted fossil record documenting the last six million years of human evolutionary history with more than 12 early hominin species recovered thus far, some of them found only in the Afar region of Ethiopia.

The Woranso-Mille paleontological project is one of the relatively young projects working in the Afar region. Understanding the importance of locating new paleontological sites, one of us (YHS) initiated an extensive survey and exploration of the northern and central Afar region of Ethiopia in the fall of 2002 under a permit issued by the Authority for Research and Conservation of Cultural Heritage of the Ministry of Culture and Tourism of Ethiopia (ARCCH). Exploration and survey largely relied on aerial photographs, satellite imagery interpretations, and air survey in order to identify sediments with paleontological potential, followed by foot survey covering vast areas. It was not until the end of the 2004 survey and exploration season that the paleontological potential of the Woranso-Mille area was identified.

The Pliocene deposits of the Woranso-Mille are unique because they represent a geological time period (3.6–3.8 Ma) that is poorly sampled in the eastern African geological sequence. Moreover, recent investigations have also identified sediments within the study area that are older than 4.3 Ma, as well as sediments that are younger than 3 Ma and older than 5 Ma. Paleontological research at Woranso-Mille in the last 10 years has largely concentrated on the fossiliferous deposits dated to between 3.5 and 3.8 Ma, largely to address the ancestor-descendant relationship that has been widely hypothesized between *Australopithecus anamensis* (4.2–3.9 Ma) and *Australopithecus afarensis* (3.7–3.0 Ma). Although a number of morphological analyses support this hypothesis, the paucity of hominin fossils from the time between 3.6 and 3.9 Ma has been a major obstacle to understanding their phylogenetic relationship. The Woranso-Mille paleoanthropological research project has contributed significantly to our understanding of middle Pliocene hominin phylogeny and systematics with the recovery of hundreds of hominin fossils of the appropriate age. These fossils emphasized the mosaic nature of the dentognathic and postcranial morphology of these early hominins and showed unequivocally that the middle Pliocene of eastern Africa was populated by a diversity of hominin species.

One of the most spectacular fossil discoveries from the Woranso-Mille study area is a partial skeleton of *Au. afarensis* (KSD-VP-1/1), the subject of this volume. The first element of the specimen was found in February 2005 and the rest of its elements were recovered from excavations conducted over 4 years following the initial discovery. Partial skeletons are extremely rare in the fossil record, with only four known from the entire Pliocene hominin fossil record. One of these partial skeletons, A.L. 288-1 (Lucy), was discovered at Hadar in 1974. This 3.2 Ma specimen assigned to the species *Au. afarensis* has been a subject of intense research and used as the major source of information to understanding the paleobiology of the species. However, it has also raised a number of debates particularly in relation to early hominin body size, shoulder girdle anatomy, thoracic shape, and locomotor adaptation of *Au. afarensis*.

KSD-VP-1/1, nicknamed *Kadanuumuu*, which means "big man" in the local Afar language, not only is a much larger individual but also has almost complete elements of the shoulder and pelvic girdles, along with ribs, cervical vertebrae, and elements of both the fore- and hindlimb that, for the first time, shed light on the cervical anatomy of early hominins and allow for a deeper understanding of limb proportions and locomotion in *Au. afarensis*. A preliminary analysis of KSD-VP-1/1 was published in 2010 in the *Proceedings of the National Academy of Sciences* that highlighted the significance of the specimen for understanding the paleobiology of *Au. afarensis*. This volume, however, presents a detailed description and analyses of all of the elements recovered, with additional data derived from computed tomography (CT), along with a better understanding of its geological, taphonomical, and paleoenvironmental context. This volume will contribute greatly to our knowledge of the postcranial anatomy of *Au. afarensis* and towards a better understanding of its overall paleobiology.

Field research conducted by the Woranso-Mille paleoanthropological project was made possible under a permit from the Authority for Research and Conservation of Cultural Heritage (ARCCH) of the Ministry of Culture and Tourism of Ethiopia, the Afar Regional State, and all of its district administrations. We would like to particularly acknowledge Mohammed Bilay (Administrator of Mille district) and Habib Wogris (Chairman of the Gega and Burtele sub district) for their support of the project. Thanks to Alemayehu Asfaw who found the first skeletal element of KSD-VP-1/1 and to the colleagues and students who participated in the excavation: Mulugeta Alene, Joshua Angelini, Erin Benson, Alan Deino, Stephanie Melillo, Hailay Reda, Liz Russell, Beverly Saylor, Gary Scott, and Robin Shultz. We are also grateful to the many field assistants from Addis Ababa and the Afar people at Waki and Waylateyta without whose participation the excavation of KSD-VP-1/1 would not have been possible. The Paleoanthropology Laboratory of the ARCCH made its laboratory and fossil storage space available and we are grateful for the support provided by the curators in the laboratory. Field and laboratory research of the Woranso-Mille project was supported by The Leakey Foundation, The Wenner-Gren Foundation, The National Geographic Society, Cleveland Museum of Natural History, and National Science Foundation (BCS-0234320, BCS-0321893, BCS-0542037, BCS-1124705, BCS-1124713, BCS-1124716, BCS-1125157, BCS-1125345).

We thank all of the authors in this volume for their contribution to our understanding of *Australopithecus* paleobiology and paleoenvironment, and to the many colleagues who have provided comments, advice, and discussion to the research conducted at Woranso-Mille in general and to the research presented in this volume specifically. The chapters presented in this volume were extensively reviewed, and we are grateful to those colleagues who contributed their time and efforts to the review process. Alex Fadiga and Caitlin Schwartz provided editorial assistance. Our deepest thanks to Eric Delson and Eric Sargis, editors of the Vertebrate Paleobiology and Paleoanthropology Series, for their support throughout the process and their patience in seeing this volume to publication and to Shalini Selvam, Sherestha Saini and Jeffrey Taub for overseeing the production of this book.

Yohannes Haile-Selassie
Denise F. Su

Contents

Contributors

Mulugeta Alene Department of Geology and Geophysics, Addis Ababa University, Addis Ababa, Ethiopia

Alan Deino Berkeley Geochronology Center, Berkeley, CA, USA

Luis Gibert Departament de Geoquímica, Petrologia i Prospecció Geològica, Universitat de Barcelona, Barcelona, Spain

Yohannes Haile-Selassie Department of Physical Anthropology, Cleveland Museum of Natural History, Cleveland, OH, USA

Bruce M. Latimer Department of Orthodontics School of Dental Medicine, Case Western Reserve University, Cleveland, OH, USA; Department of Physical Anthropology, Cleveland Museum of Natural History, Cleveland, OH, USA

C. Owen Lovejoy Department of Anthropology and Division of Biomedical Sciences, Kent State University, Kent, OH, USA; Department of Physical Anthropology, Cleveland Museum of Natural History, Cleveland, OH, USA

Stephanie M. Melillo Department of Human Evolution, Max Planck Institute for Evolutionary Anthropology, Leipzig, Germany

Marc R. Meyer Department of Anthropology, Chaffey College, Rancho Cucamonga, CA, USA

Timothy M. Ryan Center for Quantitative Imaging, EMS Energy Institute, Pennsylvania State University, University Park, PA, USA; Department of Anthropology, Pennsylvania State University, University Park, PA, USA

Beverly Z. Saylor Department of Earth, Environmental and Planetary Sciences, Case Western Reserve University, Cleveland, OH, USA

Gary Scott Berkeley Geochronology Center, Berkeley, CA, USA

Linda Spurlock Department of Anthropology and Division of Biomedical Sciences, Kent State University, Kent, OH, USA; Department of Physical Anthropology, Cleveland Museum of Natural History, Cleveland, OH, USA

Denise F. Su Department of Paleobotany and Paleoecology, Cleveland Museum of Natural History, Cleveland, OH, USA

Simone Sukhdeo Department of Anthropology, Pennsylvania State University, University Park, PA, USA

Contributors

Chapter 1
Introduction to KSD-VP-1/1: The Earliest Adult Partial Skeleton of *Australopithecus afarensis*

Yohannes Haile-Selassie

Abstract Early hominin partial skeletons are extremely rare in the fossil record, particularly for the time predating 3 Ma (million years ago). As a result, most of our understanding of Pliocene hominin paleobiology and phylogenetic relationships is based on isolated and fragmentary specimens. The Middle Pliocene species *Australopithecus afarensis* is one of the best-known early hominin species, and yet only three adult partial skeletons of the species have been recovered thus far, including the one described in this volume. The best known of these is the 3.2 Ma partial skeleton of a small female (A.L. 288-1) from Hadar, Ethiopia, and much of our understanding of the paleobiology of this species has been influenced by this specimen. The newly recovered 3.6 Ma partial skeleton from the Woranso-Mille, Ethiopia (KSD-VP-1/1, aka *Kadanuumuu*) represents a much larger male individual and may provide us with fresh insights into the paleobiology of *Au. afarensis*. This specimen not only preserves elements of the forelimb and hindlimb, but also includes complete elements, such as the scapula and clavicle, which were previously known only from fragmentary specimens. This edited volume provides the taphonomy and paleoecology of the partial skeleton, as well as detailed comparative descriptions of the preserved elements of KSD-VP-1/1 and their implications for our understanding of early hominin paleobiology. This chapter will present a basic introduction to the discovery of KSD-VP-1/1 and provide a guide to the contents of the volume.

Keywords Woranso-Mille • Hominin • Paleobiology • Paleoecology • Middle Pliocene

Y. Haile-Selassie (✉)
Department of Physical Anthropology, Cleveland Museum of Natural History, 1 Wade Oval Drive, Cleveland, OH 44106, USA
e-mail: yhailese@cmnh.org

Introduction

Australopithecus afarensis is one of the best represented early hominin species in the fossil record, known from a number of cranial and postcranial elements collected from Middle Pliocene deposits of Ethiopia, Kenya, Tanzania, and possibly Chad that range in age from ca. 3.8 Ma (million years ago) to 2.9 Ma (Leakey et al. 1976; Aronson et al. 1977; White 1977, 1980; Johanson et al. 1978a, b, 1982; Harris 1987; Brunet et al. 1995, 1996; Brown et al. 2001; Alemseged et al. 2005, 2006; Campisano and Feibel 2008; Harrison 2011). A detailed review of the species is presented in Kimbel and Delezene (2009). However, since its initial naming (Johanson et al. 1978a), its taxonomy (Ferguson 1983; Richmond and Jungers 1995; Strait and Grine 2004; Grine et al. 2006) and locomotor adaptation (Stern and Susman 1983, 1991; Susman et al. 1984, 1985; Latimer et al. 1987; Lovejoy 1988; Latimer and Lovejoy 1990a, b; Latimer 1991; Stern 2000; Lovejoy et al. 2002; Ward 2002; Ward et al. 2011) have been subjects of great debate.

Due to the relatively abundant fossil remains of the species, the craniodental anatomy of *Australopithecus afarensis* has been the subject of intensive comparative studies. Most of the complete cranial and dentognathic specimens were recovered from Hadar, Ethiopia (see Kimbel et al. 2004; Kimbel and Delezene 2009, for review). The fossil remains of this species have served as the basis for what we know about the paleobiology of, and intraspecific variation in, early hominin taxa. Specimens attributed to *Au. afarensis* show substantial amount of size and shape variation related to sexual dimorphism (Johanson et al. 1978a, b; Johanson and White 1979; Kimbel et al. 1985, 2004; Kimbel and White 1988; Lovejoy et al. 1989, among many others). Reno et al. (2003) analyzed skeletal size dimorphism in *Au. afarensis* applying extensive simulations using the smallest individual A.L. 288-1, the large sample from A.L. 333, modern humans, chimpanzees, and gorillas, which indicated that skeletal size dimorphism in *Au. afarensis* was most

Yohannes Haile-Selassie and Denise F. Su (eds.), *The Postcranial Anatomy of Australopithecus afarensis: New Insights from KSD-VP-1/1*, Vertebrate Paleobiology and Paleoanthropology, DOI 10.1007/978-94-017-7429-1_1

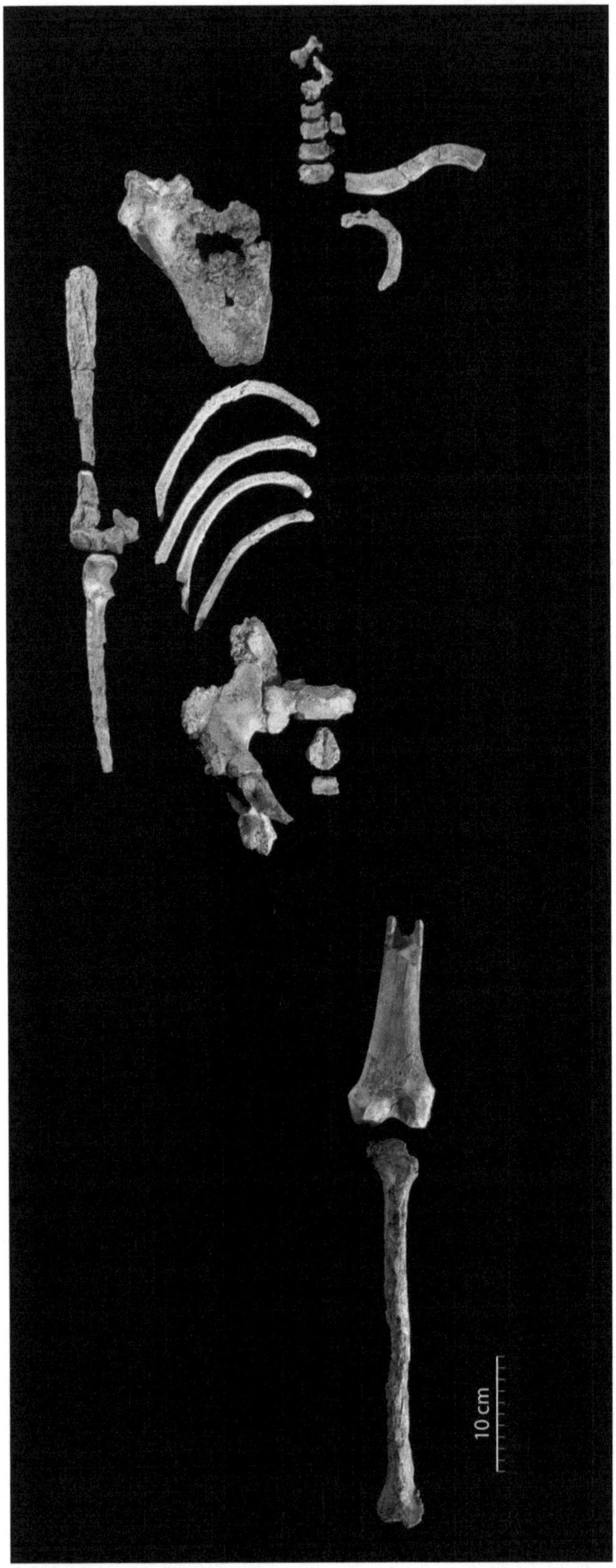

Fig. 1.1 KSD-VP-1/1 with elements arranged in their anatomical position. A complete list of elements is provided in Table 1.1. Photo courtesy of Elizabeth Russell

similar to that of modern humans. However, other comparative analyses (for example, Lague 2002) have shown that the variation falls within the range observed in gorillas and orangutans, and exceeds that of chimpanzees and humans. Some researchers have also suggested that the degree of variation seen in *Au. afarensis* is too great for a single species and proposed the possibility of multiple taxa within the hypodigm (for example, Coppens 1981, 1983; Olson 1981, 1985; Ferguson 1983, 1984; Senut 1983, 1986; Senut and Tardieu 1985; Zihlman 1985; Schmid 1989).

Until recently, only two adult partial skeletons of *Au. afarensis* with both forelimb and hindlimb elements have been reported since its initial description and naming – A.L. 288-1 (Johanson et al. 1982) and A.L. 438-1 (Drapeau et al. 2005). However, the unusually small body size of A.L. 288-1 has raised questions regarding allometry and its impact on the interpretation of *Au. afarensis* paleobiology and locomotor adaptation (e.g., Jungers 1982; Wolpoff 1983; Jungers and Stern 1983; Aiello 1992). A.L. 438-1, while a much larger individual, is missing key elements that preclude the analyses necessary to resolve the issue. Thus, there is still much to learn about the paleobiology of *Au. afarensis*. The recently announced partial skeleton, KSD-VP-1/1 (Fig. 1.1), from the Woranso-Mille study area is critically important as one of the only three known adult partial skeletons of *Au. afarensis*. KSD-VP-1/1 represents a moderately large-bodied individual, well within the range of living *Homo* in its size and certain aspects of its morphology (Haile-Selassie et al. 2010a, b); its preserved elements provide new data on limb proportions, shoulder girdle anatomy, thoracic form, and locomotor heritage in early hominins.

Background on the Woranso-Mille Study Area

The Woranso-Mille paleontological study area (WORMIL) is a relatively new hominin-bearing Mio-Pliocene site complex located in the central Afar region of Ethiopia, about 40 km north of the Hadar, Gona, and Dikika paleontological sites (Fig. 1.2). In the early 1970s, Kalb (1993, 2001) briefly visited the northern part of the study area along the old road from Mille town to Chifra and collected vertebrate fossils (mostly monkeys and pigs) from the area known as Am-Ado (Kalb's "Ahmado"). Kalb (1993, 2001) also identified some localities around the area locally known as Leadu on the Bati-Mille road further to the south. However, there is no record that exposures between Am-Ado and Leadu (ca. 12 km north–south stretch) were surveyed by Kalb or other paleontologists. A small team led by the author started intensive paleontological survey and exploration in this area in 2003. The team was able to identify a number of fossiliferous outcrops south of

Am-Ado and north of Leadu, signifying the need for long-term paleontological research in the area. In 2005, the Woranso-Mille project evolved from a small paleontological survey and exploration team to an international multidisciplinary team of scientists that began intensive and systematic research on the geology and paleontology of what is now known as the Woranso-Mille paleontological site.

A total of 85 vertebrate paleontological localities have been designated in the study area thus far, with more than 8,400 fossil specimens collected including 167 (~2% of the total fauna) hominin fossils (Haile-Selassie et al. 2007, 2010a, b, 2012; Haile-Selassie 2010). The faunal assemblage includes more than 70 mammalian species ranging in age from the Late Miocene to the Middle Pliocene. More than 95% of the specimens, including all of the hominins recovered thus far, are from sediments radiometrically dated to between 3.4 and 3.82 Ma (Deino et al. 2010; Haile-Selassie et al. 2012).

WORMIL is an important hominin-bearing site complex, sampling a time period that is poorly known in the human fossil record and providing significant fossil evidence that

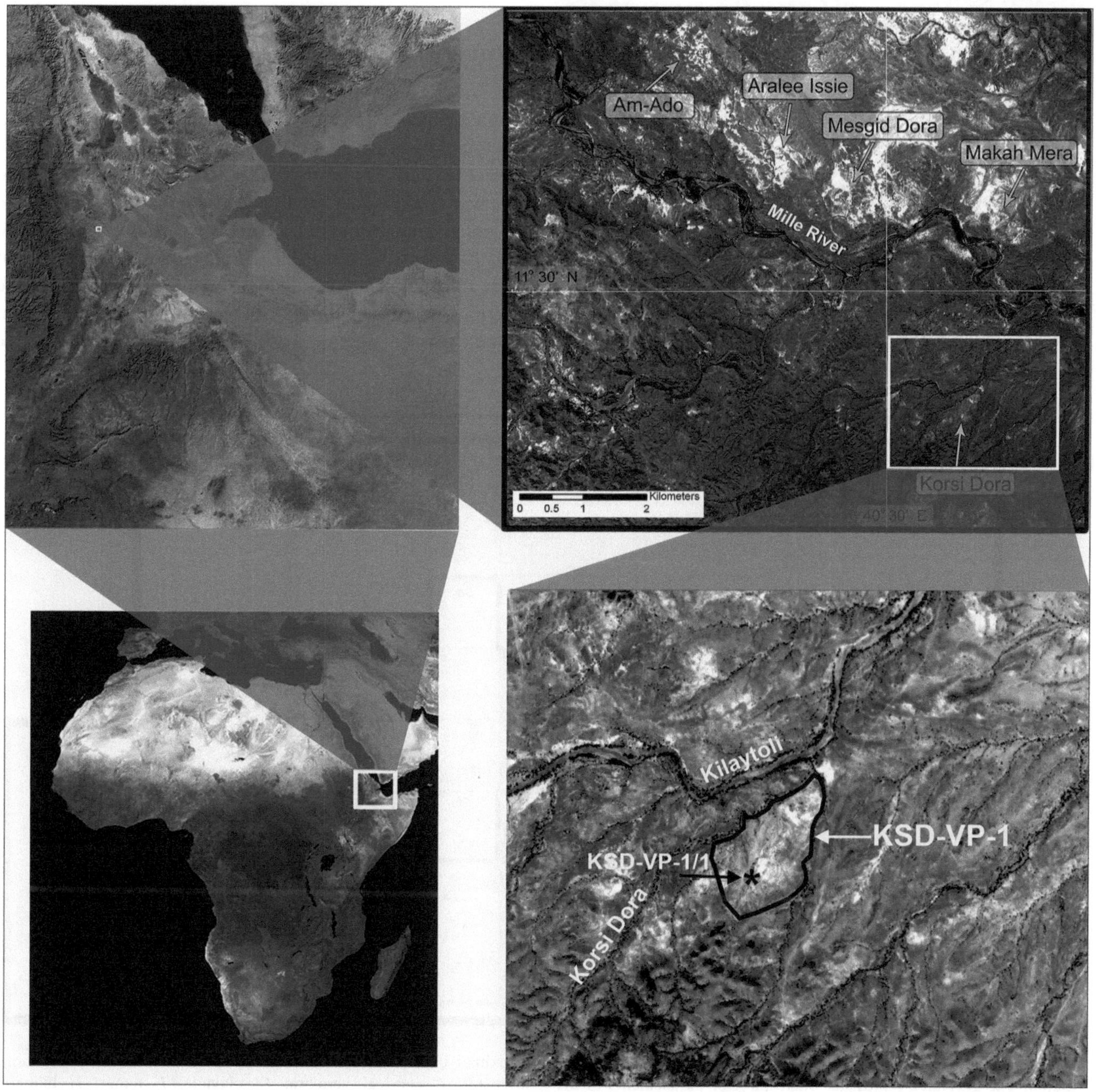

Fig. 1.2 Location map of the Woranso-Mille study area and locality boundaries of KSD-VP-1 locality (bottom right). The location of KSD-VP-1/1 is shown by a star

Fig. 1.3 Alemayehu Asfaw discovering the first element (proximal ulna) of KSD-VP-1/1 on February 10, 2005

could answer critical questions in human evolutionary studies. It is the only site, thus far, that has yielded incontrovertible evidence of multiple, contemporaneous hominin species from a single site during the Middle Pliocene (Haile-Selassie et al. 2012, 2015) and has yielded the oldest adult *Au. afarensis* partial skeleton, KSD-VP-1/1 (Haile-Selassie et al. 2010a).

Recovery, Excavation, and Preservation of KSD-VP-1/1

The partial skeleton is recovered from Korsi Dora vertebrate locality 1 (KSD-VP-1), one of the smallest and most paleontologically depauperate localities in the Woranso-Mille study area. The locality is about 300 m^2 in size, and fossiliferous sediments are exposed on the southern edge of a drainage system known as 'Kilaytoli' and east of 'Korsi Dora,' a small channel that drains to Kilaytoli (Fig. 1.2).

Recovery

The first element of KSD-VP-1/1, a proximal ulna fragment, was found by Alemayehu Asfaw (Fig. 1.3) on February 10, 2005 (11° 28′ 54.0″ N; 40° 30′ 39.3″ E [Reference Datum = WGS84]). Crawling and surface scraping in the immediate area resulted in the recovery of more parts of the ulna, fragments of the distal femur, bodies and pedicels of cervical vertebrae, humeral shaft fragments, and parts of the sacrum. Once the hominin horizon was identified, the team started excavating east and northeastward, where the surface

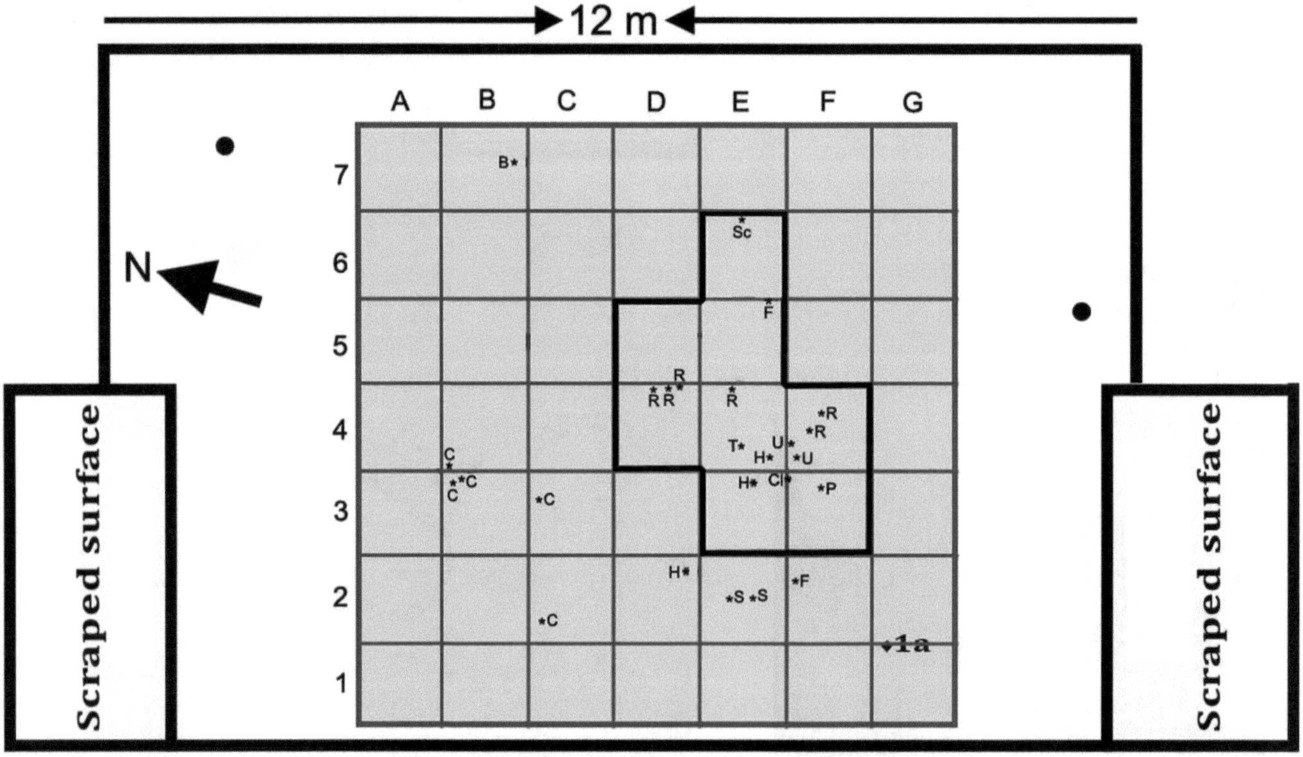

Fig. 1.4 Schematic diagram of the excavated area at KSD-VP-1/1 location. The gray area represents where most of the elements of the partial skeleton were collected. The area highlighted with the gray box is where the *in situ* specimens were recovered. 1a the first specimen (proximal ulna) that was discovered on the surface; *B* bovid molar fragment; *C* cervical vertebra; *Cl* clavicle; *F* femur; *H* humerus fragment; *R* rib; *Sc* scapula; *U* ulna fragment

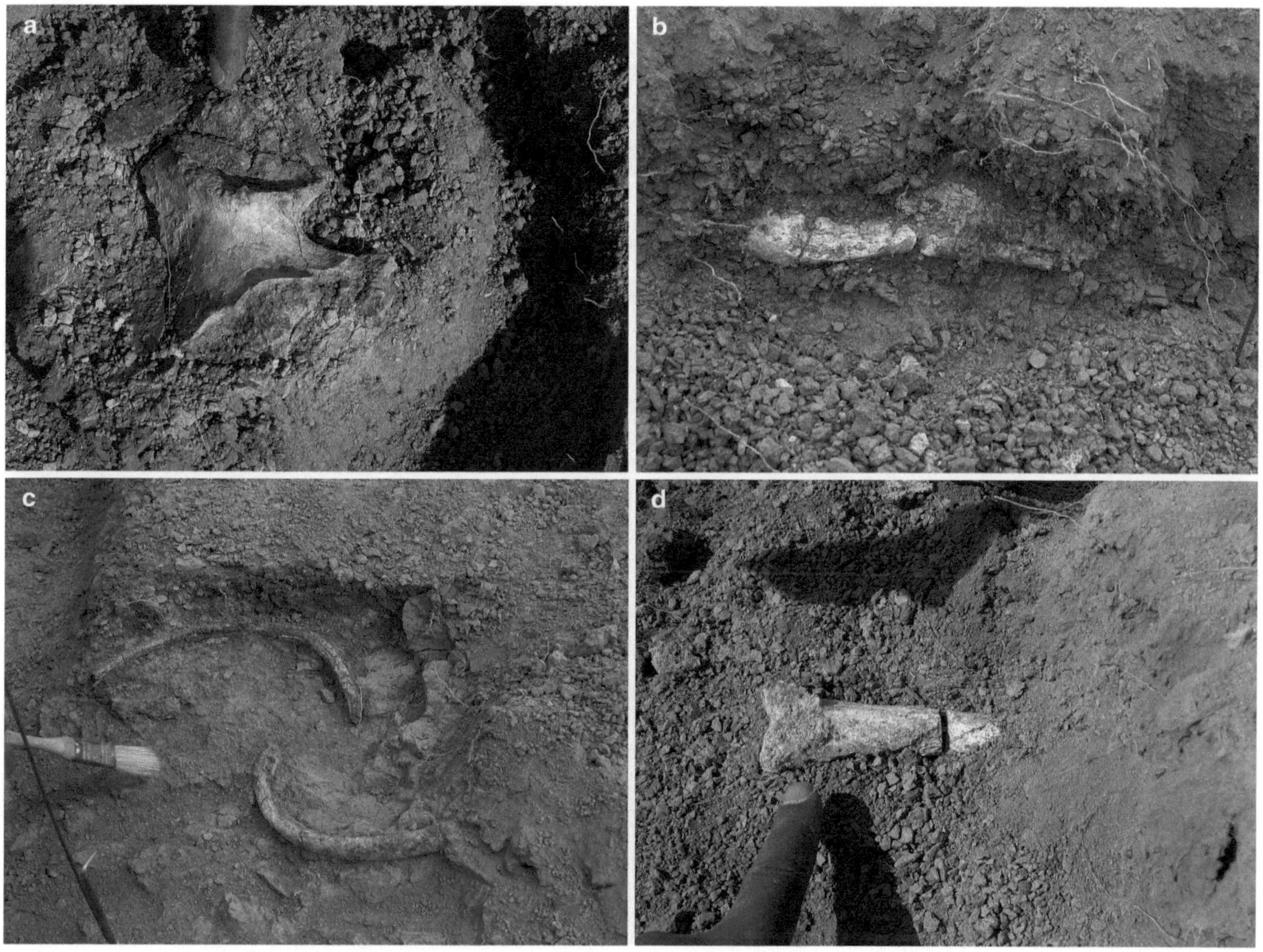

Fig. 1.5 Elements of KSD-VP-1/1 found *in situ*: **a** right innominate, **b** right clavicle, **c** lower ribs, **d** left tibia

finds were abundant, and resulted in further recovery of *in situ* skeletal elements including the left os coxa, right tibia, right clavicle, five ribs, and a left scapula (Fig. 1.4), all of which were excavated within an area of 8 m^2 (Fig. 1.5a–d) from a laminated mudstone and sandstone horizon at the base of upward-coarsening sequences that are interpreted as overbank deposits (Deino et al. 2010; Saylor et al. 2016).

Excavation

During the 2005 field season, an area of 49 m^2 was excavated, which covered the distribution of all surface fragments. A string grid was set up using metal pegs at the grid corners, and each 1 m^2 grid square was numbered with a combination of letters and numbers (Fig. 1.4). At the end of the 2005 excavation, long metal stakes were buried at all four corners of the grid perimeter, and each recovered specimen was marked with a long metal pin buried in the ground. The excavated surface and walls of the excavation were stabilized with rocks until the next field season. During the 2006 field season, the excavated area was expanded to the east and north, and numerous shaft fragments were excavated about 3 m northeast of the main bone concentration area. However, they were so fragmented that they could not be identified as belonging to KSD-VP-1/1 with any confidence. A single maxillary fragment of a bovid was also excavated about 3 m northeast of the easternmost excavated element (scapula) of KSD-VP-1/1 (Fig. 1.4).

The excavation continued for another 3 years until 2009 (Fig. 1.6a–c), with the participation of more than 45 individuals (Fig. 1.7). At the end of the 2009 field season, a total area of 85 m^2 with a vertical depth of 1.3 m on the eastern wall was excavated (Fig. 1.6d). No part of the cranium or dentition of KSD-VP-1/1 was ever recovered, and even though the enormous amount of excavated overburden was sieved, no additional bone fragments were retrieved from the sieving. Table 1.1 provides a list of the skeletal elements of KSD-VP-1/1 that are described in this monograph. It should be noted; however, that a few, mostly vertebral, fragments associated with the partial skeleton are neither listed nor described here because they cannot be identified with confidence.

Fig. 1.6 Views of the KSD-VP-1/1 excavation location. **a** Initial surface scraping immediately after the discovery of the first element (February 10, 2005). **b** Piles of excavated sediments ready for sifting. **c** Excavation in 2005. Each flag represents a specimen that was found on the surface or excavated *in situ*. **d** View of the excavation site at the end of the 2005 field season

Fig. 1.7 The excavation team at the end of the 2005 field season. More than 45 people participated in the excavation of KSD-VP-1/1 between the 2005 and 2009 field seasons

Preservation

Elements of KSD-VP-1/1 that were excavated *in situ* (the right innominate, left tibia, the ribs, the left clavicle, and the right scapula) were highly fragile and had to be consolidated using a thin mixture of vinac beads and acetone before they were extracted. Additionally, the innominate had to be plaster-jacketed before removal. All other elements were recovered in the form of small fragments (Fig. 1.8). The distribution of the skeletal elements indicates that the carcass was more likely to have been surface scattered before fossilization either by trampling or erosion. However, none of the bones shows any carnivore damage or substantial transportation (see Su 2016, for details). Compression and deformation are apparent on some of the elements, particularly on the ribs, which is likely due to the contraction and expansion of the clay sediments in which they were buried. Almost all of the *in situ* skeletal elements were covered by a thin layer of gypsum that permeated through crevices extending from clay horizons higher in the section. The surface specimens, i.e., cervical vertebrae, humerus, femur, and ulna, were fragmented into numerous pieces although they did not require significant pre-collection treatment.

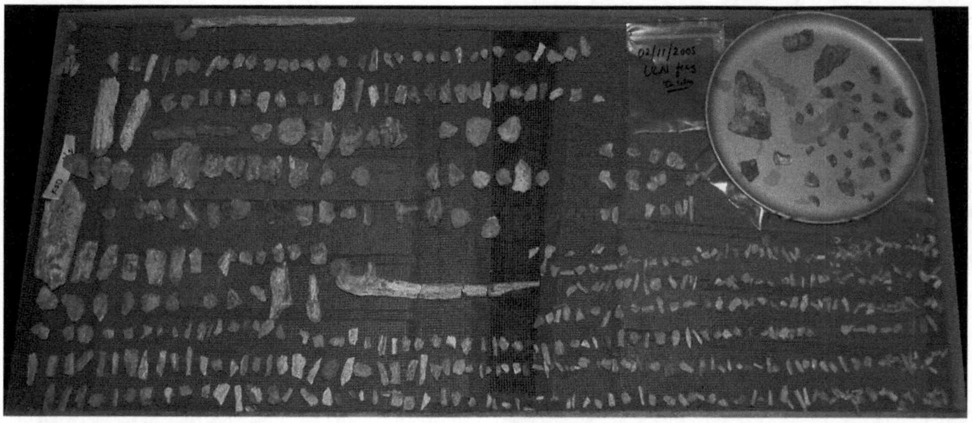

Fig. 1.8 Fragments of KSD-VP-1/1 collected by surface scraping, sieving, and excavation. More than 400 pieces were recovered representing parts of the cervical vertebrae, humerus, ulna, femur, and the sacrum. They were systematically conjoined by project scientists in the paleoanthropology laboratory of the National Museum in Addis Ababa. The fragments in the pan on top right corner belong to the innominate collected from the excavation in 2005

Table 1.1 KSD-VP-1/1 partial skeleton: inventory of preserved elements

Accession number	Element	Discovery
KSD-VP-1/1a	Right ulna	S
KSD-VP-1/1b	Right humerus	S
KSD-VP-1/1c	Left distal femur	S
KSD-VP-1/1d	Right os coxa	IS
KSD-VP-1/1e	Left tibia	IS
KSD-VP-1/1f	Right clavicle	IS
KSD-VP-1/1g	Right scapula	IS
KSD-VP-1/1h	Second cervical vertebra (C-2)	S
KSD-VP-1/1i and /1x	Third cervical vertebra (C-3)	S
KSD-VP-1/1j, /1z, and /1ac	Fourth cervical vertebra (C-4)	S
KSD-VP-1/1k, /1y and /1aa	Fifth cervical vertebra (C-5)	S
KSD-VP-1/1l	Sixth cervical vertebra (C-6)	S
KSD-VP-1/1m	Vertebral body	S
KSD-VP-1/1n	Left second rib	IS
KSD-VP-1/1o	Right lower rib (7th or 8th)	IS
KSD-VP-1/1p	Right lower rib (8th or 9th)	IS
KSD-VP-1/1q	Right upper rib (5th, 6th or 7th)	IS
KSD-VP-1/1r	Left 11th rib	IS
KSD-VP-1/1s	Middle rib fragment	S
KSD-VP-1/1t	Superior sacral body and ala	S
KSD-VP-1/1u	Posterior sacral spine fragment	S
KSD-VP-1/1v	Sacral spine fragment	S
KSD-VP-1/1w	Coccygeal body?	S
KSD-VP-1/1ab	Left articular process and articular pillar (C-7)	S
KSD-Vp-1/1ad	Right superior facet of vertebra (C-7)	S
KSD-VP-1/1ae	Right inferior facet of vertebra (C-7)	S
KSD-VP-1/1af	Posterior tubercle of the transverse process (C-7)	S
KSD-VP-1/1ag	Right lamina (C7)	S

S Surface; *IS In situ*

Fig. 1.9 Yohannes Haile-Selassie in the paleoanthropology laboratory of the National Museum of Ethiopia systematically conjoining the fragments of KSD-VP-1/1

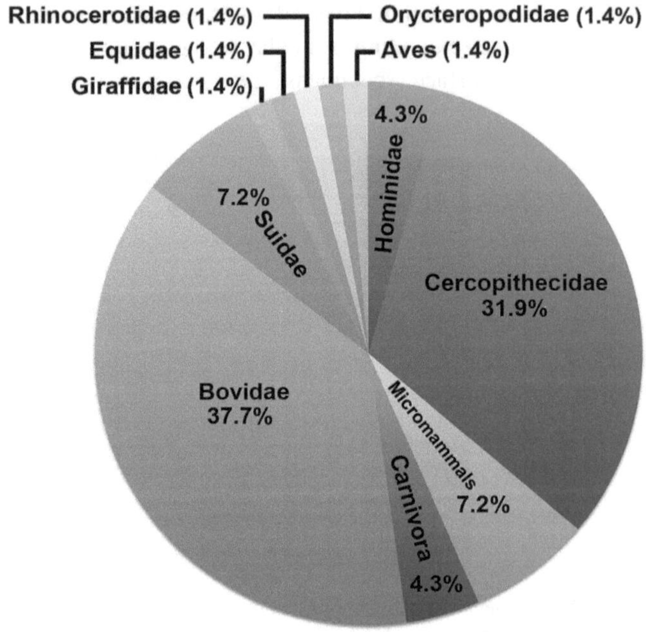

Fig. 1.10 Pie chart showing the percentage of mammalian families collected from KSD-VP-1/1 based on minimum number of individuals (MNI = 69). The micromammal percentage includes multiple families (see Table 1.2 for details)

Preparation and Curation

All original fossils collected from the study area are stored at the paleoanthropology laboratory of the National Museum of Ethiopia in Addis Ababa. Preparation of KSD-VP-1/1 included freeing elements from plaster jackets and cleaning the adhering matrix using an air scribe, at times under a microscope. Skeletal elements of KSD-VP-1/1 were joined together from over 300 bone fragments (Fig. 1.9), after which they were accessioned and photographed before being stored. The WORMIL project organizes its fossil collections by locality, taxon, and age so that project researchers can have easy access for analyses.

The Faunal Assemblage at KSD-VP-1

KSD-VP-1 is one of the smallest localities in the Woranso-Mille study area, with a total area of ca. 300 m^2. There are only 69 accessioned fossil specimens from the locality. However, a 100% collection for taphonomic analysis resulted in more than 700 mostly unidentifiable bone fragments (see Su 2016). Despite the small number of accessioned specimens, the taxonomic diversity of this locality is remarkable-23 mammalian species from 15 families are represented (Table 1.2). Bovids are the most abundant with six tribes identified, followed by cercopithecids represented by at least three species. Among the suids, *Kolpochoerus* and *Nyanzachoerus* are present. In addition to KSD-VP-1/1, two other specimens of *Au. afarensis*, an isolated lower molar (KSD-VP-1/52) and a child mandible with dentition (KSD-VP-1/39), have also been recovered from the locality. Two small carnivore species are also present: a herpestid and a lutrine (cf. *Torolutra* sp.). Among the small mammals, three genera (cf. *Serengetilagus*, *Thryonomys*, and *Hystrix*) are represented in the assemblage. Large-sized mammals are represented by Elephantidae, Rhinocerotidae, Hippopotamidae, Giraffidae, and Equidae, with at least one species identified from each family (Fig. 1.10).

Guide to Contents of the Book

This volume contains five chapters that provide detailed description and comparative analyses of KSD-VP-1/1, in addition to two chapters on its geological, taphonomic, and paleoecological context. Cumulatively, these chapters

Table 1.2 Faunal list of KSD-VP-1

- Aves
- Rodentia
  - Indet.
  - Hystricidae
    - *Hystrix* sp.
  - Thryonomyidae
    - *Thryonomys* sp.
- Lagomorpha
  - Leporidae
    - cf. *Serengetilagus* sp.
- Primates
  - Hominidae
    - *Australopithecus afarensis*
  - Cercopithecidae
    - Indet.
    - Papionini Indet.
    - *Theropithecus oswaldi* aff. *darti*
    - Colobinae Indet.
- Carnivora
  - Viverridae
    - cf. *Viverra* sp.
  - Mustelidae
    - cf. *Torolutra* sp.
- Proboscidea
  - Elephantidae
    - Indet.
- Artiodactyla
  - Bovidae
    - Indet.
    - Alcelaphini
    - Neotragini
    - Hippotragini
    - Antilopini
      - *Gazella* sp.
    - Tragelaphini
      - *Tragelaphus saraitu*
      - *Tragelaphus* sp. indet.
    - Aepycerotini
      - *Aepyceros* cf. *afarensis*
  - Suidae
    - Suinae
      - *Kolpochoerus* cf. *millensis*
    - Tetraconodontinae
      - *Nyanzachoerus kanamensis*
  - Giraffidae
    - *Giraffa* cf. *stillei*
  - Hippopotamidae
- Perissodactyla
  - Equidae
    - *Eurygnathohippus* sp.
  - Rhinocerotidae
- Tubulidentata
  - Orycteropodidae
    - *Orycteropus* sp.
- Reptilia
  - Crocodylidae

provide insights into the paleobiology and paleoecology of early *Au. afarensis* and the last chapter presents a synthesis of the broader implications of KSD-VP-1/1 for our interpretation of early hominin paleobiology.

Chapters 2 and 3 provide the geologic, taphonomic, and paleoenvironmental context of KSD-VP-1. While the fossiliferous areas within WORMIL have not yet been assigned to any specific geological formation or divided into subunits, radiometric ages have been obtained for most of the tuffs within the fossiliferous horizons (Deino et al. 2010; Haile-Selassie et al. 2010a, b, 2012, 2015). The geological age of KSD-VP-1/1 is therefore well-constrained using $^{40}Ar/^{39}Ar$ dating method. A thin tuff about 2.7 m below the partial skeleton yielded an age of 3.60 ± 0.03 Ma. Chapter 2 (Saylor et al. 2016) elaborates on the refinement of the geochronology of KSD-VP-1/1 and provides details on the sedimentological context of the locality.

Following that, Chapter 3 (Su 2016) explores the taphonomy and paleoenvironment of KSD-VP-1/1 based on faunal and geological data. The KSD-VP-1 faunal assemblage is likely autochthonous and transportation and disturbance of bones was minimal. The combined evidence indicates that the paleohabitat at KSD-VP-1 was a medium to dense woodland with some open areas of grassland or shrubland distal to the locality.

Chapters 4 through 8 provide detailed description and comparative analyses of the skeleton. One of the key pieces of information that was lacking from the preliminary description and analysis of KSD-VP-1/1 (Haile-Selassie et al. 2010a) was CT-scan data. With permission from the Ethiopian government, the original elements of the partial skeleton were transported to the United States, and each element was scanned using high-resolution computed tomography (HRCT) at the Center for Quantitative Imaging at the Pennsylvania State University. Chapter 4 (Ryan and Sukhdeo 2016) describes three-dimensional reconstructions and quantification of internal and external anatomical structures based on HRCT scan data.

KSD-VP-1/1 is the only early hominin adult specimen that preserves almost all of its cervical vertebrae and represents the oldest known cervical column in the hominin fossil record. Chapter 5 (Meyer 2016) presents a detailed description and comparative analysis of each cervical vertebra. The suite of characteristics in the KSD-VP-1/1 cervical centra is consistent with human-like orthograde posture and head carriage. However, the mosaic of derived anatomy with transitional nuchal musculature and the inchoate stage of nuchal ligament development are also consistent with the locomotor pattern of other primates without the whole-body aerial phase of humans. The KSD-VP-1/1 cervical spines suggest that the nuchal ligament, a significant aspect of the derived human axial anatomy for cursorial distance locomotion, was not fully developed as in modern humans.

The right scapula of KSD-VP-1/1 is the only known complete adult scapula of an early hominin. Chapter 6 (Melillo 2016) presents detailed description and analysis of the scapula and the clavicle to provide a more complete picture of early hominin shoulder girdle anatomy. The results corroborate Haile-Selassie et al.'s (2010a, b) conclusion that the adult *Au. afarensis* scapula, although morphologically distinct, was more similar to that of modern humans than previously recognized. Despite the morphological distinctiveness, however, the *Australopithecus* shoulder girdle in general appears to have been derived toward a morphology associated with an emphasis on the manipulatory function of the upper limb (see Schmid et al. 2013 for different opinion).

The human-like shoulder girdle anatomy of KSD-VP-1/1 described in Chapter 6 is consistent with the thorax shape of the specimen described in Chapter 7 (Latimer et al. 2016). The preserved elements of the KSD-VP-1/1 thorax comprise a number of ribs, including the second rib, and suggest that *Au. afarensis* had a human-like thoracic cage with a broad upper thorax and a deeply invaginated thoracic vertebral column. This is in stark contrast to previous reconstructions of early hominin thoracic shape (e.g., Schmid 1983, 1989, 1991) and also different from what has been inferred for *Au. sediba* where the thorax is described as mosaic with an ape-like upper thoracic shape combined with a more derived human-like shape to the lower thorax (Schmid et al. 2013).

Chapter 8 (Lovejoy et al. 2016) describes the preserved upper limb, lower limb, and pelvic elements of KSD-VP-1/1. It has been long known that understanding body proportions of early hominins is crucial for making accurate functional and phylogenetic interpretations (e.g., Robinson 1972; Walker 1973; McHenry 1978; Zihlman and Brunker 1979). Unfortunately, the paucity of partial skeletons combining upper and lower limbs, particularly those predating 3.0 Ma, has remained a major hurdle. Previous assessment of body size, limb proportions, and skeletal allometry in *Au. afarensis* relied on A.L. 288-1, which suggested that *Au. afarensis* combined human-like forelimb proportions with relatively short hindlimbs (Jungers 1982). However, the diminutive size of A.L. 288-1 led to decades of controversial speculations about the confounding effects of allometry (e.g., Wolpoff 1983; Jungers and Stern 1983). KSD-VP-1/1 belongs to an individual much larger than A.L. 288-1 and combines elements of the upper and lower limbs, allowing us to test hypotheses relating to body proportion and allometry in early hominins in Chapter 8.

Finally, Chapter 9 (Haile-Selassie et al. 2016) summarizes the broader implications of KSD-VP-1/1 for our understanding of the paleobiology of early hominins in general and *Au. afarensis* in particular, and its contributions toward reconstructing the last chimpanzee/human common ancestor (CLCA) of hominins. It also critically examines previous interpretations of early hominin shoulder girdle anatomy, thorax shape, locomotor adaptation, and skeletal bauplan in light of the new data derived from KSD-VP-1/1 compared to observations from other hominin taxa such as *Ardipithecus ramidus* and *Au. sediba*.

Acknowledgments The Paleoanthropology Laboratory of the ARCCH made its laboratory and fossil storage space available to the Woranso-Mille project. The support from the curators in the laboratory is also very much appreciated. The excavation of KSD-VP-1/1 would not have been possible without the participation of the many field assistants from Addis Ababa and the Afar people at Waki and Waylateyta. I am grateful to all of the authors in this volume for their contribution to our understanding of *Australopithecus* anatomy, paleobiology and paleoenvironment, and to the many colleagues who have provided comments, advice, and discussion to the research conducted at Woranso-Mille in general and to the research presented in this volume specifically.

References

Aiello, L. C. (1992). Allometry and the analysis of size and shape in human evolution. *Journal of Human Evolution, 22*, 127–147.

Alemseged, Z., Wynn, J. G., Kimbel, W. H., Reed, D., Geraads, D., & Bobe, R. (2005). A new hominin from the Basal Member of the Hadar Formation, Dikika, Ethiopia, and its geological context. *Journal of Human Evolution, 49*, 499–514.

Alemseged, Z., Spoor, F., Kimbel, W. H., Bobe, R., Geraads, D., Reed, D., et al. (2006). A juvenile early hominin skeleton from Dikika, Ethiopia. *Nature, 443*, 296–301.

Aronson, J. A., Schmitt, T. J., Walter, R. C., Taieb, M., Tiercelin, J.-J., Johanson, D. C., et al. (1977). New geochronologic and paleomagnetic data for the hominid-bearing Hadar Formation, Ethiopia. *Nature, 267*, 323–327.

Brown, B., Brown, F. H., & Walker, A. (2001). New hominids from the Lake Turkana basin, Kenya. *Journal of Human Evolution, 41*, 29–44.

Brunet, M., Beauvilain, A., Coppens, Y., Heintz, E., Moutaye, A. H. E., & Pilbeam, D. (1995). The first australopithecine 2,500 kilometres west of the Rift Valley (Chad). *Nature, 378*, 273–275.

Brunet, M., Beauvilain, A., Coppens, Y., Heintz, E., Moutaye, A. H. E., & Pilbeam, D. (1996). *Australopithecus bahrelgazali*, une nouvelle espèce d'Hominidé ancient de la region de Koro Toro (Tchad). *Comptes Rendus de l'Academie des Sciences IIA, 322*, 907–913.

Campisano, C. J., & Feibel, C. S. (2008). Depositional environments and stratigraphic summary of the Pliocene Hadar Formation at Hadar, Afar Depression, Ethiopia. *Geological Society of America, Special Papers, 446*, 179–201.

Coppens, Y. (1981). Le cerveau des hommes fossiles. *Comptes Rendus de l'Academie des Sciences, Paris, 292*, 3–24.

Coppens, Y. (1983). Les plus anciens fossils d'Hominides. *Pontificiae Academiae Scientiarum Scripta Varia, 50*, 1–9.

Deino, A., Scott, G., Saylor, B., Alene, M., Angelini, J. D., & Haile-Selassie, Y. (2010). ^{40}Ar/^{39}Ar dating, paleomagnetism, and tephrochemistry of Pliocene strata of the hominid-bearing Woranso-Mille area, west-central Afar Rift, Ethiopia. *Journal of Human Evolution, 58*, 111–126.

Drapeau, M. S. M., Ward, C. V., Kimbel, W. H., Johanson, D. C., & Rak, Y. (2005). Associated cranial and forelimb remains attributed to *Australopithecus afarensis* from Hadar, Ethiopia. *Journal of Human Evolution, 48*, 593–642.

Ferguson, W. (1983). An alternative interpretation of *Australopithecus afarensis* fossil material. *Primates, 24*, 397–409.

Ferguson, W. (1984). Revision of fossil hominid jaws from the Plio-Pleistocene of Hadar, in Ethiopia, including a new species of the genus *Homo* (Hominoidea: Hominidae). *Primates, 25*, 519–529.

Grine, F. E., Ungar, P. S., Teaford, M. F., & El-Zaatari, S. (2006). Molar microwear in *Praeanthropus afarensis*: Evidence for dietary stasis through time and under diverse paleoecological conditions. *Journal of Human Evolution, 51*, 297–319.

Haile-Selassie, Y. (2010). Phylogeny of early *Australopithecus*: new fossil evidence from the Woranso-Mille (Central Afar, Ethiopia). *Philosophical Transactions of the Royal Society B, 365*, 3323–3331.

Haile-Selassie, Y., Deino, A., Saylor, B., Umer, M., & Latimer, B. (2007). Preliminary geology and paleontology of new hominid-bearing Pliocene localities in the central Afar region of Ethiopia. *Anthropological Science, 115*, 215–222.

Haile-Selassie, Y., Gibert, L., Melillo, S. M., Ryan, T. M., Alene, M., Deino, A., et al. (2015). New species from Ethiopia further expands Middle Pliocene hominin diversity. *Nature, 521*, 483–488.

Haile-Selassie, Y., Latimer, B. M., Alene, M., Deino, A., Gibert, L., Melillo, S. M., et al. (2010a). An early *Australopithecus afarensis* postcranium from Woranso-Mille, Ethiopia. *Proceedings of the National Academy of Sciences, 107*, 12121–12126.

Haile-Selassie, Y., Latimer, B. M., Lovejoy, C. O., Melillo, S. M., & Meyer, M. R. (2016). Chapter 9: Conclusion: Implications of KSD-VP-1/1 for early hominin paleobiology and insights into the chimpanzee/human last common ancestor. In: Y. Haile-Selassie & D. F. Su (Eds.), *The postcranial anatomy of* Australopithecus afarensis: *New insights from KSD-VP-1/1* (pp. 179–188). Dordrecht: Springer.

Haile-Selassie, Y., Saylor, B. Z., Deino, A., Alene, M., & Latimer, B. M. (2010b). New hominid fossils from Woranso-Mille (Central Afar, Ethiopia) and taxonomy of early *Australopithecus*. *American Journal of Physical Anthropology, 141*, 406–417.

Haile-Selassie, Y., Saylor, B. Z., Deino, A., Levin, N. E., Alene, M., & Latimer, B. (2012). A new hominin foot from Ethiopia shows multiple Pliocene bipedal adaptations. *Nature, 48*, 565–569.

Harris, J. M. (1987). Summary. In M. D. Leakey & J. M. Harris (Eds.), *Laetoli: A Pliocene site in Northern Tanzania* (pp. 524–532). Oxford: Oxford University Press.

Harrison, T. (2011). Hominins from the Upper Laetolil and Upper Ndolanya Beds, Laetoli. In T. Harrison (Ed.), *Paleontology and geology of Laetoli: Human evolution in context. Volume 2: Fossil hominins and the associated fauna* (pp. 141–188). Dordrecht: Springer.

Johanson, D. C., & White, T. D. (1979). A systematic assessment of early African hominids. *Science, 203*, 321–330.

Johanson, D. C., Taieb, M., Gray, B. T., & Coppens, Y. (1978a). Geological framework of the Pliocene Hadar Formation (Afar, Ethiopia) with notes on paleontology including hominids. In W. W. Bishop (Ed.), *Geological background to fossil man* (pp. 549–564). Edinburgh: Scottish Academic Press.

Johanson, D. C., White, T. D., & Coppens, Y. (1978b). A new species of the genus *Australopithecus* (Primates: Hominidae) from the Pliocene of eastern Africa. *Kirtlandia, 28*, 1–14.

Johanson, D. C., Taieb, M., & Coppens, Y. (1982). Pliocene hominids from the Hadar Formation, Ethiopia (1973–1977): Stratigraphic, chronologic, and paleoenvironmental contexts, with notes on hominid morphology and systematics. *American Journal of Physical Anthropology, 57*, 373–402.

Jungers, W. L. (1982). Lucy's limbs: Skeletal allometry and locomotion in *Australopithecus afarensis*. *Nature, 297*, 676–678.

Jungers, W. L., & Stern, J. T. (1983). Body proportions, skeletal allometry and locomotion in the Hadar hominids: A reply to Wolpoff. *Journal of Human Evolution, 12*, 673–684.

Kalb, J. E. (1993). Refined stratigraphy of the hominid-bearing Awash Group, Middle Awash valley, Afar depression, Ethiopia. *Newsletters on Stratigraphy, 29*, 21–62.

Kalb, J. E. (2001). *Adventures in the bone trade: The race to discover human ancestors in Ethiopia's Afar depression*. New York: Copernicus Books.

Kimbel, W. H., & Delezene, L. (2009). "Lucy" redux: a review of research on *Australopithecus afarensis*. *Yearbook of Physical Anthropology, 52*, 2–48.

Kimbel, W. H., & White, T. D. (1988). Variation, sexual dimorphism and the taxonomy of *Australopithecus*. In F. E. Grine (Ed.), *Evolutionary history of the "Robust" Australopithecines* (pp. 175–192). New York: Aldine de Gruyter.

Kimbel, W. H., Rak, Y., & Johanson, D. C. (2004). *The skull of Australopithecus afarensis*. New York: Oxford University Press.

Kimbel, W. H., White, T. D., & Johanson, D. C. (1985). Craniodental morphology of the hominids from Hadar and Laetoli: evidence of "*Paranthropus*" and *Homo* in the Mid-Pliocene of Eastern Africa. In E. Delson (Ed.), *Ancestors: the hard evidence* (pp. 120–137). New York: Liss.

Lague, M. R. (2002). Another look at shape variation in the distal femur of *Australopithecus afarensis*: Implications for taxonomic and functional diversity at Hadar. *Journal of Human Evolution, 42*, 609–626.

Latimer, B. (1991). Locomotor adaptations in *Australopithecus afarensis*: The issue of arboreality. In B. Senut & Y. Coppens (Eds.), *Origine(s) de la bipèdie chez les hominidés* (pp. 169–176). Paris: CNRS.

Latimer, B., & Lovejoy, C. O. (1990a). Hallucal tarsometatarsal joint in *Australopithecus afarensis*. *American Journal of Physical Anthropology, 82*, 125–133.

Latimer, B., & Lovejoy, C. O. (1990b). Metatarsophalangeal joints of *Australopithecus afarensis*. *American Journal of Physical Anthropology, 83*, 13–23.

Latimer, B. M., Lovejoy, C. O., Johanson, D. C., & Coppens, Y. (1982). Hominid tarsal, metatarsal and phalangeal bones recovered from the Hadar Formation: 1974–1977 collections. *American Journal of Physical Anthropology, 57*, 701–719.

Latimer, B. M., Lovejoy, C. O., Spurlock, L., & Haile-Selassie, Y. (2016). Chapter 7: The thoracic cage of KSD-VP-1/1. In: Y. Haile-Selassie & D. F. Su (Eds.), *The postcranial anatomy of* Australopithecus afarensis: *New insights from KSD-VP-1/1* (pp. 143–153). Dordrecht: Springer.

Latimer, B., Ohman, J. C., & Lovejoy, C. O. (1987). Talocrual joint in African hominoids: Implications for *Australopithecus afarensis*. *American Journal of Physical Anthropology, 74*, 155–175.

Leakey, M. D., Hay, R. L., Curtis, G. H., Drake, R. E., Jackes, M. K., & White, T. D. (1976). Fossil hominids from the Laetolil Beds. *Nature, 262*, 460–466.

Lovejoy, C. O. (1988). Evolution of human walking. *Scientific American, 256*, 118–125.

Lovejoy, C. O., Kern, K., Simpson, S. W., & Meindl, R. S. (1989). A new method for estimation of skeletal dimorphism in fossil samples with an application to *Australopithecus afarensis*. In G. Giacobini (Ed.), *Hominidae: Proceedings of the 2nd International Congress of Human Paleontology* (pp. 103–108). Milan: Jaca.

Lovejoy, C. O., Latimer, B. M., Spurlock, L., & Haile-Selassie, Y. (2016). Chapter 8: The pelvic girdle and limb bones of KSD-VP-1/1. In: Y. Haile-Selassie & D. F. Su (Eds.), *The postcranial anatomy of* Australopithecus afarensis: *New insights from KSD-VP-1/1* (pp. 155–178). Dordrecht: Springer.

Lovejoy, C. O., Meindl, R. S., Ohman, J. C., Heiple, K. G., & White, T. D. (2002). The Maka femur and its bearing on the antiquity of human walking: Applying contemporary concepts of morphogenesis to the human fossil record. *American Journal of Physical Anthropology, 119*, 97–133.

McHenry, H. M. (1978). Fore- and hindlimb proportions in Plio-Pleistocene hominids. *American Journal of Physical Anthropology, 49*, 15–22.

Melillo, S. M. (2016). Chapter 6: The shoulder girdle of KSD-VP-1/1. In: Y. Haile-Selassie & D. F. Su (Eds.), *The postcranial anatomy of* Australopithecus afarensis: *New insights from KSD-VP-1/1* (pp. 113–141). Dordrecht: Springer.

Meyer, M. R. (2016). Chapter 5: The cervical vertebrae of KSD-VP-1/1. In: Y. Haile-Selassie & D. F. Su (Eds.), *The postcranial anatomy of* Australopithecus afarensis: *New insights from KSD-VP-1/1* (pp. 63–111). Dordrecht: Springer.

Olson, T. R. (1981). Basicranial morphology of the extant hominoids and Pliocene hominids: The new material from the Hadar Formation, Ethiopia and its significance in early human evolution and taxonomy. In C. B. Stringer (Ed.), *Aspects of human evolution* (pp. 99–128). London: Taylor and Francis.

Olson, T. R. (1985). Cranial morphology and systematics of the Hadar Formation hominids and "*Australopithecus*" *africanus*. In E. Delson (Ed.), *Ancestors: The hard evidence* (pp. 102–119). New York: Liss.

Reno, P. L., Meindl, R. S., McCollum, M. A., & Lovejoy, C. O. (2003). Sexual dimorphism in *Australpithecus afaranesis* was similar to that of modern humans. *Proceedings of the National Academy of Sciences, USA, 100*, 9404–9409.

Richmond, B. G., & Jungers, W. L. (1995). Size variation and sexual dimorphism in *Australopithecus afarensis* and living hominoids. *Journal of Human Evolution, 29*, 229–245.

Robinson, J. T. (1972). *Early hominid posture and locomotion*. Chicago: University of Chicago Press.

Ryan, T. M., & Sukhdeo, S. (2016). Chapter 4: KSD-VP-1/1: Analysis of the postcranial skeleton using high-resolution computed tomography. In: Y. Haile-Selassie & D. F. Su (Eds.), *The postcranial anatomy of* Australopithecus afarensis: *New insights from KSD-VP-1/1* (pp. 39–63). Dordrecht: Springer.

Saylor, B. Z., Alene, M., Deino, A., Gibert, L., Haile-Selassie, Y., Melillo, S. M. et al. (2016). Chapter 2: The geologic context of Korsi Dora and the partial skeleton KSD-VP-1/1. In: Y. Haile-Selassie & D. F. Su (Eds.), *The postcranial anatomy of* Australopithecus afarensis: *New insights from KSD-VP-1/1* (pp. 13–23). Dordrecht: Springer.

Schmid, P. (1983). Eine rekonstruktion des skelettes von A.L. 288-1 (Hadar) und deren konsequenzen. *Folia Primatologica, 40*, 283–306.

Schmid, P. (1989). How different is Lucy? In G. Giacobini (Ed.), *Hominidae: Proceedings of the 2nd International Congress of Human Paleontology* (pp. 109–114). Milan: Jaca.

Schmid, P. (1991). The trunk of the australopithecines. In: Coppens Y, Senut B, (Eds.), *Origine(s) de la bipédie chez les hominidés. Cahier de Paléoanthropologie.* (pp. 225–234). Paris: CNRS.

Schmid, P., Churchill, S. E., Nalla, S., Weissen, E., Carlso, K. J., de Ruiter, D. J., & Berger, L. R. (2013). Mosaic morphology in the thorax of *Australopithecus sediba. Science, 340*, 1234598-1–1234598-5.

Senut, B. (1983). Les hominides Plio-Pleistocènes: essai taxinomique et phylogenetique a partir de certains os longs. *Bulletin Mémoire de la Société d'Anthropologie, Paris, 10*, 325–334.

Senut, B. (1986). Long bones of the primate upper limb: monomorphic or dimorphic? In M. Pickford & B. Chiarelli (Eds.), *Sexual dimorphism in living and fossil primates* (pp. 7–22). Florence: Il Sedicesimo.

Senut, B., & Tardieu, C. (1985). Functional aspects of Plio-Pleistocene hominid limb bones. In E. Delson (Ed.), *Ancestors: The hard evidence* (pp. 193–201). New York: Alan R. Liss.

Stern, J. T. (2000). Climbing to the top: A personal memoir of *Australopithecus afarensis. Evolutionary Anthropology, 9*, 113–133.

Stern, J. T., & Susman, R. L. (1983). The locomotor anatomy of *Australopithecus afarensis. American Journal of Physical Anthropology, 60*, 279–317.

Stern, J. T., & Susman, R. L. (1991). "Total morphological pattern" versus the "magic trait": Conflicting approaches to the study of early hominid bipedalism. In B. Senut & Y. Coppens (Eds.), *Origine(s) de la bipèdie chez les hominids* (pp. 99–111). Paris: CNRS.

Strait, D. S., & Grine, F. E. (2004). Inferring hominoid and early hominid phylogeny using craniodental characters: The role of fossil taxa. *Journal of Human Evolution, 47*, 399–452.

Su, D. F. (2016). Chapter 3: The taphonomy and paleoecology of Korsi Dora vertebrate locality 1, Woranso-Mille Study Area, Ethiopia. In: Y. Haile-Selassie & D. F. Su (Eds.), *The postcranial anatomy of* Australopithecus afarensis: *New insights from KSD-VP-1/1* (pp. 25–37). Dordrecht: Springer.

Susman, R. L., Stern, J. T., & Jungers, W. L. (1984). Arboreality and bipedality in the Hadar hominids. *Folia Primatologica, 43*, 113–156.

Susman, R. L., Stern, J. T., & Jungers, W. L. (1985). Locomotor adaptations in the Hadar hominids. In E. Delson (Ed.), *Ancestors: The hard evidence* (pp. 184–192). New York: Alan R. Liss.

Walker, A. (1973). New *Australopithecus* femora from East Rudolf, Kenya. *Journal of Human Evolution, 2*, 545–555.

Ward, C. V. (2002). Interpreting the posture and locomotion of *Australopithecus afarensis*: Where do we stand? *Yearbook of Physical Anthropology, 45*, 185–215.

Ward, C. V., Kimbel, W. H., & Johanson, D. C. (2011). Complete fourth metatarsal and arches in the foot of *Australopithecus afarensis. Science, 331*, 750–753.

White, T. D. (1977). New fossil hominids from Laetolil, Tanzania. *American Journal of Physical Anthropology, 46*, 197–230.

White, T. D. (1980). Additional fossil hominids from Laetoli, Tanzania: 1976–1979 specimens. *American Journal of Physical Anthropology, 53*, 487–504.

Wolpoff, M. H. (1983). Lucy's little legs. *Journal of Human Evolution, 12*, 443–453.

Zihlman, A. (1985). *Australopithecus afarensis*: two sexes or two species? In P. Tobias (Ed.), *Hominid evolution: Past, present and future* (pp. 213–220). New York: Alan R. Liss.

Zihlman, A., & Brunker, L. (1979). Hominid bipedalism: Then and now. *Yearbook of Physical Anthropology, 22*, 132–162.

Chapter 2
The Geologic Context of Korsi Dora and the Partial Skeleton KSD-VP-1/1

Beverly Z. Saylor, Mulugeta Alene, Alan Deino, Luis Gibert, Yohannes Haile-Selassie, Stephanie M. Melillo, and Gary Scott

Abstract KSD-VP-1/1, a partial skeleton of *Australopithecus afarensis*, was excavated from Pliocene strata at Korsi Dora, 3.3 km southeast of the confluence of the Waki and Mille rivers in the northwestern part of the Woranso-Mille paleoanthropological research site. A tuff collected from ~2.7 m below the fossil horizon, at the bottom of a trench dug 25 m to the east of the fossil excavation, yielded an $^{40}Ar/^{39}Ar$ age of 3.60 ± 0.03 Ma for anorthoclase feldspar. Strata in the trench and the fossil excavation site comprise a single normal magnetozone interpreted as part of the normal subchron C2An.3n, immediately above the Gauss/Gilbert paleomagnetic transition. Geologic mapping and tephrochemical analyses combined with paleomagnetic data place the fossil horizon and the trench section into local and regional stratigraphic context by constraining the partial skeleton to be younger than the Kilaytoli tuff (KT), a ~4 m thick vitric ash with an anorthoclase feldspar age of 3.570 ± 0.014 Ma. This unit is widely recognized at Korsi Dora, in collection areas north of the Waki-Mille confluence and outside the field area. The KT correlates with the Lokochot Tuff of the Omo-Turkana Basin in Kenya. Sedimentological features of the mudstone and sandstone in and near the excavation site are consistent with deposition in a floodplain or floodplain lake proximal to a stream channel.

Keywords Ar–Ar dating • Sedimentological features • Geological correlations • Korsi Dora • Woranso-Mille

B.Z. Saylor (✉)
Department of Earth, Environmental and Planetary Sciences, Case Western Reserve University, Cleveland, OH 44106, USA
e-mail: bzs@case.edu

M. Alene
Department of Geology and Geophysics, Addis Ababa University, Addis Ababa, Ethiopia
e-mail: mulugeta_alene@yahoo.com

A. Deino · G. Scott
Berkeley Geochronology Center, 2455 Ridge Road, Berkeley, CA 94709, USA
e-mail: adeino@bgc.org

G. Scott
e-mail: gscott@bgc.org

L. Gibert
Departament de Geoquímica, Petrologia i Prospecció Geològica, Universitat de Barcelona, C/Martí i Franquès s/n, 08028 Barcelona, Spain
e-mail: lgibert@ub.edu

Y. Haile-Selassie
Department of Physical Anthropology, Cleveland Museum of Natural History, 1 Wade Oval Drive, Cleveland, OH 44106, USA
e-mail: yhailese@cmnh.org

S.M. Melillo
Department of Human Evolution, Max Planck Institute for Evolutionary Anthropology, Deutscher Platz 6, 04103 Leipzig, Germany
e-mail: stephanie_melillo@eva.mpg.de

Introduction

KSD-VP-1/1, a partial skeleton of *Australopithecus afarensis*, was excavated from the Korsi Dora collection area in the northwestern part of the Woranso-Mille paleoanthropological study area (WORMIL). WORMIL covers an area of ~60 km by ~20 km near the western margin of the Afar depression, south of the Gura'ale volcanic center and north of the road between the towns of Mille and Bati, in northern Ethiopia (Fig. 2.1). The area was first surveyed for fossils during the early 1970s, but detailed geological and paleontological research did not begin until 2004 (Haile-Selassie et al. 2007). Geological investigations to date have focused on the northern part of the WORMIL site, where Pliocene strata are exposed in the drainage basin of the Mille River (Fig. 2.1c). The Korsi Dora collection area consists of low relief exposures of fossiliferous Pliocene strata adjacent to the Kilaytoli River, an ephemeral drainage that feeds into the Mille River.

KSD-VP-1/1 was excavated from ~49 square meters of flat lying mudstone and sandstone. The fossil horizon is ~2.6 m above a lapillistone tuff with an $^{40}Ar/^{39}Ar$ age of

Yohannes Haile-Selassie and Denise F. Su (eds.), *The Postcranial Anatomy of Australopithecus afarensis: New Insights from KSD-VP-1/1*, Vertebrate Paleobiology and Paleoanthropology, DOI 10.1007/978-94-017-7429-1_2

3.60 ± 0.03 Ma (2σ). The dated tuff was collected from the bottom of a trench 25 m to the north of the excavation (Haile-Selassie et al. 2010a). We expand on the results of geologic mapping and tephrochemical studies that, together with paleomagnetic data and radiometric ages, clarify the stratigraphic and structural relationships at Korsi Dora and the position of the fossil horizon relative to marker tuffs (Haile-Selassie et al. 2010a; Deino et al. 2010). These marker tuffs enable correlation of the Korsi Dora section with a longer, well-dated composite section comprising exposures near the confluence of the Waki and Mille rivers, 3.3 km northwest of Korsi Dora, and also with comparably

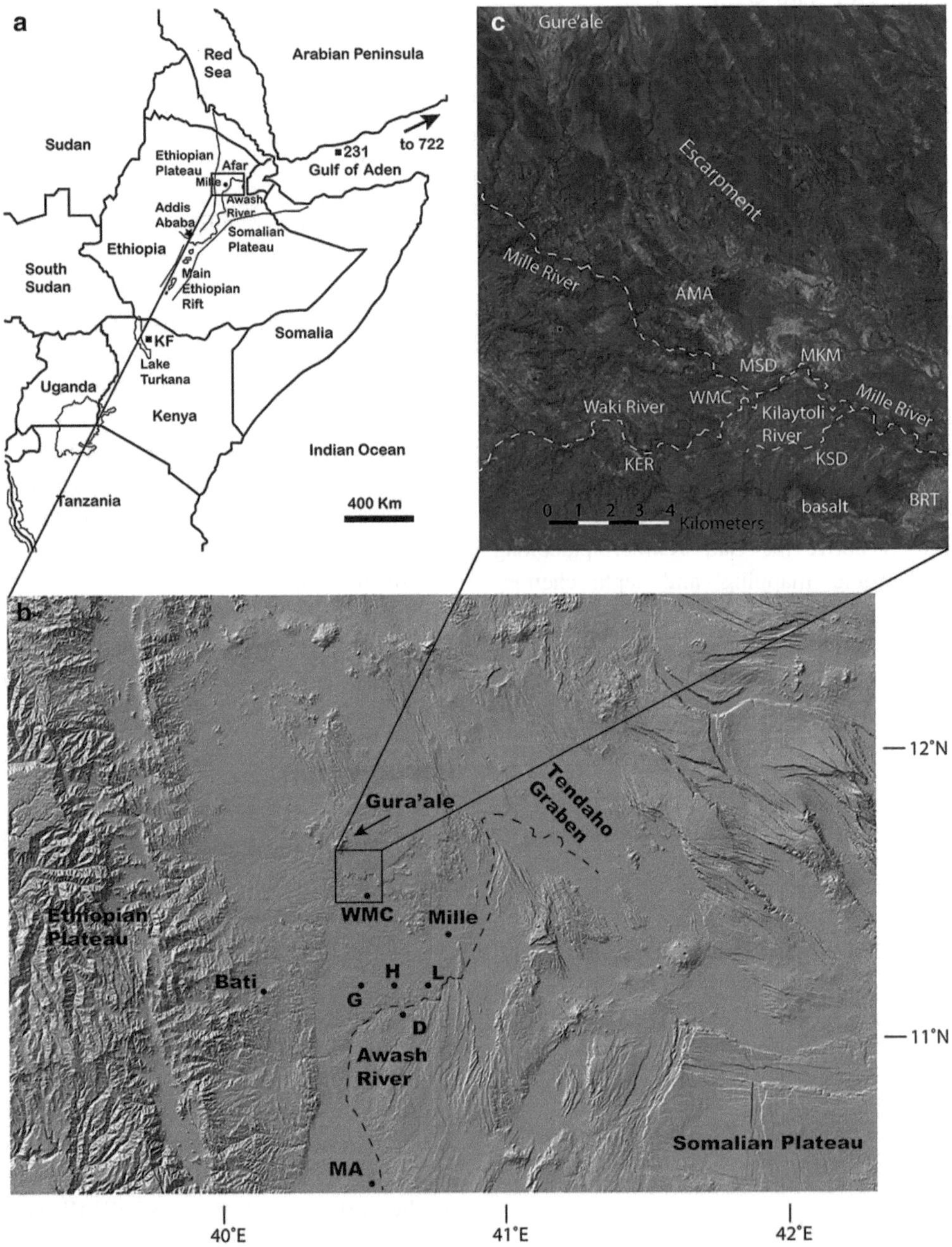

Fig. 2.1 **a** Map of the Horn of Africa and locations discussed in the text. *KF* Koobi Fora Formation; 722 and 231 are locations of deep sea cores where correlates of the Lokochot Tuff have been identified. **b** Relief map of the southwestern region of the Afar depression in Ethiopia showing the location of the Waki-Mille Confluence (WMC) in the Woranso-Mille research area relative to physiographic and tectonic features and relative to paleoanthropological research areas of the Lower and Middle Awash Valley. Base map generated from GeoMapApp (http://www.geomapapp.org) using the Global Multi-Resolution Topography Synthesis (Ryan et al. 2009). *D* Dikika; *G* Gona; *H* Hadar; *L* Ledi-Geraru; *MA* Middle Awash. **c** Map of the area around the Waki-Mille confluence (WMC) in the northern part of the Woranso-Mille study area showing the location of Korsi Dora (KSD) relative to other collection areas discussed in the text. *AMA* AmAdo; *MKM* Makah Mera; *MSD* Mesgid Dora; *KER* Kerare. Base map is a grayscale ASTER image

aged fossiliferous strata and deep-sea cores around the Horn of Africa. We also present details of the sedimentology of the excavation site and nearby exposures that provide insight into the environment of deposition and preservation of the partial skeleton.

Geologic Setting of the Woranso-Mille Study Area

WORMIL lies near the western margin of the Afar triangle (Fig. 2.1a), near the triple junction where the Main Ethiopian Rift meets the on-land propagators of the Red Sea and Gulf of Aden spreading ridges (Beyene and Abdelsalam 2005). WORMIL lies 75 km southwest of the Tendaho Graben (Fig. 2.1b), the current locus of the Red Sea propagator, where NNE-oriented normal faults associated with the Main Ethiopian Rift intersect NW-oriented faults associated with the Red Sea Rift (Acocella 2010). The Main Ethiopian Rift reached the Afar about 11 Ma. Extension along the Red Sea Rift, which began about 30 Ma, propagated into the Tendaho Graben at about 4 Ma, accompanied by a resurgence of volcanism central to rift axes, including the development of silicic volcanoes, and the emplacement of the Afar Series of basalts across much of the Afar (Acocella et al. 2008; Lahitte et al. 2003).

WORMIL shares its southern border, the Mille-Bati road, with paleoanthropological sites of the Lower Awash Valley (Fig. 2.1b). Decades of work by interdisciplinary field teams south of this border have extracted a rich fossil record from the Late Miocene to Pleistocene strata of the Awash Group (Kalb et al. 1982; Renne et al. 1999; WoldeGabriel et al. 2009). These strata, consisting of the Adu Asa (6.4 to 5.2 Ma), Sagantole (>4.6 to 3.9 Ma), Hadar (>3.8 to 2.9 Ma), and Busidima (2.7 to 0.16 Ma) Formations, filled half grabens that developed in response to the Main Ethiopian Rift extension and may also have been influenced by the Red Sea Rift (Quade et al. 2008; Wynn et al. 2008). Basin-bounding faults, inferred for the early stages of rift evolution, were likely discontinuous, forming structurally isolated basins. By the time of deposition of the Busidima Formation, however, faults had linked and localized along the western edge of a contiguous basin leading to axial drainage and the

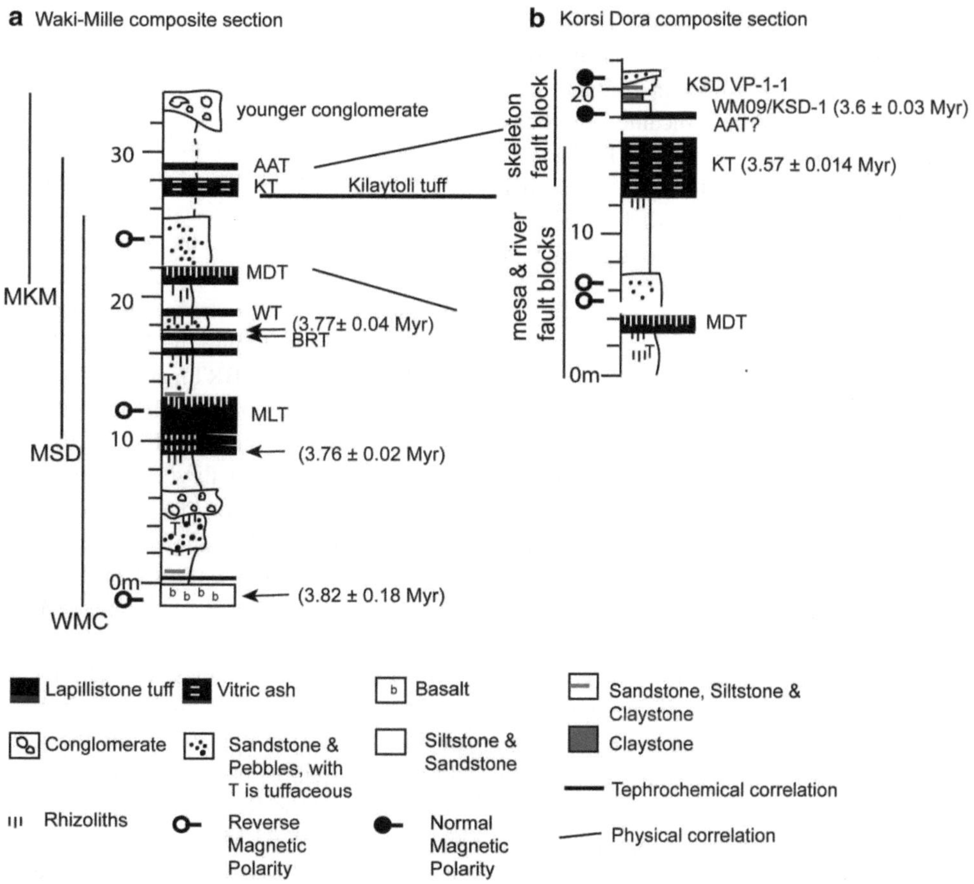

Fig. 2.2 Composite reference sections for the stratigraphy at (**a**) the Waki-Mille area, including the Waki-Mille confluence (WMC), Mesgid Dora (MSD) and Makah Mera (MKM), and (**b**) Korsi Dora (KSD). *MLT* Mille tuff sequence; *BRT* basalt-rhyolite tuff; *WT* Waki tuff; *MDT* Mesgid Dora tuff; *KT* Kilaytoli tuff; *AAT* AmAdo tuff sequence

development of the proto-Awash River (Quade et al. 2008). Thus, in addition to their rich record of Miocene–Pliocene and Pleistocene vertebrate evolution and environmental change, the formations of the Awash Group also document the volcanic and tectonic history of rifting and landscape evolution near the Afar triple junction.

The skeleton excavation site at Korsi Dora is part of a Pliocene age volcanic and sedimentary succession, exposed most extensively north of the Mille River (Deino et al. 2010). Exposures at the Waki-Mille confluence and nearby, at the Mesgid Dora and Makah Mera collection areas, constitute a well-dated composite section (Fig. 2.2), >30 m thick, of sandstone, mudstone, and conglomerate, interbedded with decimeter- to meter-scale volcanic tuffs ranging in age from ≥3.77 Ma to <3.57 Ma (Deino et al. 2010). At the confluence, the sedimentary section overlies 3.82 ± 0.18 Ma basalt (Deino et al. 2010), but more recent fieldwork has demonstrated that the tuffaceous and sedimentary strata continue below the basalt west of the confluence. Recent work has also documented the presence of additional layers of basalt within the sedimentary succession, above the confluence basalt (Alene et al. 2012). The sedimentary succession in the vicinity of the Waki-Mille confluence overlaps in age with the Basal Member of the Hadar Formation, especially as defined at Dikika (Wynn et al. 2006, 2008), but differs from the Basal Member in containing an abundance of basaltic and rhyolitic volcanic interbeds. Although tentatively considered as part of the Hadar Formation (Deino et al. 2010), it remains to be determined if strata in the Waki-Mille area, more than 40 km north of Hadar and Dikika, accumulated in a basin contiguous with that of the Basal Member.

Korsi Dora Collection Area

The Korsi Dora collection area borders the Kilaytoli River, ~2.8 km southeast of the Waki-Mille confluence. The area is named for the Korsi Dora drainage (Fig. 2.3), one of several drainages that initiate along an east–west oriented ridge of basalt, south of the Kilaytoli River, and feed northward into the Kilaytoli River or directly into the Mille River. The basaltic ridge, which extends for >10 km laterally, forms a southern boundary to exposures of tephra-rich strata typical of the Waki-Mille confluence and the surrounding area (Fig. 2.1c). The basalt is partially covered by younger conglomerate, which also covers much of Korsi Dora, obscuring the contact between basalt and topographically lower fossiliferous strata (Fig. 2.3).

For the most part, exposures of fossiliferous strata at Korsi Dora are limited to low relief outcrops, <2 m thick, where erosion along the Korsi Dora drainage and its tributaries removed younger conglomeratic cover (Figs. 2.3 and 2.4). Thicker sections, up to 6 m, are exposed along the banks of the Kilaytoli River and on a small mesa near the river.

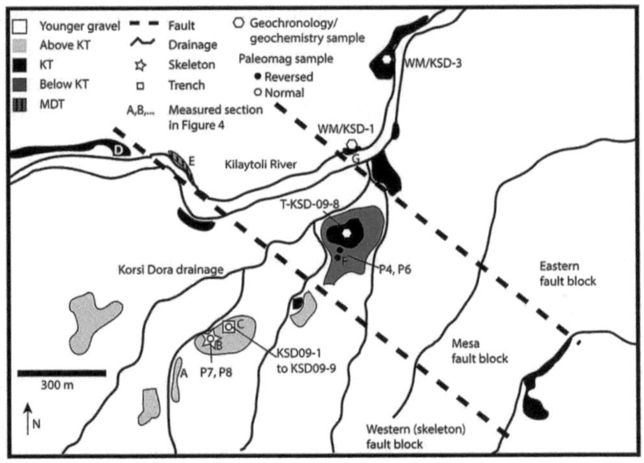

Fig. 2.3 Geologic map of the Korsi Dora area showing locations of samples and sections discussed in the text and shown in Fig. 2.4

Stratigraphic Position of the Basalt Ridge

The approximate stratigraphic position of the basalt ridge at the southern boundary of Korsi Dora can be determined from relationships at Korsi Dora and at other nearby collection areas. Basalt in the ridge extends westward to the Kerare collection area (Fig. 2.1c) where it lies above >20 m of volcaniclastic-rich strata similar to the Waki-Mille composite section, including the Mille tuff sequence (MLT), the Basalt Rhyolite tuff (BRT), and the Waki tuff (WT) or higher (Fig. 2.2). Stratigraphically higher tuffs that are not present in the Kerare section may have been removed by erosion along a >10 m deep fluvial channel, which underlies the basalt and cuts down from a position above, at least, the WT to the level of the MLT sequence. At the eastern limits of the ridge, at Burtele, the exposure of basalt curves to the north and dips eastward under a younger (≤3.47 Ma) fossiliferous succession containing the hominin partial foot fossil BRT-VP-2/73 (Haile-Selassie et al. 2012).

These relationships constrain the basalt at Korsi Dora to a position above the WT and below the section at Burtele, but do not constrain its position relative to the Korsi Dora strata, which contain the Kilaytoli tuff (KT) and thus correlate with the part of the Waki-Mille composite section that is above the WT (Deino et al. 2010). There is, however, no evidence for strata comparable to the Korsi Dora section above the basalt ridge at Burtele or elsewhere (Haile-Selassie et al. 2012).

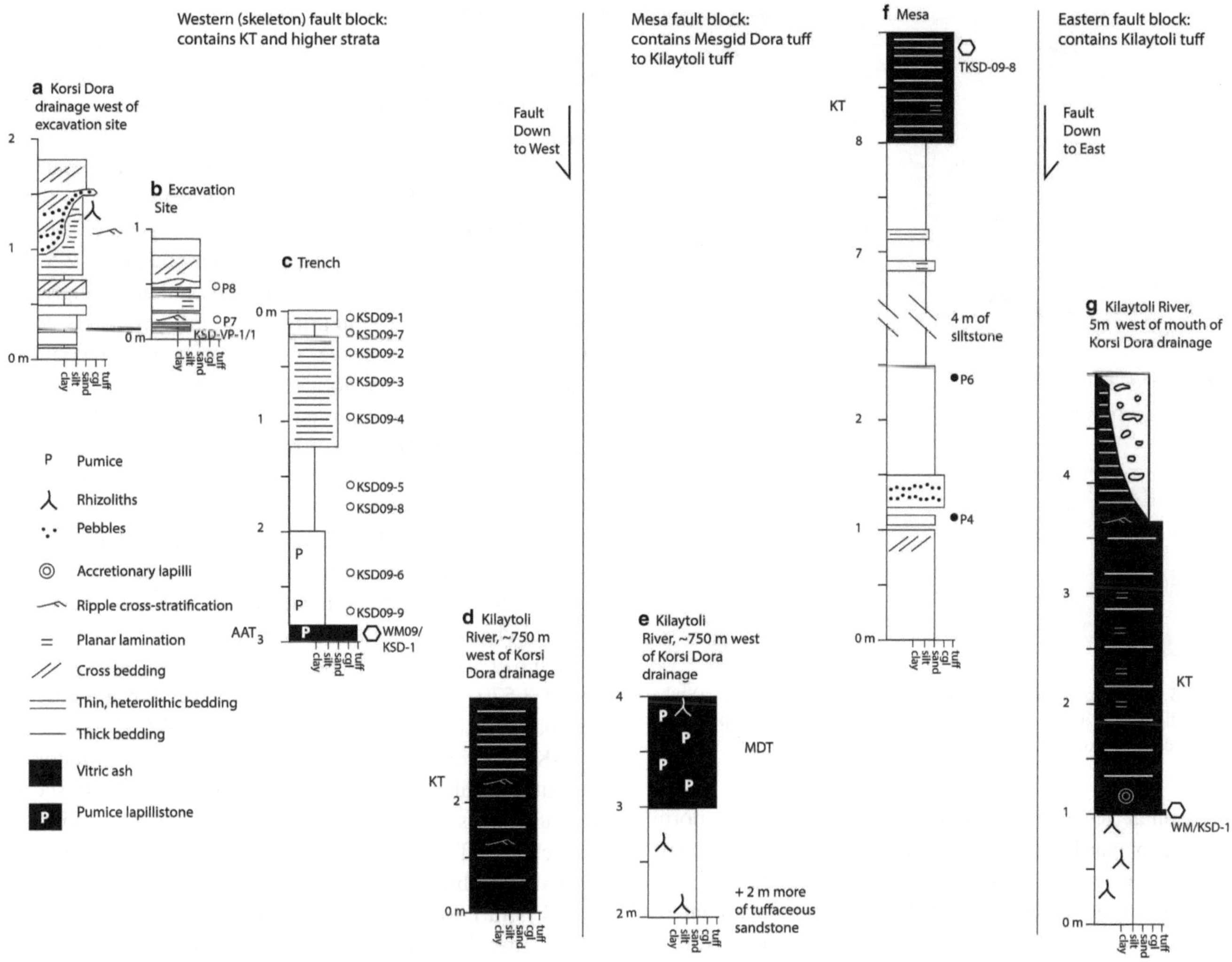

Fig. 2.4 Measured stratigraphic sections at Korsi Dora showing positions of the samples discussed in the text. For locations of sections and samples see Fig. 2.3

Also, although faults are inferred along offset linear features that define the edge of the basalt ridge, their vertical offsets appear small. The simplest interpretation is that the Korsi Dora section and the Waki-Mille composite section lie stratigraphically below the level of the Kerare-to-Burtele ridge of basalt and are older than the dated tuff at Burtele (3.460 ± 0.016 Ma, 2σ) (Haile-Selassie et al. 2012).

Kilaytoli Tuff

The Kilaytoli tuff (KT) is a white vitric ash named for exposures along the Kilaytoli River and present at several locations around Korsi Dora (Figs. 2.3 and 2.4). It is also recognized in collection areas north of the Mille River (Fig. 2.2) based on physical characteristics, stratigraphic position, $^{40}Ar/^{39}Ar$ geochronology, and glass geochemistry (Deino et al. 2010).

Where it was first described in an outcrop along the northern bank of the Kilaytoli River, 5 m west of the mouth of the Korsi Dora drainage, the KT sits on siltstone with rhizoliths (Figs. 2.4g and 2.5a). It is as much as 4 m thick but decreases to 2.6 m over the span of a few meters, truncated from above by an erosional unconformity with a younger conglomerate. A coarse crystal concentrate layer, 1 cm thick, at the base of the tuff is overlain by ~2.6 m of thin- to thick-bedded, medium to fine vitric ash, with concentrations of accretionary lapilli, followed by ~1.4 m of recessive weathering, thin- to medium-bedded fine ash with silt. Beds

Fig. 2.5 Outcrop photos of the Kilaytoli tuff on Korsi Dora (**a**) where it was first described in the Kilaytoli River and (**b**) on top of mesa. Photos courtesy of Elizabeth Russell

are tabular to locally trough shaped and exhibit normal grading, planar horizontal laminations, climbing ripple cross stratification, and local soft sediment deformation.

North of the Mille River, the KT has been documented at Mesgid Dora, Makah Mera, Aralee Issie, and AmAdo, where it varies from 1 to 3 m in thickness and is similar in characteristics to outcrops at Korsi Dora. It sits on siltstone that has been pedogenically modified to varying degrees. Stratigraphically, it ranges from 5 to 15 m above the MDT, a 1 to 3 m thick pumice lapillistone tuff, which is recognized in multiple collection areas and, depending on location, accumulated as either an airfall tuff or a fluvially reworked pumiceous deposit (Deino et al. 2010) (Fig. 2.2a).

At the Waki-Mille confluence and across much of Mesgid Dora, the section is truncated below the level of the KT, whereas at Aralee Issie and across much of Makah Mera, the KT is the highest preserved tuff in the section. At AmAdo, however, the section continues higher to include a pink pumice, lithic lapillistone that is 25 cm thick and 5 m above the top of the KT (Haile-Selassie et al. 2007, 2010b). Later work showed that the tuff reached as much as a meter thick in the AmAdo area and a second similar tuff was identified between it and the KT. These tuffs, which we will here refer to as the AmAdo tuff (AAT) sequence, were initially recognized only at AmAdo and were not named. More recently, however, a similar tuff has been identified 1–2 m above the KT in flats and hilltop exposures between Mesgid Dora and Makah Mera (Fig. 2.2a) and a similar pair of tuffs, which is present above the KT at the confluence of the Kilaytoli River and the Mille River. Furthermore, expansion of fieldwork has identified a physically similar pink, lithic, pumice lapillistone tuff above the KT in exposures along a laterally extensive, basalt-capped escarpment, ~2.5 km northeast of AmAdo (Fig. 2.1c). Based on their stratigraphic position and physical similarity, these tuff exposures correlate with the AAT sequence, which can now be recognized as a widespread marker horizon above the KT.

Deino et al. (2010) reported electron probe microanalysis (EPMA) data for glass shards in samples of the KT from the Kilaytoli River, Makah Mera, and Aralee Issie (Table 2.1), all of which exhibit bimodal distributions of Fe_2O_3 and Al_2O_3 concentration (Fig. 2.6). This distinctive distribution, along with a close similarity in all other measured oxides, supports the correlations made among sections based on physical characteristics and stratigraphic position.

Deino et al. (2010) reported a laser fusion $^{40}Ar/^{39}Ar$ mean age of 3.570 ± 0.014 Myr (2σ) for the KT, for anorthoclase feldspar crystals from two samples (WM-KSD-1 and WM-KSD-3) of the crystal concentrate layer at the bottom of the outcrops along the Kilaytoli River (Fig. 2.3). Based on age and glass geochemistry, the KT correlates with the Lokochot Tuff (Deino et al. 2010) and has extensive extra-basinal distribution. The Lokochot Tuff, defined in the Koobi Fora Formation east of Lake Turkana, has been correlated geochemically to tuffs at multiple sites of the Omo-Turkana Basin of Kenya and southern Ethiopia (Cerling and Brown 1982; Brown and Fuller 2008), and has also been recognized in the Gulf of Aden and the Arabian Sea (Fig. 2.1a) (Brown et al. 1992; DeMenocal and Brown 1999; Feakins et al. 2007). It is one of several tuffs from the Omo-Turkana Basin that are recognized in the Afar and in deep-sea drill cores around the Horn of Africa as the products of massive volcanic eruptions.

Here we report additional EPMA data for vitric ash (T-KSD-9-8) from the top of the mesa at Korsi Dora. Sample preparation and analysis followed Deino et al. (2010). The white, planar laminated, fine to medium grained ash, which is >1 m thick, is similar in physical characteristics to the KT (Figs. 2.4f and 2.5b) and, like the previously studied exposures, exhibits the characteristic bimodal distributions of Fe and Al oxide abundances. These patterns, along with similarities of other oxide abundances and physical characteristics, support identification of the vitric ash on top of the mesa and elsewhere at Korsi Dora as different exposures of the KT.

Table 2.1 Normalized wt% and 1σ Normalized wt% and 1 ms compositions for the KT

	N	Na_2O	MgO	Cl	CaO	K_2O	SiO_2	Al_2O_3	TiO_2	MnO	Fe_2O_3	Total	Total O*
WM-KSD-1													
Low Fe	11	2.67	0.04	0.14	0.19	5.11	77.14	11.34	0.19	0.07	3.11	99.88	92.99
1σ		0.52	0.02	0.02	0.01	0.30	1.03	0.12	0.06	0.03	0.17		
WM-KSD-3													
Low Fe	8	2.42	0.02	0.13	0.20	4.53	78.04	11.27	0.17	0.09	3.13	99.90	92.66
1σ		0.34	0.01	0.02	0.02	0.36	0.66	0.21	0.03	0.03	0.14		
High Fe	11	2.30	0.04	0.19	0.21	4.41	77.38	10.59	0.28	0.15	4.45	99.67	92.76
1σ		0.33	0.03	0.02	0.02	0.44	0.97	0.15	0.06	0.04	0.31		
ARI-08-5													
Low Fe	13	2.34	0.03	0.13	0.20	5.01	77.32	11.24	0.22	0.12	3.40	99.82	92.82
1σ		0.44	0.01	0.02	0.02	0.56	0.78	0.25	0.08	0.04	0.41		
High Fe	5	1.77	0.04	0.17	0.22	4.40	77.86	10.64	0.27	0.14	4.48	99.63	91.57
1σ		0.25	0.02	0.05	0.04	0.39	0.46	0.29	0.02	0.03	0.30		
MSD-08-13													
Low Fe	9	2.02	0.04	0.12	0.19	4.84	77.92	11.41	0.20	0.09	3.16	99.89	92.76
1σ		0.30	0.02	0.02	0.03	0.38	0.65	0.19	0.07	0.03	0.27		
High Fe	3	1.53	0.04	0.19	0.19	4.21	78.75	10.19	0.22	0.15	4.53	100.00	93.26
1σ		0.37	0.03	0.02	0.02	0.60	0.97	0.25	0.05	0.02	0.08		
AMA-09-4													
Low Fe	4	2.26	0.03	0.14	0.20	6.13	76.57	11.39	0.17	0.13	2.99	100.63	93.41
1σ		0.68	0.01	0.01	0.03	0.31	0.44	0.10	0.05	0.05	0.22		
High Fe	5	2.21	0.03	0.18	0.19	5.54	76.41	10.71	0.21	0.15	4.36	100.12	92.61
1σ		0.75	0.01	0.01	0.02	0.66	0.79	0.22	0.04	0.05	0.31		
T-KSD 09-8													
Low Fe	2	2.63	0.03	0.14	0.20	4.52	78.14	11.04	0.18	0.06	3.07	100.08	91.13
1σ		0.86	0.01	0.02	0.02	0.45	0.38	0.05	0.01	0.04	0.00		
High Fe	2	2.12	0.05	0.20	0.22	4.11	77.98	10.30	0.19	0.19	4.64	98.98	89.23
		0.16	0.06	0.03	0.05	0.06	0.85	0.23	0.12	0.03	0.13		

F abundances are below detection limits. Different matrix corrections for F yielded small differences from oxide totals previously reported in Deino et al. (2010). * indicates new data. AMA-09-4 and T-KSD-09-8 are new data

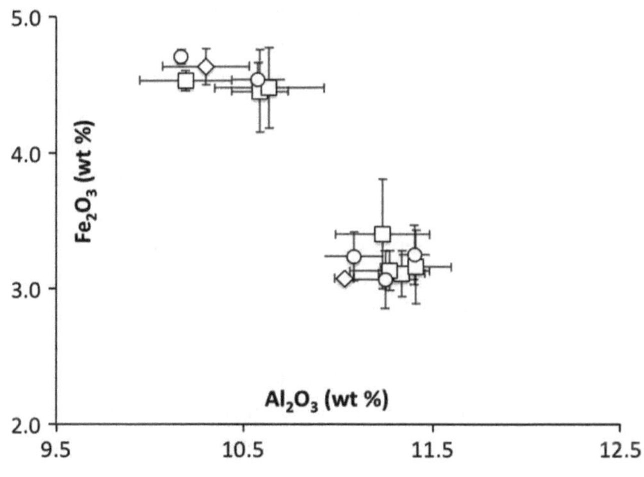

Fig. 2.6 Glass tephrochemical data (normalized, average values) for the Kilaytoli tuff and for the Lokochot Tuff of the Omo-Turkana basin showing bimodal average Fe and Al oxide abundances. Squares represent values for previously published data for samples from the Kilaytoli River and from north of the Mille River. Circles represent values for a sample of the Lokochot tuff measured at the same time as the Kilaytoli tuff. Diamonds represent new data for a sample of Kilaytoli tuff from on top of the Korsi Dora mesa

Structural Geology

The KT is an excellent chronostratigraphic marker, useful for identifying stratigraphic relationships and post-depositional structural movements of the Korsi Dora collection area (Fig. 2.3). In addition to its original description site, 5 m west of the mouth of the Korsi Dora drainage (Fig. 2.4g), the KT crops out extensively, though discontinuously, eastward, for over 1 km along the Kilaytoli River. To the west, along the Kilaytoli River, however, a span of >550 m separates the cluster of KT outcrops near the mouth of the Korsi Dora drainage from the next closest exposure of KT. There, an outcrop of KT (Fig. 2.4d) sits <5 m to the west of a short section containing a 1 m thick pumice lapillistone tuff (Fig. 2.4e). The lapillistone tuff, interpreted as MDT, has been offset along a fault to the same topographic level as the KT. South of the Kilaytoli River, the KT on top of the mesa (Fig. 2.4f) is topographically above the nearby river exposures of the tuff (Fig. 2.4g), offset by more than 6 m by a fault. Isolated bedding plane exposures of white vitric ash, similar to the KT, are also present in other drainages across the Korsi Dora area, including a drainage 700 m southeast of the mesa and a small drainage between the mesa and the excavation site (Fig. 2.3).

Topographic offset of the KT from the top of the mesa to exposures in the Kilaytoli River is the result of high angle normal faults crossing the Korsi Dora locality (Fig. 2.3). These faults contain the mesa in an uplifted horst. Exposures of KT north and east of the mesa are dropped down to the northeast, relative to the horst, along a fault that must pass between the mesa and the tuff exposures near the Korsi Dora mouth. Similarly, exposures of the KT far west of the Korsi Dora mouth, along the Kilaytoli River, are dropped down to the southwest relative to the mesa block. Placement of this second, more westerly, fault is constrained to lie between exposures of KT and MDT. Bedding plane exposures of the KT in the drainage southeast of the mesa are interpreted as part of the mesa block. By connecting the fault traces between the river exposures and the limits of these bedding plane exposures we constrain the fault trace to be approximately parallel at N 57° W. This orientation also parallels, approximately, the traces of other faults in the area.

These structural relationships place the mesa in the middle of three fault blocks. The pumice lapillistone tuff in the Kilaytoli River and the adjacent >550 m tuff-free stretch of the river are part of the mesa fault block, the relative uplift of which has exposed strata beneath the KT. The pumice lapillistone tuff, thus, is in stratigraphic position comparable to that of the MDT of the Waki-Mille composite section and, given physical similarities, is interpreted as an exposure of this tuff.

The skeleton excavation site and the trench section lie in the westernmost of the three fault blocks. This block contains exposures of the KT, specifically in the Kilaytoli River, far west of the mouth of the Korsi Dora drainage, and in a small drainage east of the trench and excavation site, but the trench section and excavation site are isolated from these exposures by younger cover (Figs. 2.3 and 2.4). Still, because the skeleton fault block is dropped down relative to the mesa block, which is capped by KT, it is expected that much of the exposures in this western block, including the excavation site, are at a stratigraphic level above that of KT.

Sedimentology and Depositional Setting

The cumulative thickness of Pliocene strata at Korsi Dora is at least 21 m. The section includes three tuffs: the KT, which is a vitric ash; a pumice lapillistone tuff, more than 8 m below the KT, which is similar in character and stratigraphic position to the MDT; and an altered lithic, pumice lapillistone tuff in the trench, which is interpreted to sit above the KT and is similar in character and stratigraphic position to the AAT. Tephrochemical correlation of the KT combined with physical and stratigraphic similarity with other tuffs support correlation of the Korsi Dora stratigraphy to the upper part of the Waki-Mille composite section, from a few meters below the level of the MDT up to a few meters above the AAT (Fig. 2.2). The stratigraphic position of the skeleton corresponds to a position above the AAT, the highest recognized tuff in the Waki-Mille composite section. Because this stratigraphic level is preserved only locally in the area, at Korsi Dora and perhaps at AmAdo and along the escarpment northeast of AmAdo, it is not yet possible to document lateral variations in facies across the region. Interpretation of the depositional environment for the skeleton must rely on local sedimentology and stratigraphy.

Sedimentary, primarily nonvolcanic, lithologies of the Korsi Dora section include sandstone, heterolithic sandstone and mudstone, siltstone, and claystone. Sandstone is present near the bottom of the Korsi Dora section, above and below the MDT (Fig. 2.4e, f), and near the top of the Korsi Dora section, above the level of the skeleton (Fig. 2.4a–c). It is medium- to thick-bedded and medium- to coarse-grained, with pebbles or tuffaceous material locally. Beds exhibit planar horizontal stratification as well as tabular and trough cross-stratification. Heterolithic sandstone and mudstone is best developed in the upper part of the Korsi Dora section, above and below the claystone layer that yielded the skeleton. It is very thin- to medium-bedded, consisting of claystone, siltstone, and fine to medium sandstone, and exhibits planar lamination and ripple cross lamination. Siltstone with minor interbeds of sandstone predominates in the mesa section between the MDT and the KT. Claystone without heterolithic interbeds is present only in the trench section.

The skeleton was excavated from a ~10 cm thick claystone layer within a section of heterolithic sandstone, siltstone, and claystone. In the trench, the heterolithic section extends about 1 m below the skeleton horizon to overlie a layer of claystone, approximately 75 cm thick. Below this claystone is about 80 cm of siltstone with dispersed pumice, underlain by altered lapillistone tuff, at the bottom of the trench. Above the skeleton horizon is about 1 m of heterolithic sandstone and mudstone that is exposed extensively in the Korsi Dora drainage and other feeder drainages west of the excavation site. Sandstone and siltstone beds have flat, non-erosive bases and, in some cases, ripple and dune forms on their tops, preserved at the base of overlying beds of claystone. Rhizoliths are rare, but locally present in sandstone beds near the top of the heterolithic section. Claystone and siltstone in the fossil horizon and elsewhere in the heterolithic assemblage contain dispersed, ~1 cm diameter, nodular, displacive evaporite minerals, likely to be gypsum. A channel surface with 75 cm of erosional relief cuts through the heterolithic sandstone and mudstone and is filled in by sandstone and pebble conglomerate, which forms an upward-fining succession composed of amalgamated trough cross-sets.

The interbedding of claystone, siltstone, and sandstone in the heterolithic section that contains the fossil horizon is indicative of episodic flow events followed by periods of quiet settling of sediment and non-accumulation. The presence of rhizoliths, although rare, indicates that between at least some of the flow events the area was colonized by plants. Whereas, the flat bed bases and the preservation of ripple and dune forms on bed tops are evidence that the episodic flows were waning, as typically happens where channelized flows spread out as they top the channel banks or otherwise enter a less restricted body. The displacive character of probable gypsum nodules in the skeletal horizon and other mudstone beds requires a diagenetic origin, rather than precipitation from the water column or at the sediment–water interface. It may be a modern diagenetic artifact, but it also could have formed early, by evaporative pumping of saline pore water through the sediment. The channelized, cross-bedded, pebbly sandstone above the heterolithic section is typical of fluvial deposition. The thick claystone below the heterolithic section, however, requires a sustained period of quiet accumulation more consistent with a lake or pond, or the distal part of a floodplain. Taken together, the sedimentological features of the mudstone and sandstone in and near the excavation site are interpreted as evidence for deposition in a floodplain or ephemeral floodplain lake, proximal to a stream channel.

The Age of KSD-VP-1/1

KSD-VP-1/1 was excavated from claystone at the bottom of a 0.9 m thick section of interbedded claystone, siltstone, and sandstone (Fig. 2.4b). The fossil horizon correlates with claystone near the top of a 3 m deep trench dug 25 m to the east of the excavation site (Fig. 2.4c). Haile-Selassie et al. (2010a) reported a laser fusion $^{40}Ar/^{39}Ar$ age of 3.60 ± 0.03 Myr (2σ) for anorthoclase feldspar from a >20 cm thick tuff (WM09/KSD-1) at the bottom of the trench. The tuff is an altered, lithic, pumice lapillistone, the bottom of which was not exposed by excavation. Paleomagnetic directions reported for the trench section are normal, as are directions reported for two samples (P7 and P8) from the excavation site (Fig. 2.4b), but reverse paleomagnetic directions were reported for two samples (P4 and P6) from the lower 2.5 m of the mesa section (Fig. 2.4f) (Deino et al. 2010; Haile-Selassie et al. 2010a).

The trench section (Fig. 2.4c) and the fossil excavation site (Fig. 2.4b) are interpreted as being part of the normal subchron C2An.3n, immediately above the Gauss/Gilbert paleomagnetic transition (3.596 Ma, ATNTS2004). The reverse paleomagnetic polarities in the lower part of the mesa (Fig. 2.4f) are interpreted as the continuation of a reverse magnetozone documented for the Waki-Mille composite section (Fig. 2.2), part of the reverse subchron C2Ar below the Gauss/Gilbert transition. Paleomagnetic results for a sample of the KT on top of the mesa were variable and not useful for determining direction, but Brown et al. (1978) and Hillhouse et al. (1986) reported reverse polarities for the Lokochot Tuff in the Koobi Fora Formation and its geochemical correlate, Tuff A, in the Shungura Formation. Because the KT is correlated, based on age and geochemistry, with the Lokochot Tuff (Deino et al. 2010), it is expected that the top of the mesa also falls within the reverse subchron C2Ar.

$^{40}Ar/^{39}Ar$ geochronology is ambiguous regarding the relative placement of the KT and the tuff in the trench, as the radiometric ages are statistically indistinguishable (3.570 ± 0.014 Ma and 3.60 ± 0.03 Ma, respectively). We do not have paleomagnetic data for the trench tuff and geochemical alteration prevents glass tephrochemistry. Still, physical differences differentiate the lithic lapillistone tuff in the trench from the fine vitric ash that constitutes the KT. The simplest interpretation is that the tuff at the bottom of the trench is conformable with the rest of the trench section and that it is positioned stratigraphically above the KT. This interpretation places the trench tuff in a similar stratigraphic position to the AAT, with which it shares physical characteristics and may correlate.

The paleomagnetic evidence positioning the mesa section and the KT stratigraphically below the trench section and the fossil excavation site is consistent with the structural evidence placing the skeletal excavation location and trench in a fault block that is dropped down relative to the mesa. This placement is also consistent with stratigraphic evidence, in that beds of sandstone and pebble conglomerate in the skeleton excavation site and in laterally equivalent exposures nearby (Fig. 2.4a) are not observed above the reverse magnetozone in the mesa, or anywhere else at Korsi Dora immediately below the KT. Similarly, there is no evidence for the KT above the level of the excavation site. We conclude that a maximum age for the fossil skeleton is given by the age of the KT tuff, 3.570 ± 0.014 Ma. A minimum age is provided by the Burtele tuff (3.469 ± 0.008 Ma), which sits stratigraphically above the basalt ridge that borders Korsi Dora. Thus, the potential depositional interval for the skeleton is a time window of about 100 ka in duration.

The depositional interval can be narrowed by applying average local sediment accumulation rates to the trench section. The trench tuff, WM09/KSD-1, has a radiometric age of 3.60 ± 0.03 Ma, but the skeleton is inferred to have a position stratigraphically above the KT, with a more precise age of 3.570 ± 0.014 Ma. Using the more precise age and sediment accumulation rates of 11 cm/ka in the Waki-Mille confluence area and 30 cm/ka in the Hadar Formation (Campisano and Feibel 2007) yields an estimated age for the skeleton of 3.56 to 3.54 Ma. This age estimate differs little from a previous estimate of 3.58 Ma based on the age of the trench tuff (Deino et al. 2010), but the structural and stratigraphic relationships clarify the position of the skeleton relative to marker tuffs and most significantly to the KT and its regional correlate, the Lokochot Tuff.

Conclusions

Pliocene strata at Korsi Dora contain distinctive marker tuffs, including the KT, which is present extensively in the WORMIL collection areas along the Mille River, near the mouth of the Waki River. The KT has been correlated with the Lokochot Tuff of the Turkana Basin based on radiometric ages and glass geochemistry. Topographic offset of the KT enables identification and mapping of northwest–southeast oriented normal faults that transect Korsi Dora, forming a horst. The partial skeleton, KSD-VP-1/1, is contained within a fault block to the west of the horst and is dropped down relative to it. Strata in the fossil excavation site and a nearby trench, which extends more than 2.7 m below the excavation horizon, comprise a normal magnetozone interpreted as part of the normal subchron C2An.3n. Based on structural, stratigraphic, and paleomagnetic evidence these sections are interpreted to lie stratigraphically above, though topographically below, a reversed polarity magnetozone that constitutes the mesa at Korsi Dora and most likely extends through the level of the KT. The reverse-polarity magnetozone is part of the reverse subchron C2Ar. The fossil skeleton is younger than the KT and younger than the Gauss/Gilbert paleomagnetic transition, both regionally identifiable stratigraphic levels. Applying regional average sediment accumulation rate to the trench section yields an estimate of 3.56 to 3.54 Ma for the age of the skeleton. KSD-VP-1/1 was excavated from a 10 cm thick bed of claystone within a ~0.9 m thick section of a heterolithic claystone, siltstone, and mudstone, which is interpreted to have been deposited by episodic flows in a floodplain or floodplain lake environment that is proximal to a stream channel.

Acknowledgments We thank the Authority for Research and Conservation of Cultural Heritage of the Ministry of Culture and Tourism of Ethiopia and the National Museum of Ethiopia for field permits and general support, the Mille District Administration for facilitating our work in the area, and the Afar people of Mille, Waki, and Waytaleyta areas for their participation in fieldwork. This project was funded by the National Science Foundation (Grant # BCS-1124716, BCS-1124705, and BCS-1125157). This paper greatly benefited from helpful comments by C. Campisano and an anonymous reviewer.

References

Acocella, V. (2010). Coupling volcanism and tectonics along divergent plate boundaries: Collapsed rifts from central Afar, Ethiopia. *Geological Society of America Bulletin, 122*, 1717–1728.

Acocella, V., Abebe, B., Korme, T., & Barberi, F. (2008). Structure of Tendaho Graben and Manda Hararo Rift: Implications for the evolution of the southern Red Sea propagator in Central Afar. *Tectonics, 27*, Tc 4016.

Alene, M., Saylor, B., Mertzman, S., Deino, A., & Haile-Selassie, Y. (2012). $^{40}Ar/^{39}Ar$ dating and geochemistry of the Woranso-Mille Pliocene basalts, central Afar, Ethiopia. *34th International Geological Congress Brisbane*, Australia.

Beyene, A., & Abdelsalam, M. G. (2005). Tectonics of the Afar depression: A review and synthesis. *Journal of African Earth Sciences, 41*, 41–59.

Brown, F., & Fuller, C. (2008). Stratigraphy and tephra of the Kibish Formation, southwestern Ethiopia. *Journal of Human Evolution, 55*, 366–403.

Brown, F. H., Sarna-Wojcicki, A. M., Meyer, C. E., & Haileab, B. (1992). Correlation of Pliocene and Pleistocene tephra layers between the Turkana Basin of East Africa and the Gulf of Aden. *Quaternary International, 13–14*, 55–67.

Brown, F. H., Shuey, R. T., & Croes, M. K. (1978). Magnetostratigraphy of the Shungura and Usno Formations, southwestern Ethiopia: New data and comprehensive reanalysis. *Geophysical Journal of the Royal Astronomical Society, 54*, 519–538.

Campisano, C., & Feibel, C. (2007). Connecting local environmental sequences to global climate patterns: Evidence from the hominin-bearing Hadar Formation, Ethiopia. *Journal of Human Evolution, 53*, 515–527.

Cerling, T. E., & Brown, F. H. (1982). Tuffaceous marker horizons in the Koobi Fora region and the Lower Omo Valley. *Nature, 299*, 216–221.

Deino, A., Scott, G., Saylor, B., Alene, M., Angelini, J., & Haile-Selassie, Y. (2010). $^{(40)}Ar/^{(39)}Ar$ dating, paleomagnetism, and tephrochemistry of Pliocene strata of the hominid-bearing Woranso-Mille area, west-central Afar Rift, Ethiopia. *Journal of Human Evolution, 58*, 111–126.

DeMenocal, P., & Brown, F. (1999). Pliocene tephra correlations between East African hominid localities, the Gulf of Aden, and the Arabian Sea. *Hominid Evolution and Climatic Change in Europe, 1*, 23–54.

Feakins, S. J., Brown, F. H., & DeMenocal, P. B. (2007). Plio-Pleistocene microtephra in DSDP site 231, Gulf of Aden. *Journal of African Earth Sciences, 48*, 341–452.

Haile-Selassie, Y., Deino, A., Saylor, B., Umer, M., & Latimer, B. (2007). Preliminary geology and paleontology of new hominid-bearing Pliocene localities in the central Afar region of Ethiopia. *Anthropological Science, 115*, 215–222.

Haile-Selassie, Y., Latimer, B. M., Alene, M., Deino, A. L., Gibert, L., Melillo, S. M., et al. (2010a). An early *Australopithecus afarensis* postcranium from Woranso-Mille, Ethiopia. *Proceedings of the National Academy of Sciences of the United States of America, 107*, 12121–12126.

Haile-Selassie, Y., Saylor, B. Z., Deino, A., Alene, M., & Latimer, B. M. (2010b). New Hominid Fossils From Woranso-Mille (Central Afar, Ethiopia) and Taxonomy of Early Australopithecus. *American Journal of Physical Anthropology, 141*, 406–417.

Haile-Selassie, Y., Saylor, B., Deino, A., Levin, N., Alene, M., & Latimer, B. (2012). A new hominin foot from Ethiopia shows multiple Pliocene bipedal adaptations. *Nature, 483*, 565–569.

Hillhouse, J. W., Cerling, T. E., & Brown, F. H. (1986). Magnetostratigraphy of the Koobi Fora Formation, Lake Turkana, Kenya. *Journal of Geophysical Research, 91*, 11581–11595.

Kalb, J. E., Oswald, E. B., Mebrate, A., Tebedge, S., & Jolly, C. J. (1982). Stratigraphy of the Awash Group, Middle Awash Valley, Afar, Ethiopia. *Newsletters on Stratigraphy, 11*, 95–127.

Lahitte, P., Gillot, P. Y., & Courtillot, V. (2003). Silicic central volcanoes as precursors to rift propagation: The Afar case. *Earth and Planetary Science Letters, 207*, 103–116.

Quade, J., Levin, N. E., Simpson, S. W., Butler, R., McIntosh, W. C., Semaw, S., et al. (2008). The geology of Gona, Afar, Ethiopia. In J. Quade & J. Wynn (Eds.), *The geology of early humans in the horn of Africa* (pp. 1–32). Boulder: The Geological Society of America.

Renne, P. R., WoldeGabriel, G., Hart, W. K., Heiken, G., & White, T. D. (1999). Chronostratigraphy of the Miocene-Pliocene of the Sagantole Formation, Middle Awash Valley, Afar Rift, Ethiopia. *Geological Society of America Bulletin, 111*, 869–885.

Ryan, W. B. F., Carbotte, S. M., Coplan, J. O., O'Hara, S., Melkonian, A., Arko, R., et al., (2009). Global multi-resolution topography synthesis. *Geochemistry Geophysics Geosystems, 10*(3). doi:10.1029/2008GC002332.

WoldeGabriel, G., Hart, W. K., Renne, P. R., Haile-Selassie, Y., & White, T. D. (2009). Stratigraphy of the Adu-Asa Formation. In: Y. Haile-Selassie & G. WoldeGabriel (Eds.), Ardipithecus kadabba: *Late Miocene evidence from the Middle Awash*, Ethiopia (pp. 27–61). Berkeley: University of California Press.

Wynn, J., Alemseged, Z., Bobe, R., Geraads, D., Reed, D., & Roman, D. (2006). Geological and palaeontological context of a Pliocene juvenile hominin at Dikika, Ethiopia. *Nature, 443*, 332–336.

Wynn, J. G., Roman, D. C., Alemseged, Z., Reed, D., Geraads, D., & Munro, S. (2008). Stratigraphy, depositional environments, and basin structure of the Hadar and Busidima Formations at Dikika, Ethiopia. In J. Quade & J. Wynn (Eds.), *The geology of early humans in the horn of Africa* (pp. 87–118). Boulder: The Geological Society of America.

Chapter 3
The Taphonomy and Paleoecology of Korsi Dora Vertebrate Locality 1, Woranso-Mille Study Area, Ethiopia

Denise F. Su

Abstract Korsi Dora Vertebrate Locality 1 (KSD-VP-1), located in the Woranso-Mille paleontological study area in Ethiopia, records one of the least-sampled temporal periods of *Australopithecus* and is one of the few to sample the earlier period of the known time range of *Au. afarensis*. It has yielded one of the most complete skeletons of *Au. afarensis* known thus far. In this paper, the taphonomy and paleoenvironment of KSD-VP-1 are explored through the collected faunal specimens. In addition to descriptive accounts of the surface modifications on the bones, data on several taphonomic factors were also collected and analyzed, the results of which suggest that the KSD-VP-1 faunal assemblage is autochthonous and that there was minimal transport and disturbance of bones. The focus of the paleoenvironmental analysis is the use of the presence and relative abundances of indicator taxa along with other lines of evidence, such as ruminant dietary adaptation and geology. The combined evidence suggests that the paleohabitat at KSD-VP-1 was likely medium to dense woodland with some open areas of grassland or shrubland distal to the locality.

Keywords KSD-VP-1/1 • Taphonomy • Paleoenvironment • Indicator species

Introduction

The Woranso-Mille paleontological study area has yielded more than 8400 fossil specimens to date, among them is a partial skeleton (KSD-VP-1/1) of *Australopithecus afarensis* from Korsi Dora Vertebrate Locality 1 (KSD-VP-1; Fig. 3.1). The age of the locality is well constrained to 3.60–3.58 Ma (Haile-Selassie et al. 2010), one of the least-sampled temporal periods of *Australopithecus*. While the collection from KSD-VP-1 is small (69 specimens cataloged), the fauna sampled is relatively diverse (Table 3.1) and, combined with geological data, can give us a better understanding of the taphonomy and paleoecology of KSD-VP-1/1, as well as add to our knowledge of the types of habitats in which *Australopithecus* lived.

Taphonomy

KSD-VP-1 Fauna

Taphonomic analyses were undertaken to better understand the processes that lead to the accumulation of the KSD-VP-1 fauna. Due to the small number of collected and cataloged specimens (N = 69), 100% surface collection was instituted at KSD-VP-1 to capture taphonomic information preserved on specimens that would not have ordinarily been collected because they are not identifiable by taxon and/or element. Thus, the total number of specimens on which taphonomic observations were made was 730, but cataloged specimens were analyzed separately from 100% surface collection specimens. KSD-VP-1 is about 210 m^2 in size, with specimens distributed across the entire locality (Fig. 3.1). The partial skeleton is excavated from a small area that measures about 13 × 6 × 1.5 m (L × W × D). Other than two bovid specimens, all specimens recovered from the excavated area are elements of the partial skeleton KSD-VP-1/1. All collected specimens from KSD-VP-1 (both cataloged and 100% surface collection) are stored at the National Museum of Ethiopia. Information on animal damage, weathering, breakage, transport, and degree of association (Behrensmeyer 1978; Hill 1980; Lyman 1994; Njau and Blumenschine 2006; Su and Harrison 2008) was collected. All bones were examined under 10× magnification. In addition to descriptive accounts of the surface modifications on the

D.F. Su (✉)
Department of Paleobotany and Paleoecology, Cleveland Museum of Natural History, 1 Wade Oval Drive, Cleveland, OH 44106, USA
e-mail: dsu@cmnh.org

Yohannes Haile-Selassie and Denise F. Su (eds.), *The Postcranial Anatomy of Australopithecus afarensis: New Insights from KSD-VP-1/1*, Vertebrate Paleobiology and Paleoanthropology, DOI 10.1007/978-94-017-7429-1_3

Table 3.1 Faunal list of KSD-VP-1

Order	Family	Subfamily/tribe	Genus and species
Primates	Hominidae	Hominini	*Australopithecus afarensis*
	Cercopithecidae	Papionini	*Theropithecus oswaldi aff. darti*
		Colobinae	
Rodentia	Hystricidae		*Hystrix* sp.
	Thryonomyidae		*Thryonomys* sp.
Lagomorpha	Leporidae		
Carnivora	Viverridae	Viverrinae	cf. *Viverra* sp.
	Mustelidae		cf. *Torolutra* sp.
Proboscidea	Elephantidae		*Elephas reckii*
Artiodactyla	Bovidae	Alcelaphini	
		Neotragini	
		Hippotragini	
		Antilopini	*Gazella* sp.
		Tragelaphini	*Tragelaphus saraitu*
			Tragelaphus sp. indet.
		Aepycerotini	*Aepyceros* cf. *afarensis*
	Suidae	Suinae	*Kolpochoerus* cf. *millensis*
		Tetraconodontinae	*Nyanzachoerus kanamensis*
	Giraffidae	Giraffinae	*Giraffa* cf. *stillei*
	Hippopotamidae		
Perissodactyla	Equidae		*Eurygnathohippus* sp.
	Rhinocerotidae		
Tubulidentata	Orycteropodidae		*Orycteropus* sp.
Reptilia	Crocodylidae		
Aves			

bones, data on several taphonomic factors were also collected and analyzed, the results of which are described below.

Weathering. Bone weathering is the "process by which the original microscopic organic and inorganic components of bone are separated from each other and destroyed by physical and chemical agents operating on the bone *in situ*, either on the surface or within the soil zone" (Behrensmeyer 1978:153). Six weathering stages (WS) are defined that are correlated with the number of years since death of the organism (see Table 3.2; Behrensmeyer 1978) and used to score the fossil specimens from KSD-VP-1. The resulting weathering profiles (Fig. 3.2) show opposite patterns for cataloged and 100% surface collection specimens. For cataloged specimens, most of the specimens are either not weathered (WS 0) or lightly weathered (WS 1), with decreasing numbers of specimens that are moderately weathered (WS 2 and 3) and heavily weathered (WS 4 and 5). For specimens from 100% surface collection, those that are heavily weathered (WS 4 and 5) dominate with decreasing numbers that are moderately weathered (WS 2 and 3) and very few show no weathering (WS 0). The observed reverse trend demonstrates the bias that is intrinsic in collections of cataloged specimens. Specimens that are heavily weathered tend not to be collected and cataloged because they are not identifiable; thus, any inferences made about the duration of accumulation based on cataloged specimens may underestimate the length of time prior to fossilization. While there may be an artificial inflation in WS 5 specimens from 100% surface collection because the majority are often nothing more than shards of bone, it still suggests that many of the bones of KSD-VP-1 were not likely to have been quickly buried and fossilized, even if there is an artificial inflation of WS 4 and WS 5 specimens.

Breakage. Data were collected on the different types of breaks observed on KSD-VP-1 specimens, including spiral, transverse, and combination breaks (Lyman 1994). Except for some elements of KSD-VP-1/1 and isolated teeth, specimens were rarely complete. The most common breakage type is transverse break for both collected and 100% surface specimens, although the latter has about three times more specimens coded with transverse break than the former; spiral fractures are much less common (Fig. 3.3). This suggests that most of the bones were broken when they were no longer fresh (Lyman 1994).

Transport. Transport was examined by scoring the degree of abrasion and association of specimens, as well as element susceptibility to fluvial transport. Behrensmeyer (1975) and Korth (1979) have suggested that one of the causes for abrasion on bone is fluvial transport, although other

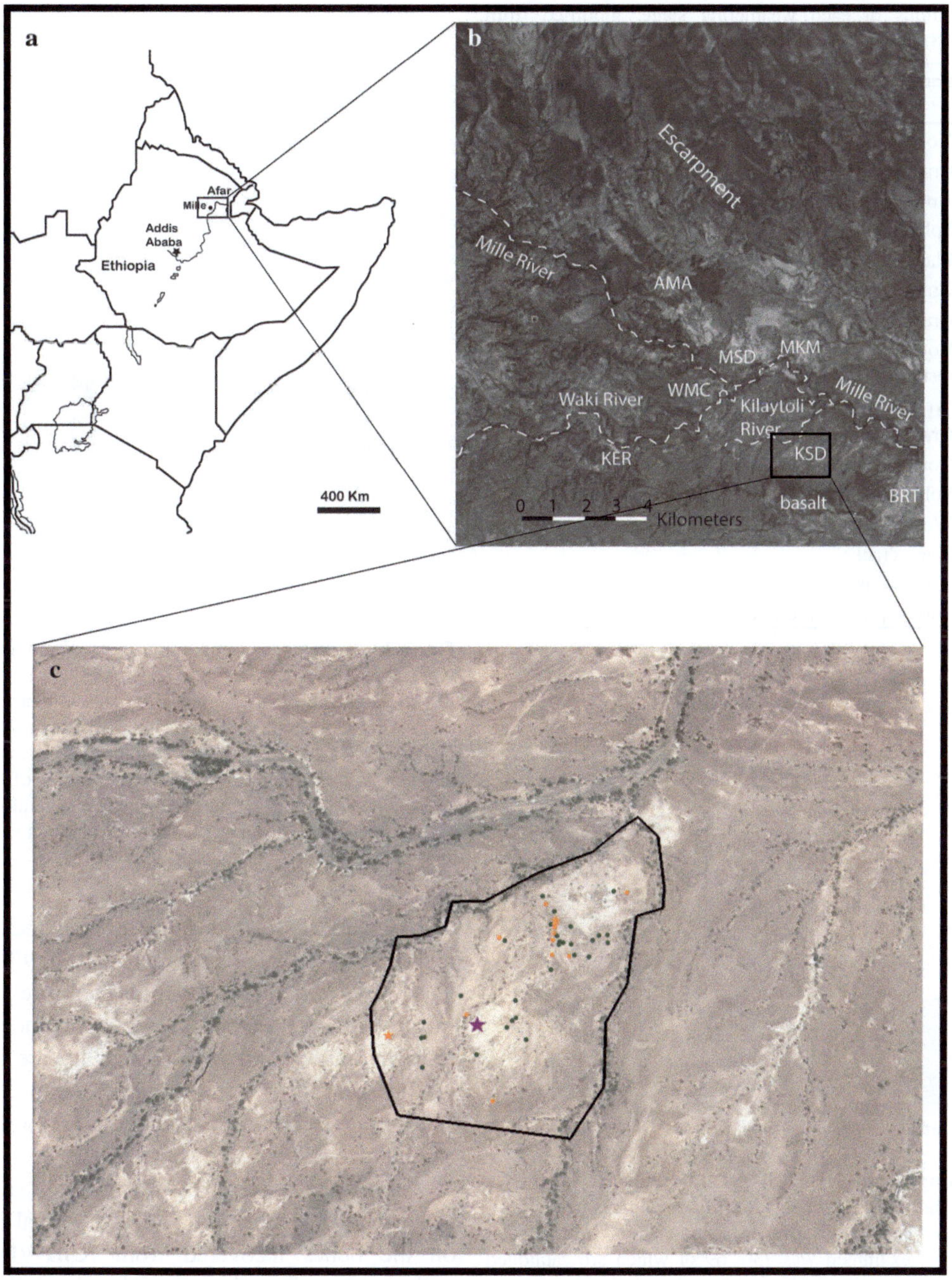

Fig. 3.1 Locator map of KSD-VP-1. **a** Map of the Horn of Africa; the box indicates the region shown in (**b**). **b** Map of the area in the northern part of the Woranso-Mille study area showing the location of Korsi Dora (KSD) in relation to other collection areas, AmAdo (AMA), Makah Mera (MKM), Mesgid Dora (MSD), Kerare (KER); the box indicates the region shown in (**c**). **c** Map that shows the boundaries of KSD-VP-1, outlined in black, and the distribution of fossil specimens recovered from the locality. Green circle = non-primate specimen, orange circle = non-hominin primate specimen, orange star = non-hominin primate partial skeleton, purple star = hominin partial skeleton, KSD-VP-1/1. **a** and **b** are modified from Fig. 2.1 of Saylor et al. (2016). Base map of **c** is from Google Earth

processes, such as trampling and aeolian activity, are also possible causes (Brain 1967; Shipman and Rose 1988). However, fluvial transport of bone abrades the entire specimen, while trampling creates deep scratches in addition to abrasion and aeolian activity abrades only the exposed surface (Lyman 1994). These differences allow for the identification of specimens with abrasion due to fluvial transport rather than other origins and only those specimens are recorded for this analysis. Not surprisingly, cataloged specimens have a much lower proportion of abraded specimens than those from 100% surface collection (Fig. 3.3), as highly abraded specimens are often lacking the morphological features necessary for

Table 3.2 Descriptions of each weathering stage, modified from Behrensmeyer (1978)

Weathering stage	Large mammal categories (>5 kg)
0	No cracking or flaking
1	Cracking parallel to fiber structure (longitudinal)
	Articular surfaces may have mosaic cracking
	Split lines begin to form
2	Flaking of outer surface (exfoliation)
	Cracks are present
	Crack edge is angular
3	Rough homogeneously altered compact bone resulting in fibrous texture
	Weathering penetrates 1–1.5 mm maximum
	Cracked edge is rounded
4	Coarsely fibrous and rough surface
	Splinters of bone loose on surface
	Weathering penetrates inner cavities
	Open cracks
5	Bone falling apart
	Large splinters present

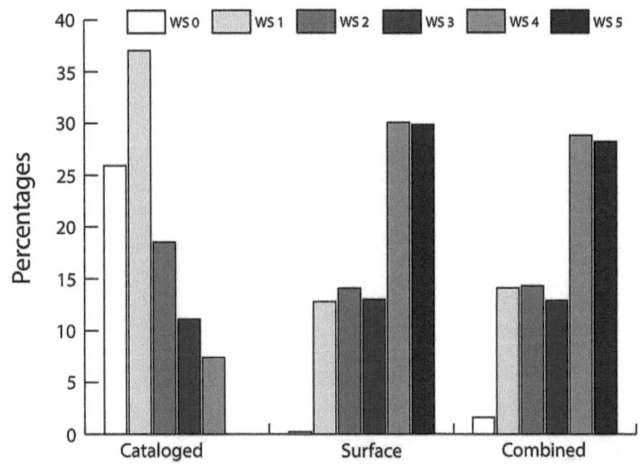

Fig. 3.2 Weathering profiles of KSD-VP-1. Cataloged specimens mostly exhibit weathering stages 1 and 2, while specimens from 100% collection are heavily skewed toward weathering stages 4 and 5. Total number of specimens examined was 730, 69 of which are cataloged specimens. Cataloged = specimens that were cataloged; Surface = specimens from 100% surface collection; Combined = Cataloged + Surface specimens. See text for discussion

taxonomic and element identification. The results indicate that there are very few abraded bones in the collection; even when the combined cataloged and 100% surface datasets are considered, abraded specimens comprise only about 10% of the collection (Fig. 3.3). This suggests that there has not been very much fluvial transportation of bones in general at KSD-VP-1. This is supported by the recovery of multiple partial skeletons, including KSD-VP-1/1 and two cercopithecid individuals,

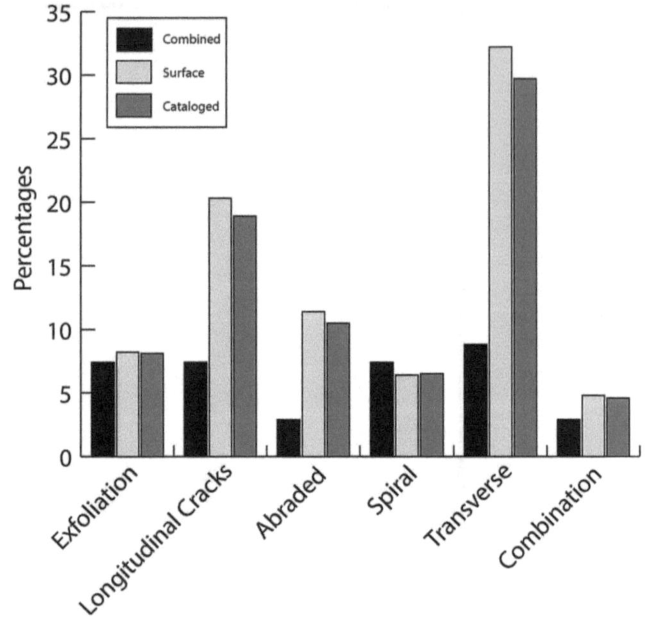

Fig. 3.3 Percentages of non-animal surface modification and breakage types at KSD-VP-1. Almost all specimens exhibit some type of surface modification and breakage. Of note is the low proportion of specimens that exhibit abrasion, suggesting the low incidence of fluvial transport of bones. The most common breakage type is transverse break. Total number of specimens examined was 730, 69 of which are cataloged specimens. Cataloged = specimens that were cataloged; Surface = specimens from 100% surface collection; Combined = Cataloged + Surface specimens. See text for discussion

whose elements (includes dentition, elements of the forelimb and hindlimb, phalanges) represent 20.9% of the total collected *individual* specimens. Furthermore, the preserved elements of these partial skeletons include ribs, vertebrae, sacrum, scapula, and ulnae, bones that are among those most easily transported by fluvial action (Voorhies 1969; Behrensmeyer 1975) and suggest a non-fluvially impacted assemblage (Behrensmeyer 1975).

Animal Damage. Data on carnivore, rodent, and insect activities were collected from the animal damage observed on the bones recovered from KSD-VP-1. Observations indicate that there is minimal animal damage; only five percent of all bones examined (cataloged and 100% surface collection) have signs of carnivore, rodent, or insect damage. There is a higher incidence of carnivore and rodent activities on cataloged specimens than those from 100% surface collection (Fig. 3.4), possibly due to the highly weathered state of most of the specimens from 100% surface collection, such that weathering activity has obliterated any signs of animal damage on them. Nevertheless, the minimal occurrences of animal damage suggest that carnivores and rodents did not have a major role in the accumulation of the KSD fauna.

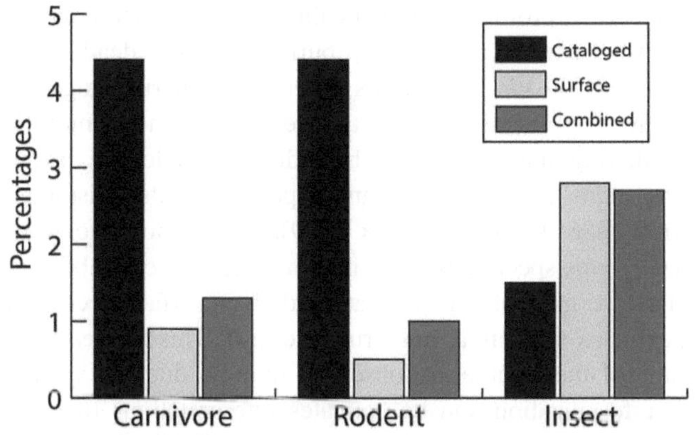

Fig. 3.4 Percentages of animal damage at KSD-VP-1. There is low incidence of animal damage on the specimens of KSD-VP-1. Cataloged specimens had higher proportions of carnivore and rodent damage than those from 100% surface collection, possibly due to the fact that most of the 100% surface collection specimens are highly weathered, which removed any original surface modifications. Total number of specimens examined was 730, 69 of which are cataloged specimens. Cataloged = specimens that were cataloged; Surface = specimens from 100% surface collection; Combined = Cataloged + Surface specimens. See text for discussion

KSD-VP-1/1

The partial skeleton, KSD-VP-1/1, exhibits different weathering and breakage patterns for each of the elements preserved; the major elements are described separately below.

a. *R. Ulna.* The ulna shows differential weathering; the ulnar head is much less weathered (WS 2) than the shaft (WS 4). It was originally found in three pieces, all exhibiting transverse breaks and longitudinal cracks.

b. *R. Humerus.* The humerus exhibits differential weathering on the anterior (WS 2) and posterior sides (WS 4). There is a transverse break on the distal end of the shaft, which likely happened after fossilization because of the presence of a cast of the medullary cavity. There are also many longitudinal cracks.

c. *L. Femur.* The femur is moderately weathered (WS 2) and shows no differential weathering. It exhibits a spiral break and longitudinal cracks.

d. *R. innominate.* As a whole, the innominate is not very weathered and ranges from WS 1-3. The heaviest weathering is on the ischium and posterior side of the pubis, which were surface collections while the rest of innominate was recovered via excavation. The pelvic body was found intact, but with large cracks between the blade and the body.

e. *L. Tibia.* The proximal tibia is missing and the remaining tibia is heavily covered with gypsum, which makes the determination of weathering stage difficult. At mid-shaft, there is only a small point of contact between the proximal and distal ends. The cortical bone on the anterior and medial side of the distal piece is gone, exposing the cancellous bone beneath. There are longitudinal cracks throughout.

f. *L. Clavicle.* The clavicle is broken into two pieces at mid-shaft, with both ends missing; the broken edges are rounded. Both fragments exhibit pitting and exfoliation on their inferior surfaces and are moderately weathered (WS = 3). There are transverse open cracks at the midpoint of both fragments that likely formed prior to fossilization.

g. *R. Scapula.* The scapula is lightly weathered (WS 1) and lacking only the acromion and coronoid processes.

h–l. *Cervical vertebrae.* Fragments of cervical vertebrae 2–7 (C2–C7) show light weathering (WS 1), except for the vertebral body of C3 and C5, which are more heavily worn (WS 3). C3 also exhibits a mosaic-cracking pattern. The vertebral body of C6 is heavily abraded on the superior proximal edge.

t. *Sacrum.* There is moderate exfoliation of the sacrum and it is moderately weathered (WS 3). The anterior surface also exhibits a mosaic-cracking pattern.

n. *2nd rib.* The 2nd rib is broken in five pieces with both longitudinal and transverse cracks with both ends of the rib absent. It is moderately weathered (WS 3).

r. *11th rib.* The 11th rib is complete, absent both ends. It is moderately weathered (WS 2).

o. *Rib.* This unidentified rib is lacking both ends. It is moderately weathered (WS 2).

p. *Rib.* This unidentified rib is lacking both ends. It is moderately weathered (WS 2).

q. *Rib.* This unidentified rib is lacking both ends. It is highly weathered (WS 4); the rib was falling apart into bone splinters but was buried and mineralized before complete disintegration. It is also covered by gypsum.

Discussion of the Taphonomy of the Site and the Partial Skeleton

The KSD-VP-1 assemblage is likely autochthonous given the low proportion of abraded bones and the presence of several associated specimens, including KSD-VP-1/1. The presence of aquatic animals (i.e., *Torolutra* and uncataloged crocodylian teeth) suggests that a large body of water was in the vicinity of KSD-VP-1. This is compatible with the sedimentological data collected thus far that indicate KSD-VP-1/1 is from claystone within a thick section of heterolithic claystone, siltstone, and mudstone, which may have been deposited by episodic flows in a floodplain or ephemeral floodplain lake proximal to a stream channel (Saylor et al. 2016).

The relative completeness of KSD-VP-1/1 and its limited area of distribution (77 m^2; see Fig. 3.5) indicate that there was very little to no transport of the individual after death and that it is unlikely that the depositional environment was high energy, at least during the time when KSD-VP-1/1 was deposited, as is also suggested by sedimentological evidence (see Saylor et al. 2016).

Weathering profiles can be difficult to interpret, particularly as they relate to time since death, due to the many factors involved in the rate of weathering, including element, taxon, exposure duration, and depositional environment (Behrensmeyer 1978; Lyman and Fox 1989). In a paleontological assemblage, it is often not possible to analytically control all of these factors (Lyman and Fox 1989); however, as a partial skeleton, KSD-VP-1/1 provides the opportunity to examine bone-weathering patterns at the locality in more depth as many of these factors are controlled. The overall weathering profile for the locality suggests that bones at KSD-VP-1 were not quickly buried after the death of an animal; KSD-VP-1/1 exhibits different weathering stages for different elements and certain elements show multiple weathering stages that may be indicative of long exposure time and/or multiple burial and exposure events (Frison and Todd 1986; Lyman and Fox 1989). It is of note, however, that *in situ* specimens (i.e., innominate, clavicle, ribs, scapula) are generally less weathered than surface recovered specimens (i.e., ulna, humerus, sacrum). This suggests that some of the weathering observed may be due to exposure after fossilization and complicates interpretations of exposure time prior to fossilization for KSD-VP-1/1 and its associated fauna. However, it is possible to conclude, due to the range of weathering stages exhibited throughout the skeleton and the presence of bones with multiple weathering stages, that the skeleton was not completely buried and fossilized soon after death, and was instead exposed for some time and then possibly partially buried, exposed, and reburied multiple times before fossilization.

The minimal carnivore damage observed on the bones suggest that it is unlikely carnivores played a significant role in the accumulation of the KSD-VP-1 assemblage, although some of the animals preserved were probably prey that were killed onsite. There are also minimal signs of animal damage on KSD-VP-1/1, and, in particular, there are no signs of

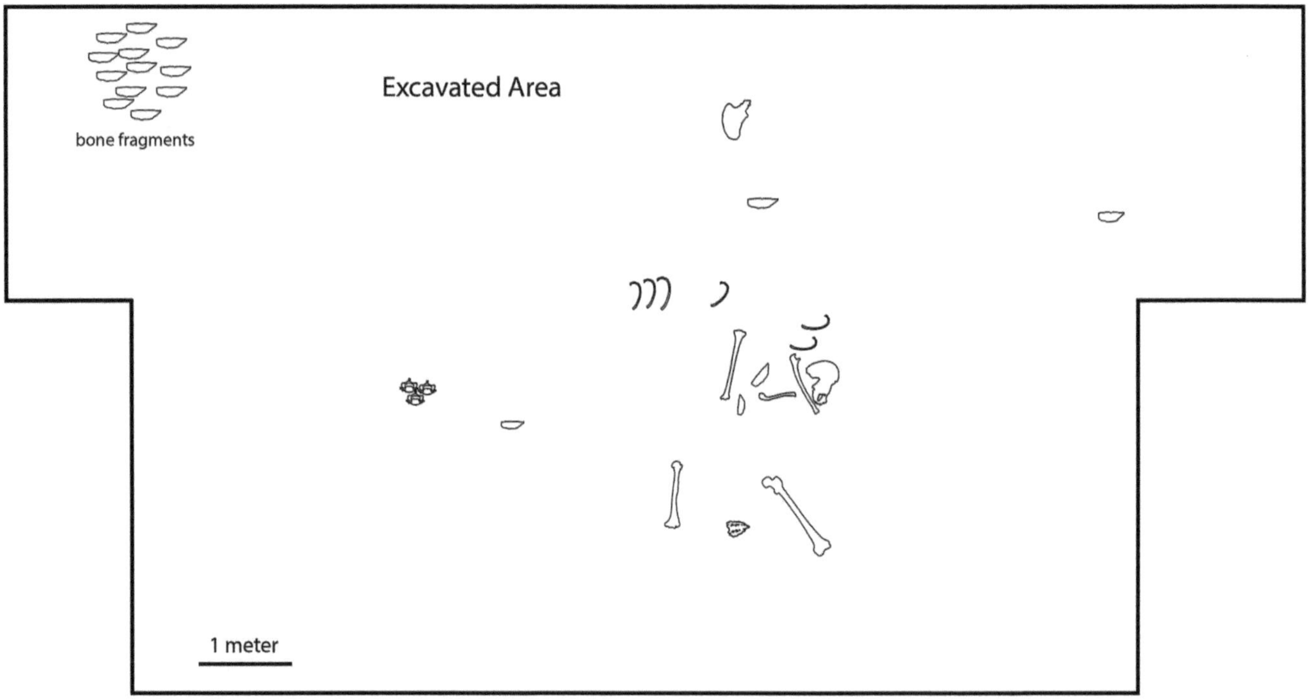

Fig. 3.5 Distribution of KSD-VP-1/1, partial skeleton of *Australopithecus afarensis*. Elements recovered are represented in the figure by a line drawing depicting that element. Note that not all elements were completely preserved, see text for description of the portions preserved and state of preservation

carnivore activity. This, in combination with associated elements of the skeleton, suggests that carnivore predation was an unlikely cause of death for KSD-VP-1/1. The totality of evidence suggests that there was differential burial of the various elements of KSD-VP-1/1. Bones that are complete and lightly weathered, i.e., scapula, were more likely to have been buried completely soon after death with minimal subsequent exposure and reburial, while those that are fragmented and weathered were exposed for longer periods of time and may have been covered and uncovered multiple times for variable lengths of time before they were finally completely buried and fossilized.

Paleoenvironment

Specimens of *Au. afarensis* have been recovered from many sites in eastern Africa, including Hadar, Laetoli, Middle Awash, Koobi Fora, West Turkana, and possibly Fejej (Fleagle et al. 1991; Kappelman et al. 1996; Kimbel and Delezene 2009; Harrison 2011). Other than Laetoli ($\sim$3.6–3.85 Ma), Woranso-Mille is the only site to sample the earlier period of the known time range of *Au. afarensis* (from ca. 3.0–3.8 Ma; Kimbel et al. 2004; Haile-Selassie et al. 2010; Harrison 2011). Thus, an analysis of the paleoenvironment of KSD-VP-1 will add to our understanding of the types of habitats that were available to *Au. afarensis* in the early part of their evolutionary history for which we have little data.

KSD-VP-1 paleoenvironment. The focus of this study will be the use of the presence and relative abundances of indicator taxa. Indicator species analysis is based on the principle of taxonomic uniformitarianism such that habitat preferences of living members of a taxon can be extended to its fossil relatives. Thus, the presence of a fossil taxon with extant members that are habitat-specific can be used to infer its paleohabitat (e.g., Gentry 1978; Vrba 1980; Shipman and Harris 1988). Recent ecomorphological and stable carbon isotopic studies show that certain fossil taxa generally share similar diets with their extant relatives (Sponheimer et al. 1999; Kingston and Harrison 2007), suggesting that indicator species can be used with reasonable confidence in reconstructing paleohabitats. Species presence, however, can be based on single or rare specimens that are ecologically or stratigraphically intrusive and, thus, erroneous indicators of habitat. This can be ameliorated by the consideration of relative abundance data and the correlation between taxon abundance and habitat preferences given proper taphonomic considerations (Bobe et al. 2007; Su et al. 2009; Su 2011). Every cataloged specimen is included in the analyses that follow and each cataloged specimen is counted as an identifiable specimen (NISP); associated specimens, such as teeth in a mandible or maxilla or partial skeleton, are counted as a single NISP.

Bovids are ideal for this type of analysis due to their abundance in the Plio-Pleistocene hominin fossil record and specificity in habitat preferences (Bobe et al. 2007; Reed 2008; Su et al. 2009; Su 2011). Bovids are among the most common taxon at KSD-VP-1 and their relative abundances are compared to those from other Plio-Pleistocene sites using Correspondence Analysis (R statistical program version 2.13.0 running on Rstudio version 0.97.551). The count data for comparative fossil sites are derived from the published literature (see Table 3.2 for a list of comparative sites and references).

Bovids constitute 36.8% of the cataloged fauna at KSD-VP-1 (Fig. 3.6a); the most common bovid tribe is Tragelaphini (36.4%, Fig. 3.6b). As a group, extant tragelaphins are adapted to complex, mosaic habitats; they require moderate to dense cover and are dependent on water (Kingdon 1997). They are heavily reliant on browse and are known, as a group, to be dedicated browsers (Kingdon 1997; Gagnon and Chew 2000; Sponheimer et al. 2003). Antilopini (includes gazelles and impalas) is the second most common group of bovids (27.3%; Fig. 3.6b) at KSD-VP-1. Modern gazelles are mostly found in less wooded habitats such as short- to medium-grasslands and open bushlands (Kingdon 1974, 1997; Estes 1991) and have significant browse in their diet such that they range from mixed-feeders to browsers in their dietary preferences (Cerling et al. 2003). Extant impalas generally prefer more wooded habitats than gazelles and are mostly found in abundance in ecotonal woodland/grassland habitats (Western 1973), similar to gazelles, however, they are mixed feeders (Cerling et al. 2003). Alcelaphins and neotragins are rare (each is 4.5% of the bovid fauna, Fig. 3.6b) and reduncins are absent from KSD-VP-1. Bovid tribe relative frequencies suggest that densely wooded habitats were dominant and grassland habitats were available in the vicinity of KSD-VP-1, but not proximal to the locality, and floodplain grassland and open to dense bushland habitats preferred by reduncins and neotragins (Kingdon 1974, 1997), respectively, were probably rare or absent from the local region.

Bovid diets can also reveal information regarding the surrounding habitat. Dietary adaptations for bovid individual specimens are assigned based on results from mesowear and enamel carbon stable isotope analyses (Su, unpublished; Levin et al. 2015). Fifty percent of the bovid specimens at KSD-VP-1 are browsers, followed by 44.4% of mixed feeders; only 5.6% of bovids are dedicated grazers. Combined with bovid tribe frequencies, this suggests that woody or C_3 plants were in much higher abundance than C_4 plants at KSD-VP-1.

Due to modern primates' reliance and dependence on trees, the presence and abundance of their fossil relatives are often used to indicate the availability of wooded vegetation. Cercopithecids are among the most common faunal element at KSD-VP-1; they comprise 32.4% of the total fauna and are

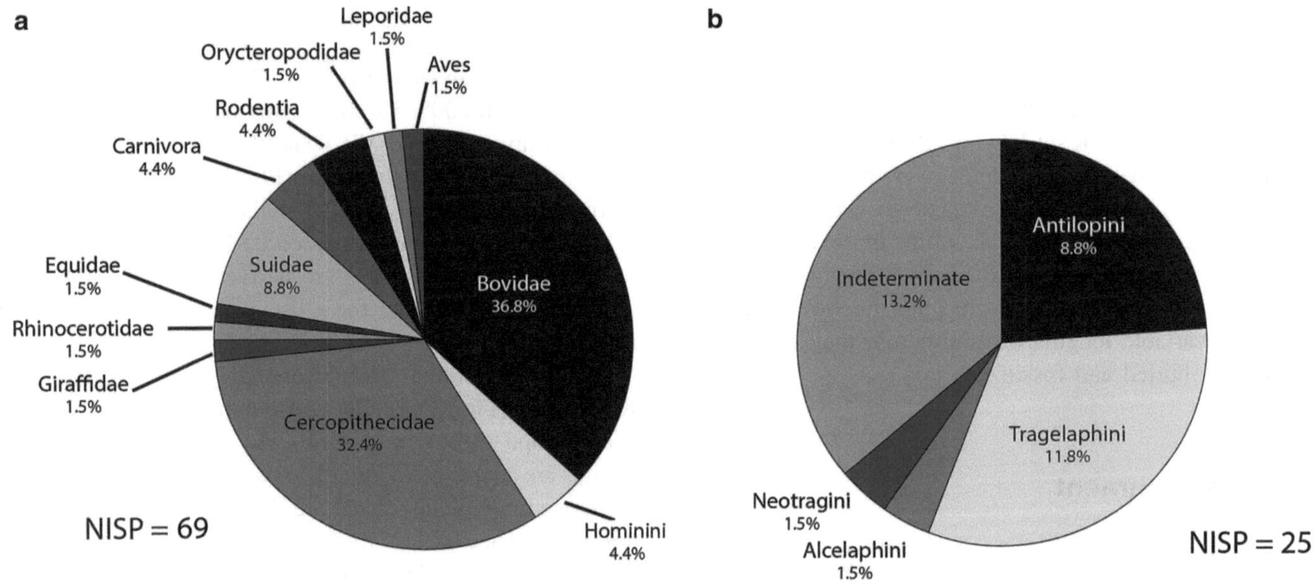

Fig. 3.6 Pie charts of the KSD-VP-1 faunal composition based on cataloged specimens. **a** Proportions of major taxonomic groups at KSD-VP-1. Bovids and Cercopithecids are the two most common taxa at the locality. **b** Proportions of Bovidae at KSD-VP-1. Tragelaphins are the most common bovid at the locality. Indeterminate specimens are postcranial elements that cannot be assigned to bovid tribes

almost as abundant as bovids (Fig. 3.6a), which in combination with the bovid and dietary data, are suggestive of the wide availability of trees at KSD-VP-1. While not the most common elements of the KSD-VP-1 fauna, aquatic animals are present. There is a single specimen of *Torolutra* identified, as well as dental specimens of Crocodylia, indicating there was water in the vicinity of KSD-VP-1 and that the locality may be sampling margins along that body of water.

The results from the analyses presented here suggest that wooded habitats were common at KSD-VP-1. Specimens belonging to taxa that are closely associated with wooded vegetation such as tragelaphins and cercopithecids dominate the fauna; dedicated grazers are a small component of the fauna while those that rely on browse either exclusively or partially are dominant. The presence of aquatic animals and the sedimentological data suggest that there was a large body of water in the vicinity of KSD-VP-1 and that it was slow moving when it deposited sediments at KSD-VP-1. The dominant vegetation was unlikely to have been a floodplain grassland, however, as there have been no reduncins or hippopotamids recovered from the locality. Thus, the combined evidence suggests that KSD-VP-1 was distal to a large body of water; the paleohabitat was likely medium to dense woodland with grassy and bushed ground cover. A caveat of this interpretation is the small sample size of the fossil assemblage. An examination of the number of genera identified and the number of individual specimens collected at each locality at Woranso-Mille shows that genus diversity increases with larger sample sizes until the number of individual specimens reaches about 200, at which point, the number of genera identified plateaus (Fig. 3.7). This suggests that only the most common faunal elements have been sampled at KSD-VP-1 and that rare elements were likely to have been lost from the fossil assemblage. It is likely that many of the rarer elements of the fauna that are indicative of more open habitats may have, in life, lived mostly in other areas of Woranso-Mille that were less vegetated and are indeed ecologically intrusive elements at KSD-VP-1, which is constrained to an area of about 210 m^2 and sampling a very small area of Woranso-Mille. This is further emphasized by the high abundance of primates, which are usually one of the

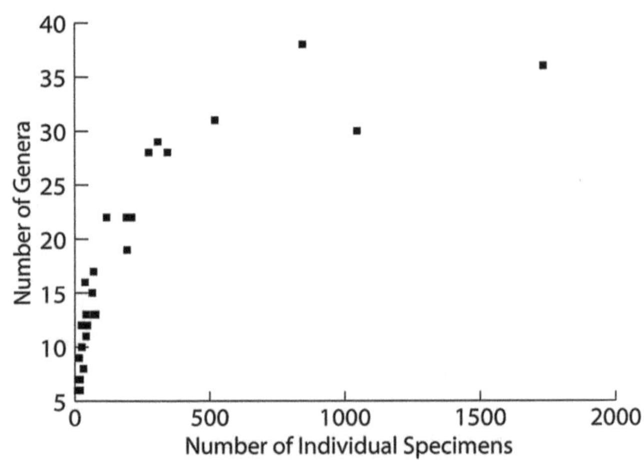

Fig. 3.7 Genus richness versus sampling size at Woranso-Mille. Each square represents a locality in the Woranso-Mille Study Area. Genus richness increases with increasing sample sizes. See text for further discussion

more rare components of a fossil assemblage, and is likely a true reflection of the abundance of trees at KSD-VP-1.

Eastern Africa Plio-Pleistocene paleoenvironments. There have been numerous studies on the paleoenvironments of Pliocene hominins (Woldegabriel et al. 1994; Wynn 2000; Harris and Leakey 2003; Wynn et al. 2006; Alemseged et al. 2007; Bobe et al. 2007; Reed 2008; White et al. 2009a, b; Harrison 2011; Su 2011; Bedaso et al. 2013; Behrensmeyer and Reed 2013) that suggest they were found in complex, mosaic habitats and that there was a trend toward more open habitats through the Pliocene in eastern Africa. In order to place the paleoenvironmental interpretation of KSD-VP-1 into this context, those of other major eastern Africa Plio-Pleistocene sites are briefly summarized here.

Aramis, Ethiopia. Remains of *Ardipithecus ramidus,* including a partial skeleton, have been recovered from Aramis (4.4 Ma) in the Middle Awash, Ethiopia (White et al. 1994, 2009a). Geological, isotopic, and faunal evidence indicate that Aramis was densely wooded with areas of wooded grassland during the Pliocene (Woldegabriel et al. 1994; White et al. 2009b), although other interpretations of the isotopic evidence suggest that Aramis had much higher proportions of wooded grassland than reconstructed (Cerling et al. 2010).

Kanapoi, Kenya. *Australopithecus anamensis* is known from Kanapoi in the Turkana Basin and dated to about 4.2 Ma (Leakey et al. 1995; Harris and Leakey 2003). The associated fauna suggests that Kanapoi was a complex mosaic of woodland and grassland with a predominance of woodland habitats (Harris and Leakey 2003). However, Wynn (2000) infers that *Au. anamensis* was more closely associated with the more open habitats present at Kanapoi based on paleosol stable isotopes.

Laetoli, Tanzania. Laetoli is a Pliocene (3.6–3.8 Ma) site in northeastern Tanzania from which specimens of *Australopithecus afarensis* have been recovered. Unlike many other Plio-Pleistocene eastern African sites, Laetoli lacked a permanent large body of water (Su and Harrison 2008; Ditchfield and Harrison 2011). The cumulative geological, faunal, and stable isotopic evidence suggests that Laetoli was a mosaic of habitats, likely dominated by bushlands and grasslands with areas of woodland and riparian woodland along ephemeral river courses (Su and Harrison 2008; Harrison 2011; Su 2011); other interpretations based on bovid ecomorphology and ruminant mesowear suggest, however, that Laetoli was more densely vegetated with a dominance of woodland habitats (Kovarovic and Andrews 2007; Bishop et al. 2011; Kaiser 2011; Kovarovic and Andrews 2011).

Hadar, Ethiopia. Specimens of *Au. afarensis* are recovered from the Sidi Hakoma, Denen Dora, and Kada Hadar Members (3.42–2.94 Ma) of the Hadar Formation; while none are known from the Basal Member (3.42–3.45 Ma; Campisano and Feibel 2008), it is still included here as the oldest Member of the Formation. Reconstructions of the paleoenvironment at Hadar indicate vegetation regime shifts during the 500 ka of these Members. The fauna of the Basal Member indicate a mosaic of woodland and shrubland habitats, although pollen data suggest that there was a forested area near a river channel. The Sidi Hakoma Member spans from 3.42–3.26 Ma (Campisano and Feibel 2008). It was an ecotonal environment at its base that then shifted between various proportions of woodland, shrubland, and grassland habitats with a riparian component until the expansion of paleolake Hadar, which resulted in lake margin wetlands at the top of the sequence (Reed 2008; Behrensmeyer and Reed 2013). The Denen Dora Member, dated to ~3.2 Ma, spans only about 50 kyr (Campisano and Feibel 2008); faunal evidence suggests that it shifted from a mosaic woodland habitat to one with extensive wetlands or floodplain grasslands, after which open woodland or wooded grassland habitats dominated (Reed 2008; Behrensmeyer and Reed 2013). The Kada Hadar Member is dated to ~3.2–2.94 Ma (Campisano and Feibel 2008); its paleohabitat changes from open woodland with some edaphic grassland to an arid, open woodland or shrubland habitat around 3.12 Ma (Reed 2008).

Dikika, Ethiopia. Numerous specimens of *Au. afarensis* have been recovered from the Basal and Sidi Hakoma Members (3.42–3.26 Ma; Campisano and Feibel 2008) of the Hadar Formation at Dikika (Alemseged et al. 2005, 2006). The Dikika fauna represents those found in mesic deltaic habitats with some grasslands (Wynn et al. 2006). Stable isotopic data suggest that grassland, wooded grassland, and woodland habitats persisted throughout both Members, albeit in different proportions through time (Bedaso et al. 2013).

Omo, Ethiopia. While the isolated teeth from the Shungura Formation at Omo are not confidently attributed to *Australopithecus* (Suwa et al. 1996); this is an important site for understanding eastern African Pliocene paleoecology and is thus included in this summary. Fossils from the older members of the Shungura Formation of the Omo Valley in Ethiopia are generally derived from fluvial deposits (Bobe 1997; Bobe et al. 2007). During Members B and C (3.36–2.6 Ma), the habitat was likely wooded and moist as inferred from relative abundances of bovid tribes; alcelaphins and antilopins, both indicators of more open habitats, were quite rare, suggesting that grasslands were a small component of the Omo paleoenvironment (Alemseged et al. 2007; Bobe et al. 2007).

General Discussion. Given the variety of habitats available to early hominins, it is unclear if they have a preferred habitat. *Australopithecus afarensis*, in particular, is found in a broad range of paleohabitats from closed woodlands to open shrubland/grassland to edaphic grasslands (Bobe et al. 2007; Reed 2008; Su 2011; Behrensmeyer and Reed 2013).

Table 3.3 Comparative hominin-bearing Plio-Pleistocene sites used in the bovid relative abundance analysis

Locality	Age (Ma)	References
Lothagam, Kenya		Leakey and Harris (2001)
Apak member	~5–4.2	
Kaiyumung member	<3.9	
Kanapoi, Kenya	~4.2	Harris and Leakey (2003)
Laetoli, Tanzania	3.8–3.6	Su (2011)
Hadar, Ethiopia		Reed (2008)
Basal member	3.8–3.4	
Sidi Hakoma member	3.42–3.26	
Denen Dora member	3.26–3.18	
Kada Hadar member	3.2	
Omo, Ethiopia		Bobe (1997), Alemseged (2003)
Shungura member B	3.36–2.95	
Shungura member C	2.95–2.6	
Middle Awash, Ethiopia		
Aramis	4.4	White et al. (2009a, b)

In order to better understand the differences and similarities in habitat between KSD-VP-1 and other Plio-Pleistocene sites, the relative abundances of the KSD-VP-1 bovids are compared to the above sites using Correspondence Analysis. The first and second dimensions account for 56.8 and 26.5% of the inertia, respectively. Three distinct clusters can be seen in the plot of Dimensions 1 and 2 – (1) KSD-VP-1, Aramis, and Kanapoi (2) Laetoli localities, and (3) all other Plio-Pleistocene sites (Fig. 3.8; see Table 3.3 for the list of comparative sites, Dikika is not included in the analysis due to the lack of comparable published data). This suggests that the relative abundances of bovid tribes at KSD-VP-1 are more similar to those found at Aramis and Kanapoi than at other Plio-Pleistocene sites and is likely driven by the high proportion of tragelaphins at KSD-VP-1, Aramis (White et al. 2009a, b) and Kanapoi (Harris and Leakey 2003). It is clear that KSD-VP-1 is unlike Pliocene Laetoli, which lacked a permanent large body of water (Su and Harrison 2008; Ditchfield and Harrison 2011) or many of the other Plio-Pleistocene sites, most of which had varying proportions of floodplain or edaphic grasslands as part of their paleohabitats (Bonnefille and DeChamps 1983; Wesselman 1985; Reed 1997, 2008; Bobe and Eck 2001; Bobe et al. 2002; Bonnefille et al. 2004), and suggestive of similarities in ecological structure among KSD-VP-1, Aramis, and Kanapoi. It is important to point out, however, that KSD-VP-1 is sampling not just a much smaller area than any of the comparative sites, but also a single point in time. It is, effectively, sampling a microhabitat that members of a population of *Au. afarensis* visited and/or utilized at a specific time, such that the relatively homogeneously wooded habitat reconstruction for KSD-VP-1 is partly a reflection of its restricted area and time. It is likely that if the entire Woranso-Mille Study Area was considered at the same geographic and temporal scale as those of the comparative sites, the reconstruction would be much more mosaic in nature. This emphasizes the importance of the consideration of both temporal and geographic scale in considering hominin habitat preferences and exploitation.

Australopithecus *paleoenvironment and diet*. One of the main goals of paleoenvironmental reconstruction is to provide researchers with the range of possible dietary resources available to early hominins. Stable carbon isotopic signatures of *Au. afarensis* teeth from Hadar and Woranso-Mille

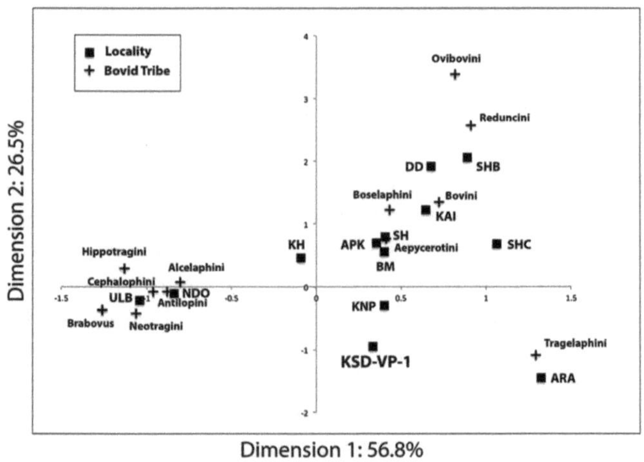

Fig. 3.8 Plot of correspondence analysis on bovid tribes from KSD-VP-1 and other Plio-Pleistocene fossil hominin sites. A plot of Dimensions 1 and 2 of a Correspondence Analysis show that there are three general groups of hominin sites – (1) KSD-VP-1, KNP, and ARA, (2) ULB and NDO, (3) all other Plio-Pleistocene hominin sites. See text for further discussion. Abbreviations: *KSD-VP-1* = Korsi Dora Vertebrate Locality 1, *KNP* Kanapoi; *ARA* Aramis; *APK* Apak Member (Lothagam, Kenya); *KAI* Kaiyumung Member (Lothagam, Kenya); *ULB* Upper Laetolil Beds (Laetoli, Tanzania); *NDO* Upper Ndolanya Beds (Laetoli, Tanzania); *BM* Basal Member (Hadar, Ethiopia); *SH* Sidi Hakoma Member (Hadar, Ethiopia); *DD* Denen Dora Member (Hadar, Ethiopia); *KH* Kada Hadar Member (Hadar, Ethiopia); *SHB* Shungura Member B (Omo, Ethiopia); *SHC* Shungura Member C (Omo, Ethiopia)

are indicative of a wide range of C_3 and C_4 food items in their diet (Wynn et al. 2013; Levin et al. 2015), which suggest that *Au. afarensis* was exploiting the mosaic nature of their environments to their greatest potential. However, microwear studies show that not only do *Au. afarensis* specimens show remarkable similarity in their occlusal microwear (Grine et al. 2006) that belies the broad range of their $\delta^{13}C$ values, they share similar occlusal microwear signatures with *Au. anamensis* (Ungar et al. 2010) even though *Au. anamensis* did not incorporate C_4 or crassulacean acid metabolism (CAM) resources into its diet (Cerling et al. 2013). This implies that any inferences that might be made regarding the ecologies of early hominins based on proxy dietary signals are fraught with complications and uncertainties.

Nevertheless, given the current evidence, it is reasonable to suggest that *Au. afarensis* may have been generalists who consumed an isotopically varied diet from a variety of resources in its diverse environment. This is in direct contrast with *Au. anamensis*, who ate mostly C_3 food resources (Cerling et al. 2013) even though they were also found in complex mosaics of woodland and grassland habitats (Harris and Leakey 2003). Unfortunately, it is not possible at this point to postulate how *Au. afarensis* utilized their environments and which specific resources they exploited, but continued research into their habitats, paleobiology, and inferred dietary behavior at finer resolutions will help to advance current hypotheses and formulate new ones.

Conclusions

Korsi Dora vertebrate locality 1 (KSD-VP-1) samples one of the lesser-known time periods of hominin evolution and has yielded one of the most complete *Australopithecus afarensis* skeletons known (KSD-VP-1/1; Haile-Selassie et al. 2010). Results of taphonomic analyses indicate that the KSD-VP-1 assemblage is autochthonous; there was very little carnivore damage and minimal transport of bones. Weathering profile of the assemblage suggests that bones were generally not buried quickly after the death of an animal, but the lack of transport and carnivore damage and higher proportion of partial skeletons and associated elements indicates that the assemblage was also not disturbed prior to burial.

The ecological proxies examined in this study suggest that while some shrubland and grassland was likely present in the general area of the Woranso-Mille Study Area, wooded habitats were dominant at KSD-VP-1. There is further sedimentological and faunal evidence that there was likely a large body of water, but that KSD-VP-1 was sampling its shallow margin where water was slow moving.

KSD-VP-1 is only the first of the Woranso-Mille localities to be subjected to detailed taphonomic and paleoecological analyses. Future research on other localities will provide a detailed picture of the spatial and temporal distribution of vegetation types at Woranso-Mille and provide a better understanding of *Au. afarensis* resource use at Woranso-Mille. The paleoenvironmental reconstruction of KSD-VP-1 as a medium to densely wooded habitat indicates that *Au. afarensis* had access to and may have utilized significant proportions of C_3 resources at Woranso-Mille, although their enamel isotopic signature indicates the incorporation of significant C_4 resources in their diet (Levin et al. 2015) and a generalist strategy in acquiring dietary resources. Recent analyses of *Au. afarensis* paleoenvironments (Bobe et al. 2007; Reed 2008; Su 2011; Behrensmeyer and Reed 2013) and enamel carbon isotope composition (Wynn et al. 2013) corroborate this and suggest that populations of *Au. afarensis* were generalist omnivores who exploited the wide range of resources that their mosaic environment had to offer and was likely the first early hominin to incorporate significant C4/CAM resources into its diet (Wynn et al. 2013; Levin et al. 2015), at as early as 3.76 Ma (Levin et al. 2015). Ongoing research into the paleoenvironment and functional morphology of *Au. afarensis* at Woranso-Mille and other *Au. afarensis* sites can help us better appreciate the relationship between diet, behavior, morphology, and environment, as well as refine our understanding of the evolutionary history of *Au. afarensis*.

Acknowledgments I am grateful to Y. Haile-Selassie for inviting me to study the taphonomy and paleoecology of KSD-VP-1 and to the members of the Woranso-Mille Paleontological project who collected the fossil material analyzed in this study. I thank the Authority for Research and Conservation of Cultural Heritage and the National Museum of Ethiopia of the Ministry of Culture and Tourism for permission to study the fossil material under their care. The Afar Regional Government, its local administrative units, and the Afar people of the Mille District are thanked for facilitating and participating in the fieldwork that resulted in the recovery of the material discussed here. Thanks to S. Curran, K. Reed, and an anonymous reviewer for helpful comments. Field research was supported by Grants BCS-0234320, BCS-0542037, and BCS-0321893 from the National Science Foundation and by the Leakey Foundation, the Wenner-Gren Foundation, and the National Geographic Society. Laboratory research for this study was supported by the Leakey Foundation.

References

Alemseged, Z. (2003). An integrated approach to taphonomy and faunal change in the Shungura Formation (Ethiopia) and its implication for hominid evolution. *Journal of Human Evolution*, *44*, 451–478.

Alemseged, Z., Wynn, J. G., Kimbel, W. H., Reed, D., Geraads, D., & Bobe, R. (2005). A new hominin from the basal member of the Hadar Formation, Dikika, Ethiopia, and its geological context. *Journal of Human Evolution*, *49*, 499–514.

Alemseged, Z., Spoor, F., Kimbel, W. H., Bobe, R., Geraads, D., Reed, D., & Wynn, J. G. (2006). A juvenile early hominin skeleton from Dikika, Ethiopia. *Nature, 443*, 296–301.

Alemseged, Z., Bobe, R., & Geraads, D. (2007). Comparability of fossil data and its significance for the interpretation of hominin environments. In A. K. Berhrensmeyer, Z. Alemseged, & R. Bobe (Eds.), *Hominin environments in the East African Pliocene: An assessment of the faunal evidence* (pp. 159–181). Dordrecht: Springer.

Bedaso, Z. K., Wynn, J. G., Alemseged, Z., & Geraads, D. (2013). Dietary and paleoenvironmental reconstruction using stable isotopes of herbivore tooth enamel from middle Pliocene Dikika, Ethiopia: Implication for *Australopithecus afarensis* habitat and food resources. *Journal of Human Evolution, 64*, 21–38.

Behrensmeyer, A. K. (1975). The taphonomy and paleoecology of Plio-Pleistocene vertebrate assemblages east of Lake Rudolf, Kenya. *Bulletin of the Museum of Comparative Zoology, 146*, 473–578.

Behrensmeyer, A. K. (1978). Taphonomic and ecologic information from bone weathering. *Paleobiology, 4*, 150–162.

Behrensmeyer, A.K. & Reed, K.E. (2013). Chapter 4 Reconstructing the habitats of australopithecus: Paleoenvironments, site taphonomy, and faunas. *The Paleobiology of Australopithecus* (pp. 41–60). Dordrecht, Heidelberg, London, New York: Springer.

Bishop, L. C., Plummer, T. W., Hertel, F., & Kovarovic, K. (2011). Chapter 17 Paleoenvironments of Laetoli, Tanzania as determined by antelope habitat preferences. In K. E. Reed, J. G. Fleagle, & R. E. Leakey (Eds.), *Paleontology and geology of laetoli: Human evolution in context. Volume 1: Geology, geochronology, paleoecology and paleoenvironment* (pp. 355–366). Dordrecht: Springer.

Bobe, R. (1997). *Hominid environments in the Pliocene: An analysis of fossil mammals from the Omo Valley, Ethiopia*. Ph.D., University of Washington.

Bobe, R., & Eck, G. G. (2001). Responses of African bovids to Pliocene climatic change. *Paleobiology, 27*, 1–47.

Bobe, R., Behrensmeyer, A. K., & Chapman, R. E. (2002). Faunal change, environmental variability and late Pliocene hominin evolution. *Journal of Human Evolution, 42*, 475–497.

Bobe, R., Behrensmeyer, A. K., & Eck, G. G. (2007). Patterns of abundance and diversity in late Cenozoic bovids from the Turkana and Hadar Basins, Kenya and Ethiopia. In A. K. Behrensmeyer, Z. Alemseged, & R. Bobe (Eds.), *Hominin environments in the East African Pliocene: An assessment of the faunal evidence* (pp. 129–157). Dordrecht: Springer.

Bonnefille, R., & DeChamps, R. (1983). Data on Fossil Flora. In J. de Heinzelin (Ed.), *The Omo Group: Archives of the International Omo Research Expedition, Annales, S. 8, Sciences Geologiques* (pp. 191–207). Tervuren: Musée de l'Afrique Centrale.

Bonnefille, R., Potts, R., Chalié, F., Jolly, D., & Peyron, O. (2004). High-resolution vegetation and climate change associated with Pliocene *Australopithecus afarensis*. *Proceedings of the National Academy of Science, 101*, 12125–12129.

Brain, C. K. (1967). Bone weathering and the problem of bone pseudo-tools. *South African Journal of Science, 63*, 97–99.

Campisano, C. J., & Feibel, C. S. (2008). Depositional environments and stratigraphic summary of the Pliocene Hadar formation at Hadar, Afar depression, Ethiopia. *Geological Society of America Special Papers, 446*, 179–201.

Cerling, T. E., Harris, J. M., & Passey, B. H. (2003). Diets of East African bovidae based on stable isotope analysis. *Journal of Mammalogy, 84*, 456–470.

Cerling, T. E., Levin, N. E., Quade, J., Wynn, J. G., Fox, D. L., & Kingston, J. D., et al. (2010). Comment on the paleoenvironment of *Ardipithecus ramidus*. *Science, 328*, 1105.

Cerling, T. E., Manthi, F., Mbua, E., Leakey, L., Leakey, M., & Leakey, R., et al. (2013). Stable Isotope-based diet reconstructions of Turkana Basin hominins. *Proceedings of the National Academy of Sciences U S A, 110*, 10501–10506.

Ditchfield, P., & Harrison, T. (2011). Chapter 3 sedimentology, lithostratigraphy and depositional history of the Laetoli Area. In T. Harrison (Ed.), *Paleontology and geology of laetoli: Human evolution in context. Volume 1: Geology, geochronology, paleoecology and paleoenvironment* (pp. 47–76). Dordrecht: Springer.

Estes, R. D. (1991). *The behavior guide to african mammals: Including hoofed mammals, carnivores, primates*. Berkeley: University of California Press.

Fleagle, J. G., Rasmussen, D. T., Yirga, S., Bown, T. M., & Grine, F. E. (1991). New hominid fossils from Fejej, southern Ethiopia. *Journal of Human Evolution, 21*, 145–152.

Frison, G. C., & Todd, L. C. (1986). *The colby mammoth site: Taphonomy and archaeology of a clovis kill in northern Wyoming*. Albuquerque: University of New Mexico Press.

Gagnon, M., & Chew, A. E. (2000). Dietary preferences in extant African Bovidae. *Journal of Mammalogy, 81*, 490–511.

Gentry, A. W. (1978). Bovidae. In V. J. Maglio & H. B. S. Cooke (Eds.), *Evolution of African mammals* (pp. 540–572). Cambridge: Harvard University Press.

Grine, F. E., Ungar, P. S., Teaford, M. F., & El-Zaatari, S. (2006). Molar microwear in *Praeanthropus afarensis*: Evidence for dietary stasis through time and under diverse paleoecological conditions. *Journal of Human Evolution, 51*, 297–319.

Haile-Selassie, Y., Latimer, B. M., Alene, M., Deino, A. L., Gibert, L., & Melillo, S. M., et al. (2010). An early *Australopithecus afarensis* postcranium from Woranso-Mille, Ethiopia. *Proceedings of the National Academy of Sciences U S A, 107*, 12121–12126.

Harris, J. M., & Leakey, M. G. (Eds.). (2003). *Geology and vertebrate paleontology of the early pliocene site of Kanapoi, northern Kenya*. Los Angeles: Natural History Museum of Los Angeles County.

Harrison, T. (2011). Laetoli Revisited: Renewed Paleontological and Geological Investigations at Localities on the Eyasi Plateau in Northern Tanzania. In T. Harrison (Ed.), *Paleontology and geology of laetoli: Human evolution in context. Volume 1: Geology, geochronology, paleoecology and paleoenvironment* (pp. 1–15). Dordrecht: Springer Press.

Hill, A. P. (1980). Early postmortem damage to the remains of some contemporary East African mammals. In A. K. Behrensmeyer & A. P. Hill (Eds.), *Fossils in the making: Vertebrate taphonomy and paleoecology* (pp. 131–152). Chicago: The University of Chicago Press.

Kaiser, T. M. (2011). Feeding ecology and niche partitioning of the Laetoli ungulate faunas. In T. Harrison (Ed.), *Paleontology and geology of laetoli: Human evolution in context. Volume 1: Geology, geochronology, paleoecology and paleoenvironment* (pp. 329–354). Dordrecht: Springer.

Kappelman, J., Swisher III, C. C., Fleagle, J. G., Yirga, S., Bown, T. M., & Feseha, M. (1996). Age of *Australopithecus afarensis* from Fejej, Ethiopia. *Journal of Human Evolution, 30*, 139–146.

Kimbel, W. H., & Delezene, L. K. (2009). "Lucy" redux: A review of research on *Australopithecus afarensis*. *American Journal of Physical Anthropology, 140*(Suppl. 49), 2–48.

Kimbel, W.H., Rak, Y., & Johanson, D.C. (2004). *The skull of* Australopithecus afarensis Oxford: Oxford University Press.

Kingdon, J. (1974). *East African mammals: An atlas of evolution in Africa, volume IIIC (Bovids)*. Chicago: The University of Chicago Press.

Kingdon, J. (1997). *The Kingdon field guide to African mammals*. San Diego: Academic Press.

Kingston, J. D., & Harrison, T. (2007). Isotopic dietary reconstructions of Pliocene herbivores at Laetoli: Implications for early hominin paleoecology. *Palaeogeography, Palaeoclimatology, Palaeoecology, 243*, 272–306.

Korth, W. W. (1979). Taphonomy of microvertebrate fossil assemblages. *Annals of the Carnegie Museum, 48*, 235–285.

Kovarovic, K., & Andrews, P. (2007). Bovid postcranial ecomorphological survey of the Laetoli paleoenvironment. *Journal of Human Evolution, 52*, 663–680.

Kovarovic, K., & Andrews, P. (2011). Chapter 18 Environmental change within the Laetoli fossiliferous sequence: Vegetation catenas and bovid ecomorphology. In T. Harrison (Ed.), *Paleontology and geology of laetoli: Human evolution in context. Volume 1: Geology, geochronology, paleoecology and paleoenvironment* (pp. 367–380). Dordrecht: Springer.

Leakey, M. G., Feibel, C. S., McDougall, I., & Walker, A. (1995). New four-million-year-old hominid species from Kanapoi and Allia Bay, Kenya. *Nature, 376*, 565–571.

Leakey, M. G., & Harris, J. M. (2001). Lothagam: Its significance and contributions. In M. G. Leakey & J. M. Harris (Eds.), Lothagam: The dawn of humanity in eastern Africa (pp. 625–660). New York: Columbia University Press.

Levin, N.E., Haile-Selassie, Y., Frost, S.R., & Saylor, B.Z. (2015). Dietary change among hominins and cercopithecids in Ethiopia during the early Pliocene. *Proceedings of the National Academy of Sciences, 112*, 12304–12309.

Lyman, R. L. (1994). *Vertebrate taphonomy*. Cambridge: Cambridge University Press.

Lyman, R. L., & Fox, G. L. (1989). A critical evaluation of bone weathering as an indication of bone assemblage formation. *Journal of Archaeological Science, 16*, 293–317.

Njau, J. K., & Blumenschine, R. J. (2006). A diagnosis of crocodile feeding traces on larger mammal bone, with fossil examples from the Plio-Pleistocene Olduvai Basin, Tanzania. *Journal of Human Evolution, 50*, 142–162.

Reed, K. E. (1997). Early hominid evolution and ecological change through the African Plio-Pleistocene. *Journal of Human Evolution, 32*, 289–322.

Reed, K. E. (2008). Paleoecological patterns at the Hadar hominin site, Afar Regional State, Ethiopia. *Journal of Human Evolution, 54*, 743–768.

Saylor, B. Z., Alene, M., Deino, A., Gibert, L., Haile-Selassie, Y., Melillo, S., & Scott, G. (2016). The geologic context of Korsi Dora and the partial skeleton of KSD-VP-1/1. In Y. Haile-Selassie & D. F. Su (Eds.), *The postcranial anatomy of* Australopithecus afarensis*: New insights from KSD-VP-1/1* (pp. 13–23). Dordrecht: Springer.

Shipman, P., & Harris, J. M. (1988). Habitat preference and paleoecology of *Australopithecus boisei* in Eastern Africa. In F. E. Grine (Ed.), *Evolutionary history of the "robust" australopithecines* (pp. 343–381). New York: Aldine de Gruyter.

Shipman, P., & Rose, J. J. (1988). Bone tools: An experimental approach. In S.L. Olsen (Ed.), *Scanning electron microscopy in archaeology* (pp. 303–335). Oxford: British Archaeological Reports International Series.

Sponheimer, M., Reed, K. E., & Lee-Thorp, J. A. (1999). Combining isotopic and ecomorphological data to refine bovid paleodietary reconstruction: A case study from the Makapansgat Limeworks hominin locality. *Journal of Human Evolution, 36*, 705–718.

Sponheimer, M., Lee-Thorp, J. A., deRuiter, D. J., Smith, J. M., van der Merwe, N. J., & Reed, K. E., et al. (2003). Diets of southern African Bovidae: Stable isotope evidence. *Journal of Mammalogy, 84*, 471–479.

Su, D. F. (2011). Large Mammal Evidence for the Paleoenvironment of the Upper Laetolil and Upper Ndolanya Beds of Laetoli, Tanzania. In T. Harrison (Ed.), *Paleontology and geology of laetoli: Human evolution in context. Volume 1: Geology, geochronology, paleoecology and paleoenvironment* (pp. 381–392). Dordrecht: Springer.

Su, D. F., & Harrison, T. (2008). Ecological implications of the relative rarity of fossil hominins at Laetoli. *Journal of Human Evolution, 55*, 672–681.

Su, D. F., Ambrose, S. H., Degusta, D., & Haile-Selassie, Y. (2009). Paleoenvironment. In Y. Haile-Selassie & G. WoldeGabriel (Eds.), *Ardipithecus kadabba: Late Miocene evidence from the Middle Awash, Ethiopia*. Berkeley: University of California Press.

Suwa, G., White, T. D., & Howell, F. C. (1996). Mandibular postcanine dentition from the Shungura Formation, Ethiopia: Crown morphology, taxonomic allocations, and Plio-Pleistocene hominid evolution. *American Journal of Physical Anthropology, 101*, 247–282.

Ungar, P. S., Scott, R. S., Grine, F. E., & Teaford, M. F. (2010). Molar microwear textures and the diets of *Australopithecus anamensis* and *Australopithecus afarensis*. *Philosophical Transactions of the Royal Society of London, B, Biological Sciences, 365*, 3345–3354.

Voorhies, M. R. (1969). Taphonomy and population dynamics of an early Pliocene vertebrate fauna, Knox County, Nebraska. *Rocky Mountain Geology, 8*, 1–69.

Vrba, E. (1980). The significance of bovid remains as indicators of environment and prediction patterns. In A. K. Behrensmeyer & A. P. Hill (Eds.), *Fossils in the making* (pp. 247–271). Chicago: University of Chicago Press.

Wesselman, H. B. (1985). Fossil micromammals as indicators of climatic change about 2.4 Myr ago in the Omo Valley. *Ethiopia. South African journal of science, 81*, 260–261.

Western, D. (1973). The structure, dynamics and changes of the Amboseli ecosystem. Ph.D. dissertation, University of Nairobi, Nairobi.

White, T. D., Suwa, G., & Asfaw, B. (1994). *Australopithecus ramidus*, a new species of early hominid from Aramis, Ethiopia. *Nature, 375*, 306–312.

White, T. D., Asfaw, B., Beyene, Y., Haile-Selassie, Y., Lovejoy, C. O., Suwa, G., & WoldeGabriel, G. (2009a). *Ardipithecus ramidus* and the paleobiology of early hominids. *Science, 326*, 64–86.

White, T. D., Ambrose, S. H., Suwa, G., Su, D. F., DeGusta, D., & Bernor, R. L., et al. (2009b). Macrovertebrate Paleontology and the Pliocene Habitat of *Ardipithecus ramidus*. *Science, 326*, 67, 87–93.

Woldegabriel, G., White, T. D., Suwa, G., Renne, P., Deheinzelin, J., & Hart, W. K., et al. (1994). Ecological and Temporal Placement of Early Pliocene Hominids at Aramis, Ethiopia. *Nature, 371*, 330–333.

Wynn, J. G. (2000). Paleosols, stable carbon isotopes, and paleoenvironmental interpretation of Kanapoi, Northern Kenya. *Journal of Human Evolution, 39*, 411–432.

Wynn, J. G., Alemseged, Z., Bobe, R., Geraads, D., Reed, D., & Roman, D. C. (2006). Geological and palaeontological context of a Pliocene juvenile hominin at Dikika, Ethiopia. *Nature, 443*, 332–336.

Wynn, J. G., Sponheimer, M., Kimbel, W. H., Alemseged, Z., Reed, K., & Bedaso, Z. K., et al. (2013). Diet of *Australopithecus afarensis* from the Pliocene Hadar Formation, Ethiopia. *Proceedings of the National Academy of Sciences USA, 110*, 10495–10500.

Chapter 4
KSD-VP-1/1: Analysis of the Postcranial Skeleton Using High-Resolution Computed Tomography

Timothy M. Ryan and Simone Sukhdeo

Abstract The KSD-VP-1/1 partial *Australopithecus afarensis* skeleton from Woranso-Mille, Ethiopia, provides an excellent opportunity to reconstruct various aspects of the paleobiology of *Australopithecus*. High-resolution computed tomography (HRCT) scan data were collected from each skeletal element. Three-dimensional reconstructions were used to visualize and quantify internal and external anatomical structures. The skeleton was heavily mineralized and preservation of the internal bone structures, such as the endosteal borders of diaphyseal cortical bone and the trabecular structure of the epiphyses was generally poor. The three-dimensional HRCT data also permitted the reconstruction of the pelvic girdle and the fragmented distal femur using rigid transformations of isosurface reconstructions.

Keywords High-resolution computed tomography • Cortical bone • Trabecular bone

Introduction

The partial *Australopithecus afarensis* skeleton from Woranso-Mille, Ethiopia, provides an excellent opportunity to reconstruct the paleobiology of *Australopithecus* based on a generally well-preserved skeleton of a large-bodied male (Haile-Selassie et al. 2010). This fossil, KSD-VP-1/1, preserves components of the upper and lower limbs, pelvic and pectoral girdles, ribs, and cervical vertebrae. The importance of the KSD-VP-1/1 skeleton arises from the unique suite of anatomical features present indicating a relatively derived postcranial skeleton. Though this early *Australopithecus* dates to approximately 3.6 Ma, the derived features of the thorax and lower limb suggest a much earlier acquisition of critical anatomical features, such as relatively long lower limbs, that are typically associated with later hominins of the genus *Homo* (Haile-Selassie et al. 2010).

Analyses of the external anatomical features of the skeleton provide important insights into the evolutionary morphology and locomotor behavior of this specimen and *Australopithecus* in general, but further observations and understanding may be gained from analyses of the internal morphology of the skeleton. High-resolution computed tomography (HRCT) offers the ability to characterize the internal morphology of the skeleton nondestructively, and the resultant high-quality image data and metrics from three-dimensional (3D) datasets can reveal novel information regarding *Australopithecus* paleobiology. These 3D HRCT data permit the reconstruction of fragmentary and deformed elements using traditional and 3D geometric morphometric techniques, the reconstruction of paired elements through mirror imaging and rigid transformation of missing parts, and the creation of 3D visualizations of internal and external anatomical features. Functional and locomotor hypotheses can be evaluated by examining functionally relevant internal structures, preserved through virtual reconstructions and visualizations, and integrating the internal observations with those from external anatomical observations.

The goals of this chapter are to provide an overview and evaluation of the HRCT data acquired from the KSD-VP-1/1 skeleton, to present 3D visualizations of the skeletal elements with associated interpretations of anatomical features, and to use 3D reconstructions of various elements to investigate anatomical form and function. In doing so, this chapter will explore the advantages of using 3D visualizations and virtual reconstruction for fossil analysis, particularly in cases where the fossil material is partial or damaged.

T.M. Ryan (✉) · S. Sukhdeo
Department of Anthropology, Pennsylvania State University, University Park, PA 16802, USA
e-mail: tmr21@psu.edu

S. Sukhdeo
e-mail: sms6065@psu.edu

T.M. Ryan
Center for Quantitative Imaging, EMS Energy Institute, Pennsylvania State University, University Park, PA 16802, USA

Yohannes Haile-Selassie and Denise F. Su (eds.), *The Postcranial Anatomy of Australopithecus afarensis: New Insights from KSD-VP-1/1*, Vertebrate Paleobiology and Paleoanthropology, DOI 10.1007/978-94-017-7429-1_4

Most of the comparative and functional analyses of the KSD-VP-1/1 skeletal morphology will be covered in other chapters in this volume, so this chapter will be primarily descriptive in nature. Descriptions of the HRCT data from each skeletal element will be accompanied by 3D visualizations, highlighting relevant features of the external and internal anatomy, and a report of measurements collected from the HRCT data. In selected cases, reconstructions of fragmentary or paired elements, (e.g., innominate) were undertaken and presented visually for further analysis.

Materials and Methods

The individual elements of the KSD-VP-1/1 skeleton were scanned on the OMNI-X HD-600 high-resolution X-ray CT scanner (Varian Medical Systems, Lincolnshire, IL) in the Center for Quantitative Imaging (CQI) at the Pennsylvania State University. Because the fossil is relatively heavily mineralized, each element was mounted in a plastic cylinder and embedded in 0.1 mm glass beads prior to scanning as a means of reducing beam hardening and other scan artifacts. The fossil specimens were embedded directly in the glass beads with no other material surrounding the fossil. No metal filtration was used on the X-ray source to further harden the X-ray beam. The diameter of the plastic cylinder varied depending on the size of each fossil element. The plastic tube containing each fossil was positioned vertically in the HRCT system to collect transverse CT slices and to maximize scan resolution. For each element, serial cross-sectional scans were collected covering the entire bone. Source energy settings varied depending on the specimen (Table 4.1). Scans were collected using 2,800 views with 8 samples averaged per view and approximately 350 slices per rotation. Voxel sizes ranged between 0.041 and 0.129 mm depending on the size of the specimen (Table 4.1) with the principal goal of collecting the highest resolution images possible given the size of the specimen. One exception to this approach was the scan of the ribs which were mounted together in one plastic tube and scanned at approximately 0.066 mm voxel size. As a result of the specific hardware configuration, the OMNI-X HD-600 scanner does not generate isotropic voxels, resulting in slice thicknesses slightly larger than the inline pixel dimensions. All CT data were reconstructed as 16-bit TIFF grayscale images with a 1024 × 1024 pixel matrix.

The image data were processed to extract each fossil element from the surrounding glass beads and to create individual datasets for each fossil specimen. Once each element was isolated from the glass beads, three-dimensional isosurface reconstructions, and volume renderings were produced using Avizo 7.1 Standard visualization software (Visualization Sciences Group, Burlington, MA). Each element was processed in the same way to facilitate visualization of the fossils and identification of notable features preserved in the fossilized remains as revealed by the HRCT data. An initial assessment of the existing structure determined if further steps could be taken in certain cases to reconstruct the original anatomy from fragmentary elements.

Three-dimensional isosurface reconstructions are a quick and commonly used method for visualizing CT data by generating a 3D triangulated surface from the segmented image data (Lorensen and Cline 1987; Schroeder et al. 1998; Zollikofer and Ponce de Leon 2005). Colored and textured surfaces reveal details of the morphological features of each element and allow for virtual manipulation of the object. Isosurface reconstructions were produced for each individual bone using automatic histogram-based segmentation algorithms. Each specimen was inspected following automatic segmentation and edited manually when necessary to include or exclude relevant structures. For several elements including the cervical vertebrae, ribs, innominate, and sacrum, the 3D triangulated surfaces were translated in 3D space within Avizo to produce reconstructions of multiple elements in general anatomical position. The articular surfaces of the distal femur and distal tibia were defined digitally by selecting only triangles on the articular surfaces.

Three-dimensional volume renderings were produced for each element using a glow colormap with diffuse lighting

Table 4.1 High-resolution computed tomography scan settings for each specimen

Element	Specimen numbers	Field of view (mm)	Pixel size (*x*, *y*) (mm)	Slice thickness (*z*) (mm)	Energy settings
Cervical vertebrae	KSD-VP-1/1h,i,j,k,l	40.45	0.040	0.041	180 kV; 0.250 mA
Left clavicle	KSD-VP-1/1f	63.49	0.062	0.066	180 kV; 0.250 mA
Ribs	KSD-VP-1/1n,o,p,q,r	63.49	0.062	0.066	180 kV; 0.250 mA
Right scapula	KSD-VP-1/1g	132.10	0.129	0.129	250 kV; 0.200 mA
Right humerus	KSD-VP-1/1b	77.82	0.076	0.074	250 kV; 0.100 mA
Right ulna	KSD-VP-1/1a	63.49	0.062	0.066	180 kV; 0.250 mA
Innominate	KSD-VP-1/1d	132.10	0.129	0.129	250 kV; 0.200 mA
Sacral body and ala	KSD-VP-1/1t	63.49	0.062	0.066	180 kV; 0.250 mA
Left distal femur	KSD-VP-1/1c	77.82	0.076	0.074	250 kV; 0.100 mA
Left tibia	KSD-VP-1/1e	77.82	0.076	0.074	250 kV; 0.100 mA

and specularity (Schroeder et al. 1998; Zollikofer and Ponce de Leon 2005). Volume rendering is a ray-casting technique that produces a 3D rendering of an object by simulating the transmission of light through the 3D volume data. Compared to 3D surface reconstructions which rely on an isocontour approach that loses most density and some structural details of the original dataset, volume-based visualizations retain the density information contained in the original image dataset and are therefore useful for highlighting density and textural details of each fossil element. This technique uses faux shading from a glow colormap and edge coloring to create a 3D rendering. By highlighting surface features, a volume rendering makes it easier to discern subtle changes in the topography associated with muscle attachments, insertions, and the delineation of articular surfaces.

Supplementing the volume renderings, multiple cross-sectional orthoslices were extracted to reveal the internal structure of the fossil specimens. Orthoslices are two-dimensional images that can be extracted from any of the three orthogonal planes, as well as adjusted to obliquely slice through an element. The orthoslices generally showed significant damage and cortical exfoliation throughout the postcranial skeleton; in a better-preserved fossil specimen, orthoslices could be used for visual inspection of the internal trabecular bone architecture and later aid in the extraction of volumes of interest.

A small comparative sample consisting of extant hominids (*Homo sapiens*, *Pan troglodytes*) and several *Australopithecus* fossil specimens was used to illustrate anatomical features of the KSD-VP-1/1 skeleton. Three-dimensional HRCT data for the human sample were used from a separate project (Shaw and Ryan 2012; Macintosh et al. 2013). Adult chimpanzee femoral and tibial data were obtained from the Digital Morphology Museum, Kyoto University Primate Research Institute (http://www.pri/kyoto-u.ac.jp/dmm/WebGallery/index.html). Three-dimensional laser surface scans were collected for several fossil casts of original specimens from the Penn State Matson Museum of Anthropology cast collection using a NextEngine 3D Scanner HD. Spatial resolution of surface scans was approximately 0.5 mm in all dimensions.

Results

High-Resolution Computed Tomography Scans

The HRCT scan data for the various KSD-VP-1/1 elements indicate relatively heavy mineralization and many of the bones display significant exfoliation and erosion of the periosteal bone surfaces. In general, the shafts of the long bones do not preserve much of the original cortical bone, but instead they represent eroded casts of the diaphyseal medullary cavity. Dense mineral infilling is present in most of the skeletal elements, and in some of the specimens, the density of the mineral infilling matches or exceeds the density of the fossilized bone. The endosteal cortical bone in many elements is difficult to differentiate from the matrix, although this varies between elements and often within a single element. Preservation of trabecular bone structure is similarly variable within and between elements. The matrix obscures internal bony features, making visual assessment challenging. When present, trabecular bone manifests as relatively low-density regions within the denser matrix that fills intertrabecular spaces. These regions are apparent in the proximal and distal tibia, os coxa, cervical vertebral bodies, clavicle, distal humerus and distal femur. Depositional damage and diagenesis result in the lack of material contrast between struts and matrix in their intervening spaces. These preservation limitations prevent effective quantification of trabecular bone architecture.

Right Scapula (KSD-VP-1/1g)

KSD-VP-1/1g is a nearly complete right scapula (Fig. 4.1, also see Fig. 6.3 in Melillo 2016). The morphology of this specimen has already been discussed in great detail by Haile-Selassie et al. (2010). The overall structure is fairly intact, with a complete glenoid fossa and preserved axillary and vertebral borders. Few early hominin scapulae exist for comparison, but three scapulae representing *Au. afarensis* (A.L. 288-1), *Au. africanus* (Sts 7), and *Au. sediba* (MH2) can be used to compare glenoid fossa dimensions among early hominins (Vrba 1979; Johanson et al. 1982; Churchill et al. 2013). The glenoid fossa measurements are summarized in Table 4.2. The maximum glenoid length and breadth dimensions reveal that KSD-VP-1/1 has a much larger glenoid fossa than the small-bodied A.L. 288-1 specimen and MH2. The KSD-VP-1/1 glenoid size fits more closely with Sts 7 and falls within the modern human range (Vrba 1979; Churchill and Trinkaus 1990; Churchill et al. 2013).

The glenoid index, however, suggests that the shape of the KSD-VP-1/1 glenoid fossa is more similar to that of A.L. 288-1, with values of 71.6 and 70.8, respectively (Table 4.2). These index values fall within the overlapping ranges of modern human and chimpanzees. In contrast, Sts 7 and MH2 exhibit lower glenoid indices, suggesting they possess either long fossae relative to width, or narrow fossae relative to length, compared to modern humans and KSD-VP-1/1 (Churchill et al. 2013). Johanson et al. (1982) describes the A.L. 288-1 glenoid fossa as ovate, while the glenoid fossa of Sts 7 appears to have a more exaggerated pear shape, though the fossa is not complete (Vrba 1979; Larson 2013).

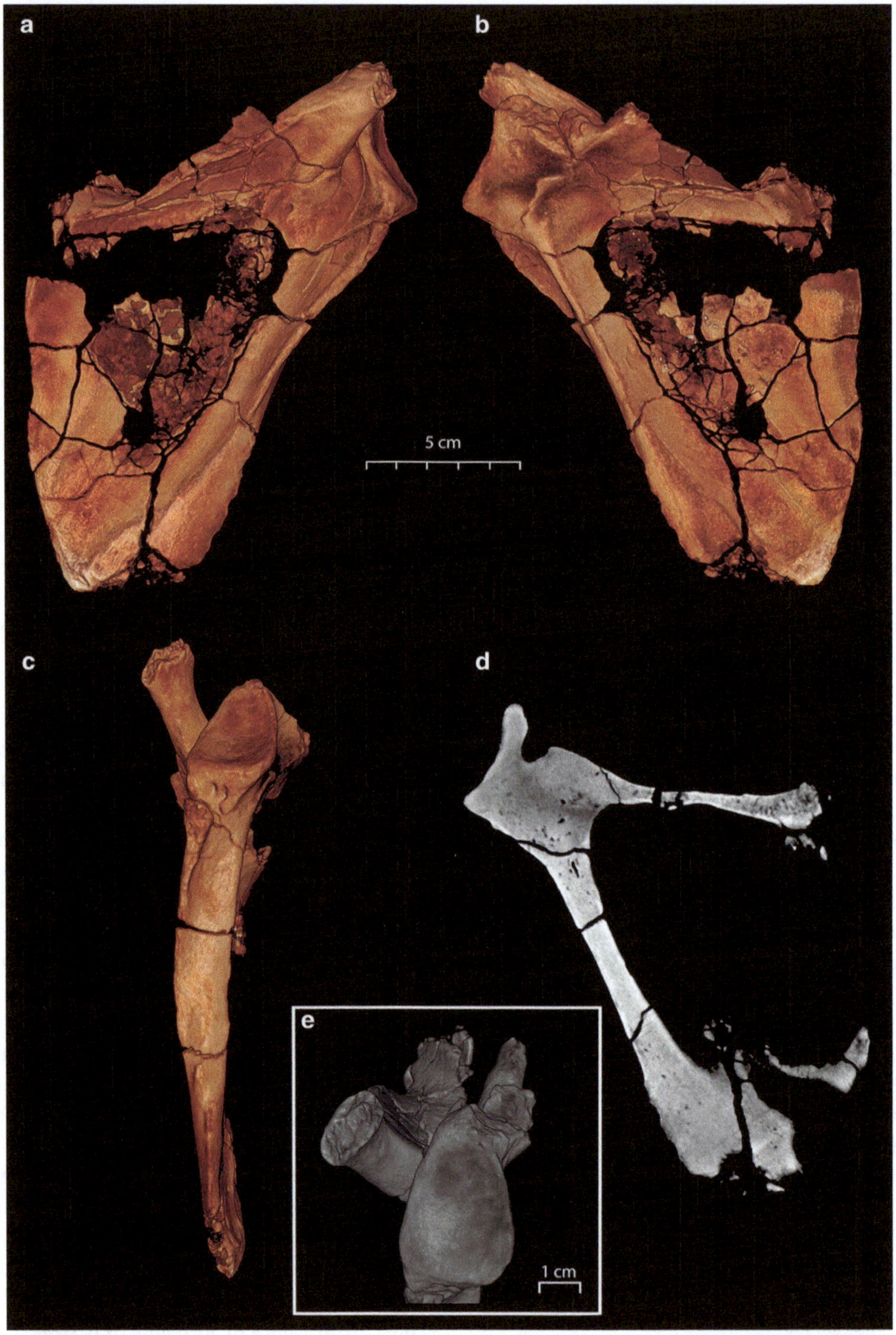

Fig. 4.1 Three-dimensional volume renderings and HRCT slice of KSD-VP-1/1g, right scapula. **a** Posterior view. **b** Anterior view. **c** Lateral view showing glenoid. **d** Two-dimensional coronal HRCT section through glenoid showing lateral and superior borders. **e** Detail of glenoid fossa

A qualitative assessment of KSD-VP-1/1 supports the glenoid index in comparison, with the glenoid fossa having a more ovate shape (Fig. 4.1).

The relatively well-preserved infraspinous fossa portion, as well as the base of the spine and portions of the supraspinous fossa, indicate that the KSD-VP-1/1 scapula has a tall scapula with a comparatively narrow mediolateral breadth, which is the characteristic of apes and modern humans (Larson 2013). A series of angular measurements from Haile-Selassie et al. (2010) describe the orientation of the shoulder joint and indicate that KSD-VP-1/1 shares some morphology with *Gorilla gorilla*, while other features match

Table 4.2 KSD-VP-1/1g scapular metrics (in mm)

Measurement	KSD-VP-1/1[a]	*Australopithecus*	*H. sapiens* (mean, SD)	*P. troglodytes* (mean, SD)	*G. gorilla* (mean, SD)	*P. pygmaeus/abelii* (mean, SD)
Glenoid Fossa						
Maximum glenoid length	35.6	[b]35.5 (*Sts 7*) [c]25.7 (*A.L. 288-1*)	[e]33.9 (3.1)			
Maximum glenoid breadth	25.5	[b]19.5 (*Sts 7*) [c]18.2 (*A.L. 288-1*)	[e]24.0 (2.2)			
Glenoid size (length × breadth$^{0.5}$)	30.1	26.3 (*Sts 7*) 21.6 (*A.L. 288-1*) 22.4 (*MH2*)	27.9 (2.4)	25.9 (2.1)	38.4 (5.2)	28.8 (2.6)
Glenoid index (100 × breadth/length)	71.6	54.9 (*Sts 7*) 70.8 (*A.L. 288-1*) 55.6 (*MH2*)	68.3 (5.2)	71.2 (4.7)	62.9 (5.7)	62.1 (4.3)
Scapular Angles						
Glenoid–Axillary	128.0	[f]117.0 (*Sts 7*) 123.4 (*MH2*) [f]120.0/121.9 (*DIK-1-1*)	137.8 (4.6)	119.0 (4.9)	122.4 (4.3)	124.1 (3.5)
Glenoid–Spinal	76.0	[b]90.0 (*Sts 7*) 82.9 (*MH2*) [f]82.9/83.4 (*DIK-1-1*)	102.4 (3.5)	76.7 (3.7)	77.8 (4.3)	79.7 (5.7)
Axillary–Vertebral	31.0	42.1 (*MH2*) [f]41.9/35.4 (*DIK-1-1*)	45.5 (3.0)	32.7 (3.6)	42.9 (3.1)	38.7 (3.3)
Spinal–Axillary	50.0	[f]29.8 (*Sts 7*) [f]38.2 (*A.L. 288-1*) 28.2 (*MH2*) [f]37.0/38.5 (*DIK-1-1*)	52.6 (3.9)	17.7 (3.3)	21.4 (3.8)	24.4 (3.7)

[a]KSD-VP-1/1 glenoid fossa values collected by authors and scapular angles are from Haile-Selassie et al. (2010)
[b]From Vrba (1979)
[c]From Johanson et al. (1982)
[d]MH2, *H. sapiens*, *P. troglodytes*, *G. gorilla*, and *P. pygmaeus/abelii* values for glenoid size, glenoid index, and the scapular angles from Churchill et al. (2013)
[e]From Churchill and Trinkaus (1990)
[f]From Green and Alemseged 2012. Note, DIK-1-1 is a juvenile and the scapular angles reflect the right/left scapulae

the *Homo sapiens* pattern more closely. The cranially-oriented glenoid fossa combined with a transversely-oriented spine is clearly seen in a coronal slice through KSD-VP-1/1g in Fig. 4.1d. The angular measurements are summarized in Table 4.2, along with comparable data taken from the literature.

Left Clavicle (KSD-VP-1/1f)

KSD-VP-1/1f is a nearly complete left clavicle with relatively good external and internal preservation (Fig. 4.2, also see Fig. 6.4 in Melillo 2016). Though the clavicle is missing a significant portion of the sternal end, it retains much of its original morphology, with muscle attachment sites and anteroposterior curvature that appears similar to other hominins and modern humans. However, without the full clavicular dimensions, estimations of the internal/external and inferior/superior curvature are not reliable for comparison with other species (Voisin 2006, 2008).

Volume renderings highlight the surface topography of the fossil, revealing a well-preserved, smooth inferior surface with one large transverse fracture near the midshaft of the specimen and several smaller fractures. The superior surface is more rugose with significant exfoliation of the cortical bone surface. In anterior view, the clavicle possesses a slight inferior curvature from the acromial to the sternal end, which is also present in African apes (Larson 2013). The complete right clavicle of MH2 (*Au. sediba*) also presents an inferiorly angled clavicle, suggesting that KSD-VP-1/1 and MH2 share a high scapular position with African apes (Churchill et al. 2013). The curvature may indicate that the pectoral girdle of KSD-VP-1/1 and MH2 was similar to the primitive hominini condition described by Larson et al. (2007), though KSD-VP-1/1f contradicts the prediction of a short clavicle with an estimated length of 156.4 mm compared to the estimated length of 107.5 mm for MH2.

Three roughly parasagittal cross sections through the clavicle, as well as a longitudinal transverse section along the clavicular shaft, reveal variation in cortical bone preservation (Fig. 4.2). Medially, the cortical bone exhibits clearly discernable internal and external margins. More significant erosion of the cortical bone structure is evident in the middle and lateral aspects of the preserved shaft region. The transverse section along the shaft exposes trabecular bone structure in some regions of the clavicle (Fig. 4.2). However, the apparent preservation of the trabecular bone appears to be primarily the retention of empty intertrabecular spaces in some locations along the length of the bone.

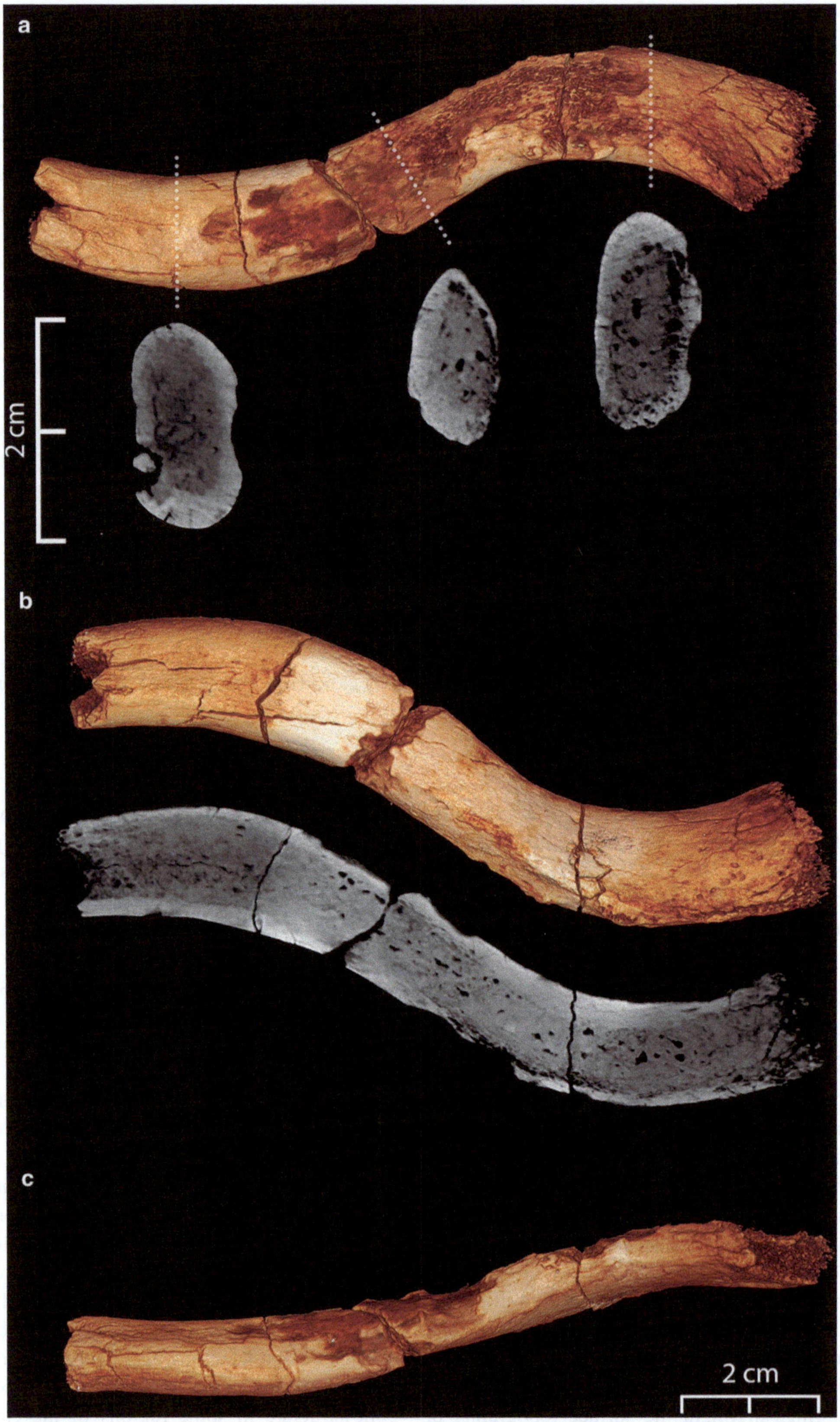

Fig. 4.2 Three-dimensional volume renderings and two-dimensional HRCT sections of KSD-VP-1/1f, left clavicle. **a** Superior view with 2D CT slices indicated by white lines. Scale refers to 2D cross section slices only. **b** Inferior view with 2D transverse CT slice showing internal structure, same scale as 3D renderings. **c** Anterior view. Medial is to the left of the image. The dotted white lines indicating location of each cross section are oriented orthogonal to the shaft in each location. Scale at bottom for all 3D renderings

Right Humerus (KSD-VP-1/1b)

KSD-VP-1/1b preserves approximately the distal two-thirds of the right humerus (Haile-Selassie 2010), including most of the trochlea and capitulum, although there is some exfoliation of the articular surfaces in this region (Fig. 4.3, also see Fig. 8.2 in Lovejoy et al. 2016). A significant portion of the olecranon fossa is missing, as well as the distal medial portion of the metaphysis and distal diaphysis. The diaphyseal cortical bone is significantly eroded along most of the shaft, and its final dimensions and appearance represent what appears to be an incomplete cast of the medullary cavity. Proximal portions of the partial humerus preserve

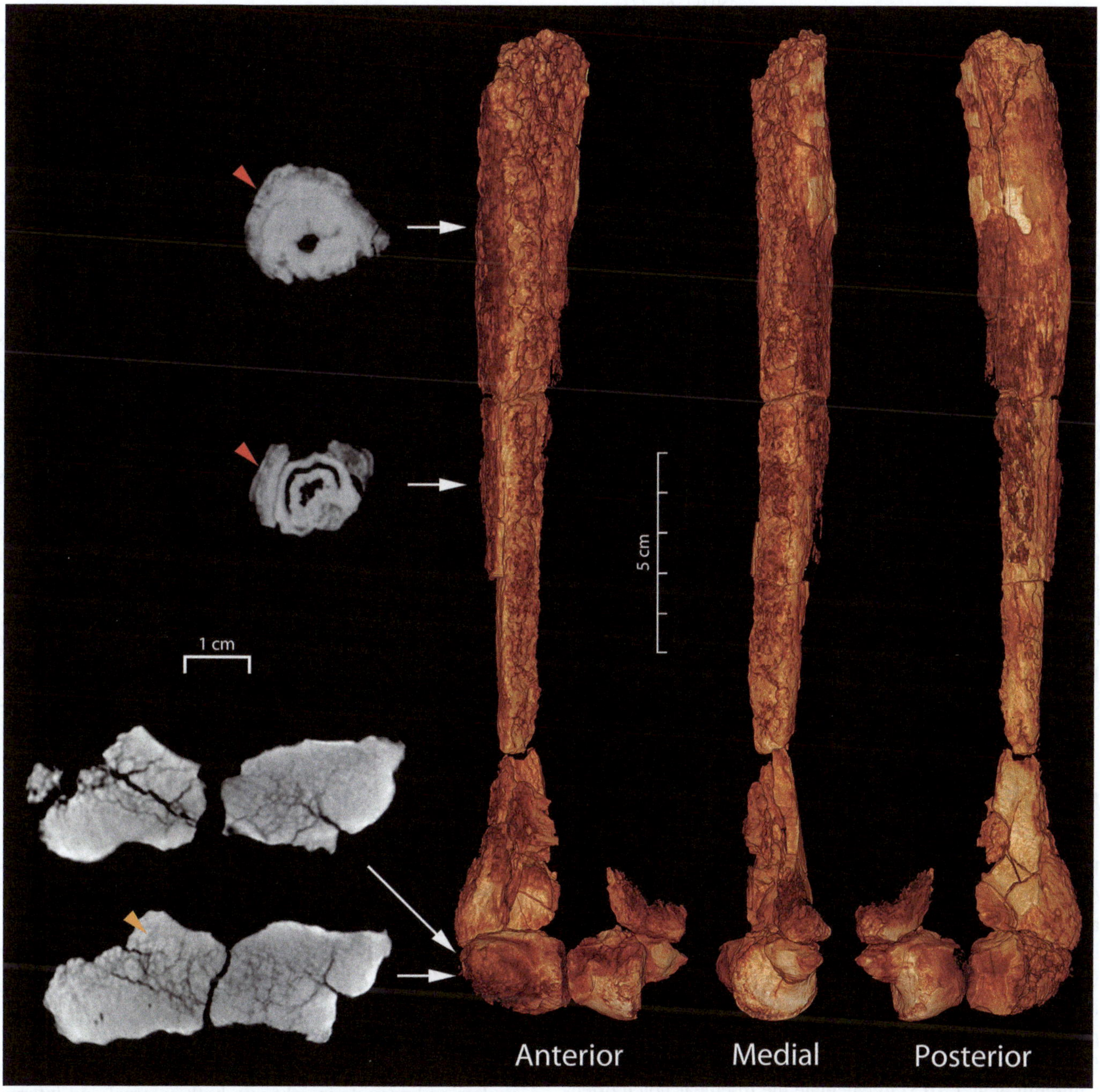

Fig. 4.3 Three-dimensional volume renderings and two-dimensional HRCT sections of KSD-VP-1/1b, right humerus. Volume renderings of the two right humerus fragments showing anterior, medial, and posterior views. Two-dimensional HRCT sections on the left reveal some preservation of cortical bone structure along diaphysis (red arrows) and possible preservation of trabecular bone in the trochlea and capitulum (orange arrow). Trabecular bone appears as discontinuous, low-density features within the higher density matrix and is clearly distinguishable from fractures

Table 4.3 KSD-VP-1/1b humeral metrics (in mm)

Measurement	KSD-VP-1/1	*Australopithecus*	Early *Homo* (mean, SD)	*H. sapiens* (mean, SD)	*P. troglodytes* (mean, SD)	*G. gorilla* (mean, SD)
Distal articular breadth	[a]48	[b]38.5 (6.3) [b]35.1 (3.7) (*MH2*) [c]30.1 (*A.L. 288-1*) [c]41.0 (*Stw 431*)	[b]46.4 (7.4)	[a]44.9 (4.6)	[a]45.7 (13.0)	[a]61.0 (8.8)
Biepicondylar breadth	58.8	[b]55.0 (11.4) [b]53.5 (4.9) (*MH2*) [c]41.0 (*A.L. 288-1*) [c]59.0 (*Stw 431*)	[b]63.0 (5.6)	[b]62.1 (5.3)	[b]62.1 (4.6)	[b]88.4 (13.2)
Trochlear breadth	[a]27	[a]19.3 (*A.L. 288-1*)	–	[a]27.3 (2.9)	[a]27.2 (6.4)	[a]27.3 (2.9)

Australopithecus sample includes: A.L. 288-1, BOU-VP-12/1, BOU-VP-35/1, KSD-VP-1/1b, A.L. 333-107, KNM-BC 1745, Omo 199-73-27, KNM-ER 1473, Sts 7, Stw 328, A.L. 137-48a, A.L. 322-1, KNM-ER 739, KNM-ER 1504, KNM-KP 271, KNM-ER 6020, SKX 10924, SK 24600, TM 1517, Stw 431. Early *Homo* sample includes: KNM-WT 15000, D4507, Gomboré IB 7594 and Kabwe E.898
[a]From Haile-Selassie et al. (2010)
[b]From Churchill et al. (2013)
[c]From Dobson (2005)

some of the cortical bone, but cross sections are not complete (Fig. 4.3). The cross sections reveal multiple cavities within the matrix of the medullary cavity. Trabecular bone structure is visible in both the trochlea and capitulum as low-density structures interspersed with small fractures through the specimen. Despite poor preservation at the distal end of the humerus, some measurements of the articular region were taken from the isosurface reconstruction, using a measurement tool that anchors its endpoints to the triangulated surface. The metrics show that KSD-VP-1/1 has a distal articular breadth that is approximately 1.6 times that of A.L. 288-1, supporting the larger estimated body size of KSD-VP-1/1 compared to A.L. 288-1. The measurements for articular and biepicondylar breadth fall within modern human and chimpanzee ranges (Dobson 2005; Haile-Selassie et al. 2010). Descriptive measurements of KSD-VP-1/1b are listed in Table 4.3 together with some comparative values from the literature.

Right Ulna (KSD-VP-1/1a)

KSD-VP-1/1a is a partial right ulna preserving approximately 60% of the total length (Haile-Selassie et al. 2010) (Fig. 4.4, also see Fig. 8.4 in Lovejoy et al. 2016). The proximal end, including both the olecranon and coronoid processes, is very well preserved. The trochlear notch is oriented anteriorly like in other early hominins, suggesting different load-bearing postures than found in humans or suspensory apes (Haile-Selassie et al. 2010). Using the method described by Churchill et al. (2013) to calculate the trochlear notch orientation, KSD-VP-1/1a has a value of 93.5, which is higher than that seen in the comparative *Australopithecus* sample, whose mean value was 82.6 (Churchill et al. 2013). This high value corroborates the results found by Haile-Selassie et al. (2010) using the method outlined in Drapeau (2004) and Drapeau et al. (2005). The anterior-facing trochlear notch is suggested as a unique trait of early hominins, now including KSD-VP-1/1, which differs significantly from the superior-facing trochlear notches found in African apes and modern humans (Haile-Selassie et al. 2010). Additionally, the dimensions of the trochlear notch and the proximal articular surface in KSD-VP-1/1a are generally close to the values found in early *Homo*, and within the ranges of *Homo sapiens* and *Pan troglodytes*. Other descriptive measurements for the ulna are provided in Table 4.4.

A longitudinal parasagittal HRCT section as well as transverse HRCT slices reveal relatively poor preservation of cortical or trabecular bone structure throughout the entire specimen (Fig. 4.4). There is no evidence of preserved trabeculae in the olecranon or coronoid processes. The HRCT scans also show very little evidence of preserved endosteal cortical surfaces along the shaft or in the coronoid and olecranon processes. It seems likely, given the preservation in other KSD-VP-1/1 specimens, that the cortical and trabecular bone are not missing from these regions, but are of the same density as the surrounding matrix and not discernable in the HRCT images.

Cervical Vertebrae (KSD-VP-1/1h, KSD-VP-1/1i, KSD-VP-1/1j, KSD-VP-1/1k, KSD-VP-1/1l)

The second through sixth cervical vertebral bodies are fairly complete and provide insight into both external and internal morphology of the *Australopithecus* cervical region. Three-dimensional surface reconstructions of each vertebra in anatomical position are shown in Fig. 4.5 (also see Figs. 5.1–5.8 in Meyer 2016). For C3–C6, only the vertebral

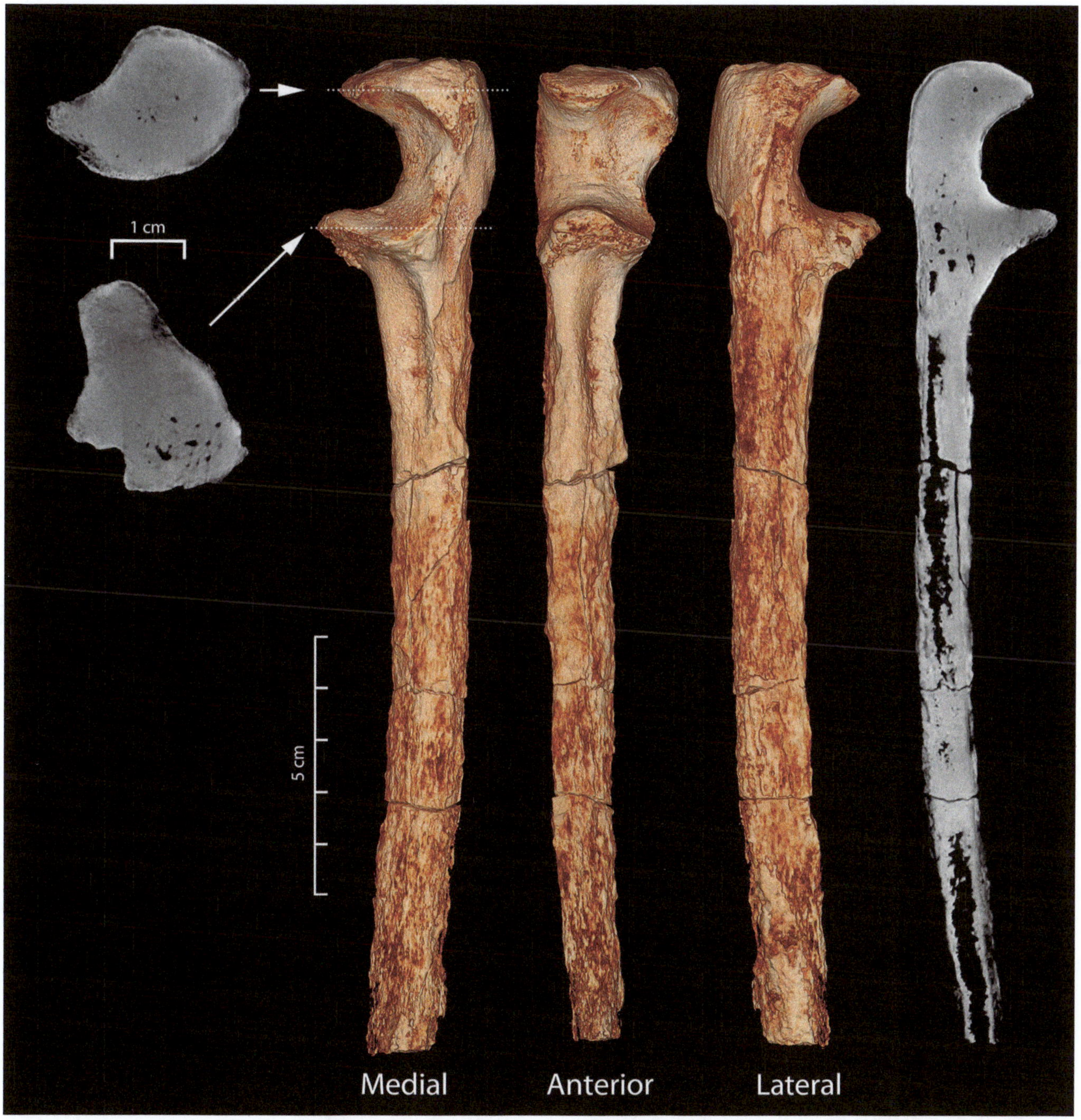

Fig. 4.4 Three-dimensional volume renderings and two-dimensional HRCT sections of KSD-VP-1/1a, right ulna. Views of ulna include medial, anterior, and lateral. White dotted lines indicate locations of two cross sections. Very little cortical bone is preserved on the diaphysis and no evidence of internal trabecular architecture is evident in the olecranon or coronoid processes as evidenced by two cross sections

body is preserved, but a portion of the lamina is preserved in C2. The C2 is missing the dens, but retains a portion of the body and almost the entire lamina on the left side, including part of the superior articular facet.

The HRCT scans of C2–C6 reveal relatively well-preserved trabecular and cortical bone structure (Fig. 4.6). The preservation in these elements demonstrates the variability in preservation possible within a site and within a skeleton. Variability within a site is most evident in C4, where the trabeculae appear as distinct high-density structures on the right side, but on the left side of the orthoslice, the trabeculae and matrix share similar density values. In direct contrast to C4, the trabeculae of C2 and C3 are preserved as distinct low-density regions in the anterior

Table 4.4 KSD-VP-1/1a ulnar metrics (in mm)

Measurement	KSD-VP-1/1	*Australopithecus*	Early *Homo*	*H. sapiens* (mean, SD)	*P. troglodytes* (mean, SD)	*G. gorilla* (mean, SD)
Trochlear notch orientation index	93.5	[a]84.0 (0.07) (*MH2*) [a]82.6 (4.3)	[a]84.5 (5.3)	[a]70.6 (4.5)	–	–
Trochlear notch[b]						
Height	25.6	15.7–23.2	23.1	23.2 (2.4)	22.0 (3.5)	28.8 (4.7)
Depth	13.2	7.9–11.3	11.9	12.1 (1.6)	11.5 (1.8)	14.0 (2.1)
Radial notch						
Proximodistal height	7.7	6.8–10.2	12.9	11.1 (2.1)	11.5 (1.5)	18.3 (2.3)
Anteroposterior length	15.3	9.8–16.4	13.8	17.4 (2.7)	15.8 (2.3)	24.8 (3.5)
Articular surface of trochlear notch						
Mediolateral breadth of proximal portion	26.5	16.1–25.7	26.2	25.1 (2.9)	22.0 (2.0)	31.5 (5.1)
Mediolateral breadth of middle portion	22.4	12.3–15.8	17.3	15.3 (2.4)	15.6 (2.9)	31.9 (5.7)
Mediolateral breadth posterior to radial notch	22.6	16.0–24.7	24.7	26.9 (3.0)	24.7 (2.6)	38.8 (5.8)
Mediolateral breadth of diaphysis distal to radial notch	16.0	15.2–19.6	23.9	22.1 (3.2)	18.2 (2.0)	28.3 (4.2)
Maximum anteroposterior distance at coronoid	32.2	22.5–36.1	33.4	34.8 (3.0)	36.5 (3.4)	45.9 (6.6)
Maximum mediolateral breadth of olecranon process	27.5	13.4–27.8	28.2	21.5 (2.5)	22.3 (2.8)	28.9 (5.4)

[a]From Churchill et al. (2013). Trochlear notch orientation index is (100*olecranon AP diameter/coronoid height). *Australopithecus* sample includes: A.L. 288-1, A.L. 438-1, OH 36, KNM-ER 1500, Omo 141–23, Stw 113, Stw 380, Stw 398, Stw 431 and SKX 8761(mean, SD). Early *Homo* sample includes: Omo L40-19 and KNM-BK 66 (mean, SD)

[b]All of the following data from Drapeau et al. (2005) unless otherwise indicated. *Australopithecus* sample includes: A.L. 288-1, A.L. 438-1, OH 36, and Stw 113. Early *Homo* sample includes: Omo L40-19. KSD-VP-1/1a measurements were collected by the authors

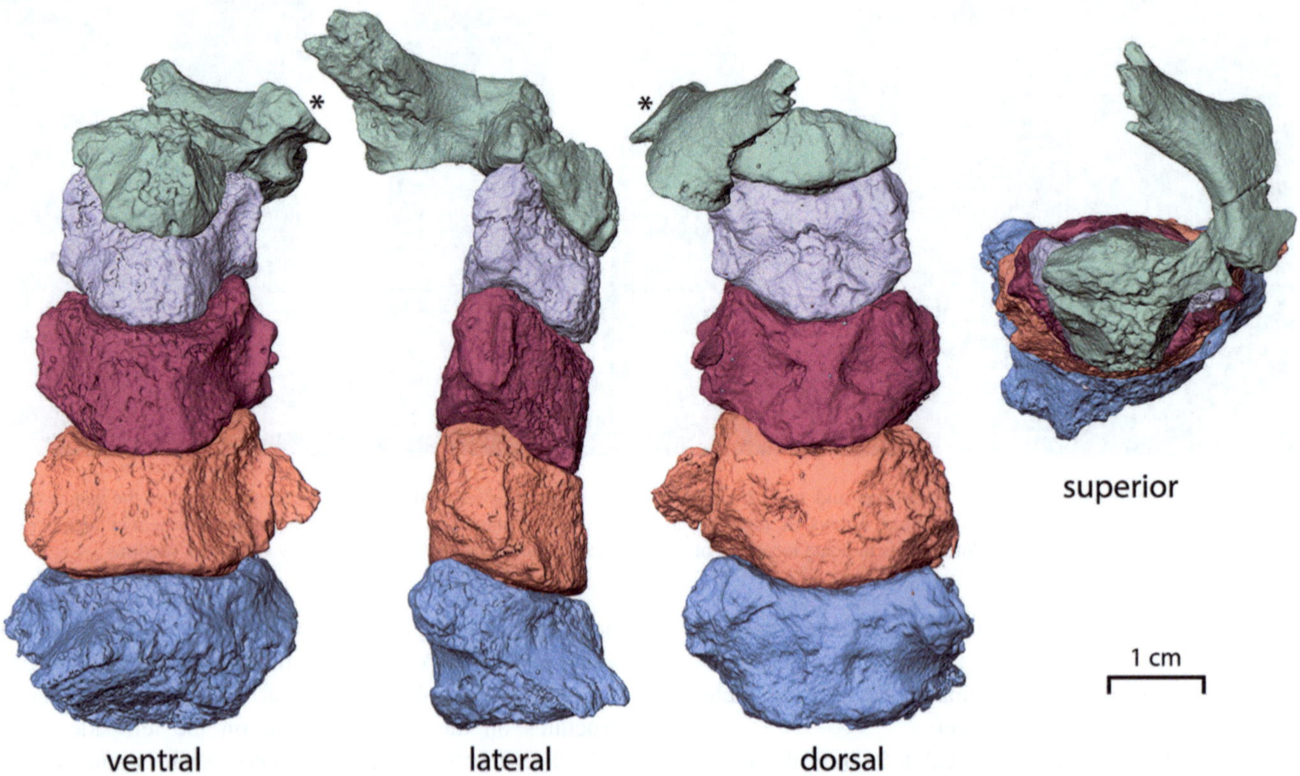

Fig. 4.5 Reconstruction of KSD-VP-1/1 cervical anatomy using 3D isosurface reconstructions. Views: anterior, lateral, posterior, and superior. The partial superior articular facet on C2 is indicated in each view (*)

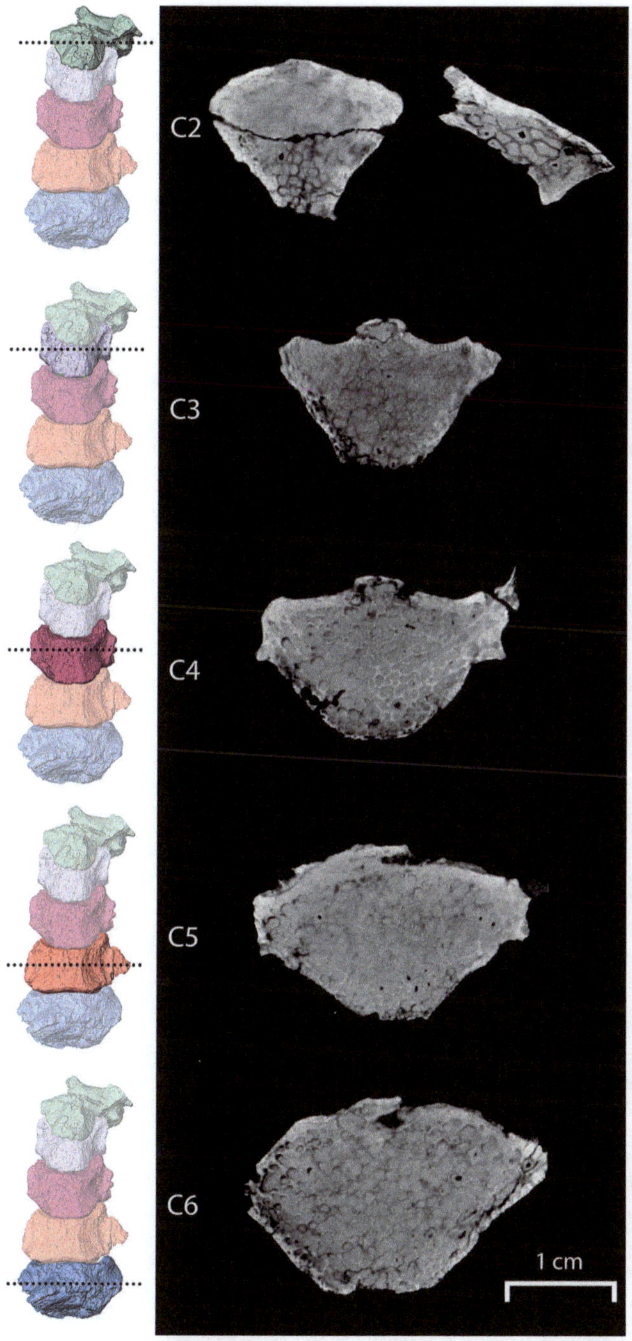

Fig. 4.6 Transverse HRCT slices through each cervical vertebral body. Trabeculae are clearly preserved in each specimen with the best preservation evident in C2 though C4

part of the body and especially in the lamina. Throughout the skeleton, trabecular preservation varies between these two extremes, with the majority of the skeleton exhibiting trabecular bone that is indistinct from the surrounding matrix because of similar density values between the bone and the matrix. The internal trabecular architecture of C5 and C6 provide examples of poorly delineated trabeculae, even though the cortical bone is relatively clear.

The cortical bone shell is not clearly discernible in the vertebral body of C2, but it does appear to be differentiated from the internal matrix of the lamina (Fig. 4.6). The cortical shell is fairly well defined in C3–C6, providing insight into cortical bone thickness in the cervical vertebral column (Fig. 4.6). In all four vertebral elements, the cortical shell appears to be thickest on the posterior surface of the body, and this is particularly apparent in C4 and C5. The lateral and anterior margins of the body in these vertebrae appear to have thin and porous cortical bone. The thin cortical shell is seen most clearly on the anterior left side of C4, in which the distinctive trabecular structure abuts a very thin cortical shell. This feature does not appear to be the result of significant exfoliation or erosion of the external cortical bone shell. A similar pattern can be seen on the anterior right side of C6, although the cortical shell is somewhat thicker in this specimen.

Ribs KSD-VP-1/1n (Left Second), KSD-VP-1/1q (Right 5/6/7), KSD-VP-1/1s (Middle Rib Frag), KSD-VP-1/1o (Right 7/8), KSD-VP-1/1p (Right 8/9), KSD-VP-1/1r (Left 11)

The 3D HRCT data of the five ribs provide insight into the shape of the thorax (Haile-Selassie et al. 2010). The significance of the preserved rib anatomy has been discussed in detail elsewhere (Haile-Selassie et al. 2010). Volume renderings reveal numerous cracks and exfoliated areas along each of the shafts, though some of the ribs exhibit a fairly smooth surface over large portions of the shaft. The internal cortical and trabecular structures of several preserved ribs are shown in Fig. 4.7. In general, the preservation is similar to that observed in other skeletal elements from this individual and especially in the second rib fragment cracks are clearly visible running through the cross sections (Fig. 4.7a, also see Fig. 7.1 in Latimer et al. 2016). The cortical bone shell is reasonably well-defined along the length of the ribs. The internal endosteal boundary is discernable in most sections along the rib shafts. Trabecular bone, however, is relatively poorly preserved internally (Fig. 4.7).

Right Os Coxa (KSD-VP-1/1d) and Superior Sacral Body (KSD-VP-1/1t)

KSD-VP-1/1d and 1/1t represent the partial right innominate and most of the superior sacral body, respectively (Fig. 4.8,

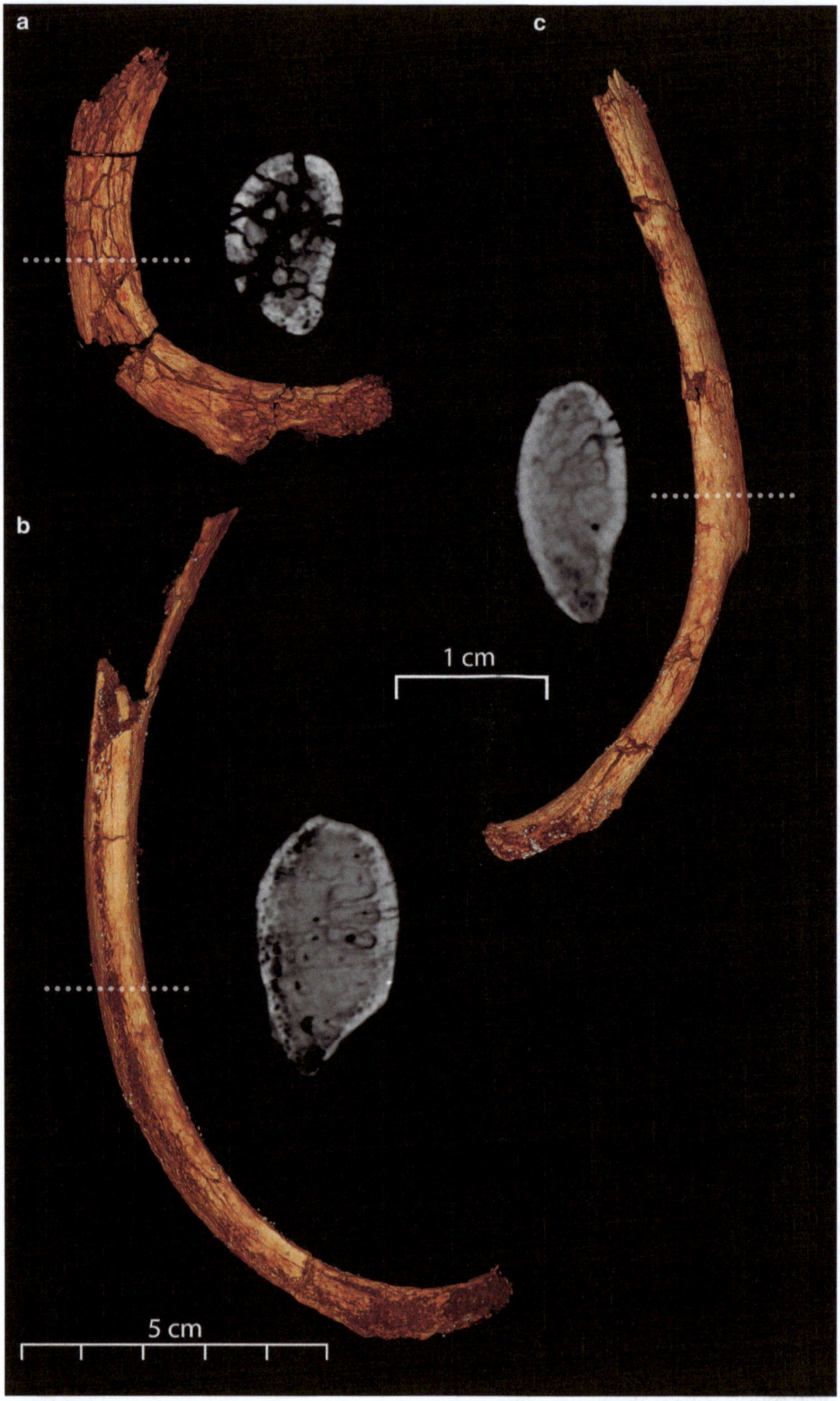

Fig. 4.7 Three-dimensional volume renderings and two-dimensional HRCT sections of ribs. **a** KSD-VP-1/1n, left second rib. **b** KSD-VP-1/1o, right 7th or 8th rib. **c** KSD-VP-1/1p, right 8th or 9th rib. Two-dimensional HRCT sections from each rib are indicated by the white dotted lines. The cross sections show the relatively clear preservation of cortical bone as well as internal trabecular bone preserved as discontinuous low-density structures. Scale in the middle is for 2D slices and scale at the bottom is for 3D renderings only

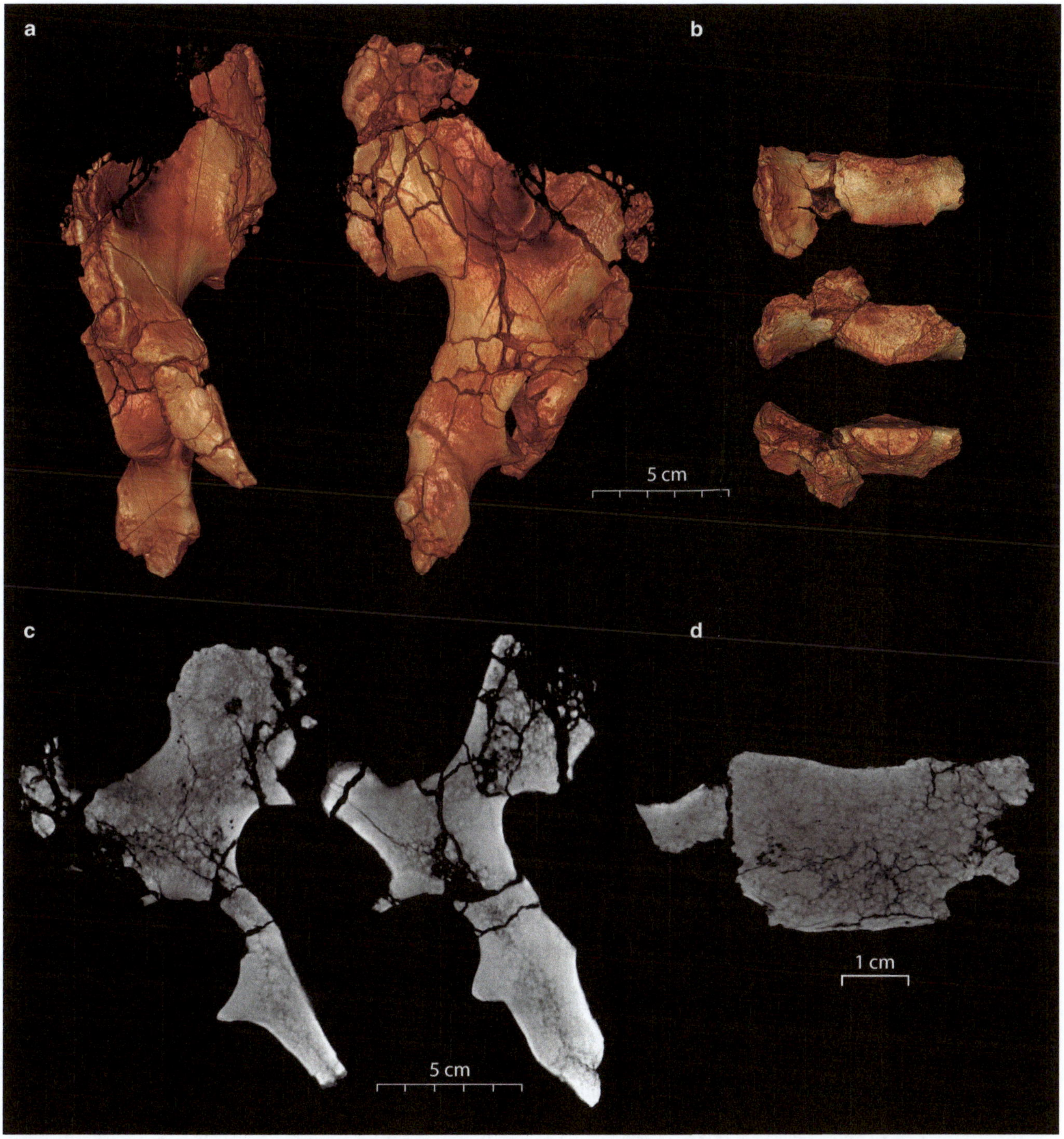

Fig. 4.8 Three-dimensional volume renderings and 2D HRCT sections of KSD-VP-1/1d (right os coxa) and KSD-VP-1/1t (superior sacral body). **a** The right os coxa in anterior and posterior views. **b** The superior sacral body in anterior, superior, and inferior views. **c** 2D HRCT coronal sections of the os coxa reveal very little differentiation of cortical and trabecular bone in the innominate. **d** Coronal section through the sacral body indicates some preservation of trabeculae when the image is enlarged for closer examination

also see Figs. 8.12 and 8.17 in Lovejoy et al. 2016). Even though the innominate has several significant fractures and is missing important features, these two elements provide insight into *Australopithecus* pelvic morphology. The superior pubic ramus was broken and translated inferiorly during preservation causing a slight superoinferior displacement of the acetabulum.

The HRCT data provide little insight into the internal morphology of the KSD-VP-1/1 innominate. The cortical bone boundaries and trabecular bone structure normally

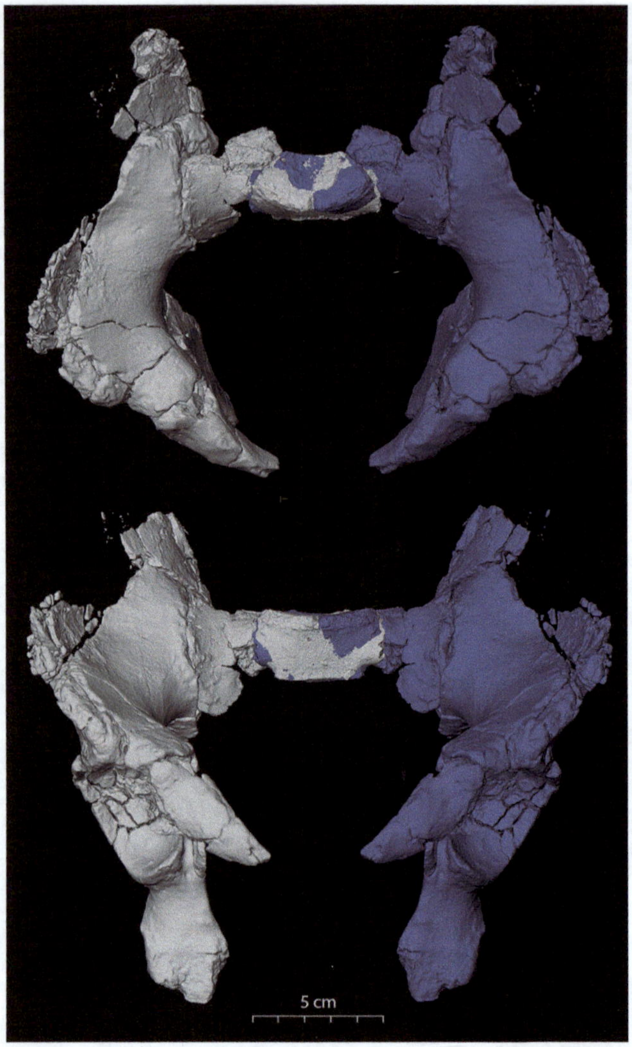

Fig. 4.9 Three-dimensional isosurface reconstruction of the KSD-VP-1/1 pelvis. The gray portions are original fossil material. The left os coxa and partial sacrum, rendered in blue, were produced by mirror imaging the original fossil and translating the pieces into anatomical position. Isosurfaces were used in this reconstruction to facilitate repositioning of the individual elements

infilling the innominate are not apparent from the HRCT images (Fig. 4.8). What is evident from the HRCT data are significant fractures running through various regions of the innominate and only discontinuous regions of trabecular bone poorly preserved as low-density regions within the bone. The sacral body is more similar to the cervical vertebrae in terms of preservation of internal bone structure (Fig. 4.8). Even though there are significant fractures crossing the bone, the cortical shell and some trabecular structure are discernable from the HRCT scans. Like the cervical vertebral bodies, however, the preservation is not such that the trabecular structure can be extracted and quantified.

The HRCT data also allow a reconstruction of the pelvic girdle using 3D triangulated isosurface data (Fig. 4.9). Using 3D surface reconstructions for the right innominate and sacrum, the two bones were articulated digitally in Avizo 7.1. Duplicates of each element were mirror imaged to generate the left side of the pelvis and translated into anatomical position. First, the missing left sacral ala was reconstructed using the mirrored right ala by overlaying the flipped KSD-VP-1/1t onto the original specimen and using the sacral body to align them accurately. Once the left ala was reconstructed, the mirror-imaged left innominate was translated and rotated into articulation with the sacrum. The pelvic reconstruction focused on rigid transformations of relatively undeformed features. The acetabulum and pubis were therefore not reconstructed here. The resulting reconstruction allows for an effective visualization of pelvic morphology in KSD-VP-1/1.

Left Femur (KSD-VP-1/1c)

KSD-VP-1/1c is a partial left distal femur, truncated somewhere in the superior popliteal surface region (Fig. 4.10, also see Fig. 8.8 in Lovejoy et al. 2016). The distal femur is relatively intact and complete, with only a few gaps in the bone surrounding the articular surface. Without the corresponding proximal femoral material from which to estimate femoral length (Lovejoy and Heiple 1970), it is difficult to determine the percentage of femoral length represented by this distal fragment. Haile-Selassie et al. (2010) estimated a length of between 418 and 438 mm based on likely crural indices, suggesting that the preserved fossil represents approximately 41–43% of the distal end of the bone preserved.

The HRCT scan data demonstrates the significant distortion of the femoral shaft and damage to the distal condyles (Fig. 4.10). It is not possible to differentiate the cortical bone structure from matrix filling the medullary cavity in these HRCT data. As with other elements, the trabecular bone structure is not well-differentiated in the distal condyles (Fig. 4.10), regions typically filled with highly organized trabeculae.

The anterior portion of the distal femur was relatively well preserved, except for a significant gap between the anterior articular, or patellar, surface, and the anterior aspect of the distal metaphysis. An examination of the patellar surface reveals a small portion of the surface on the proximal side of the gap in one of the fragments making up the anterior central metaphysis. This small section could indicate displacement of the fragments during preservation or a proximally-expanded lateral portion of the patellar surface. The preserved portion of the anterior articular surface suggests a deep patellar groove with a lateral border projecting more anteriorly than the medial border. The high lateral

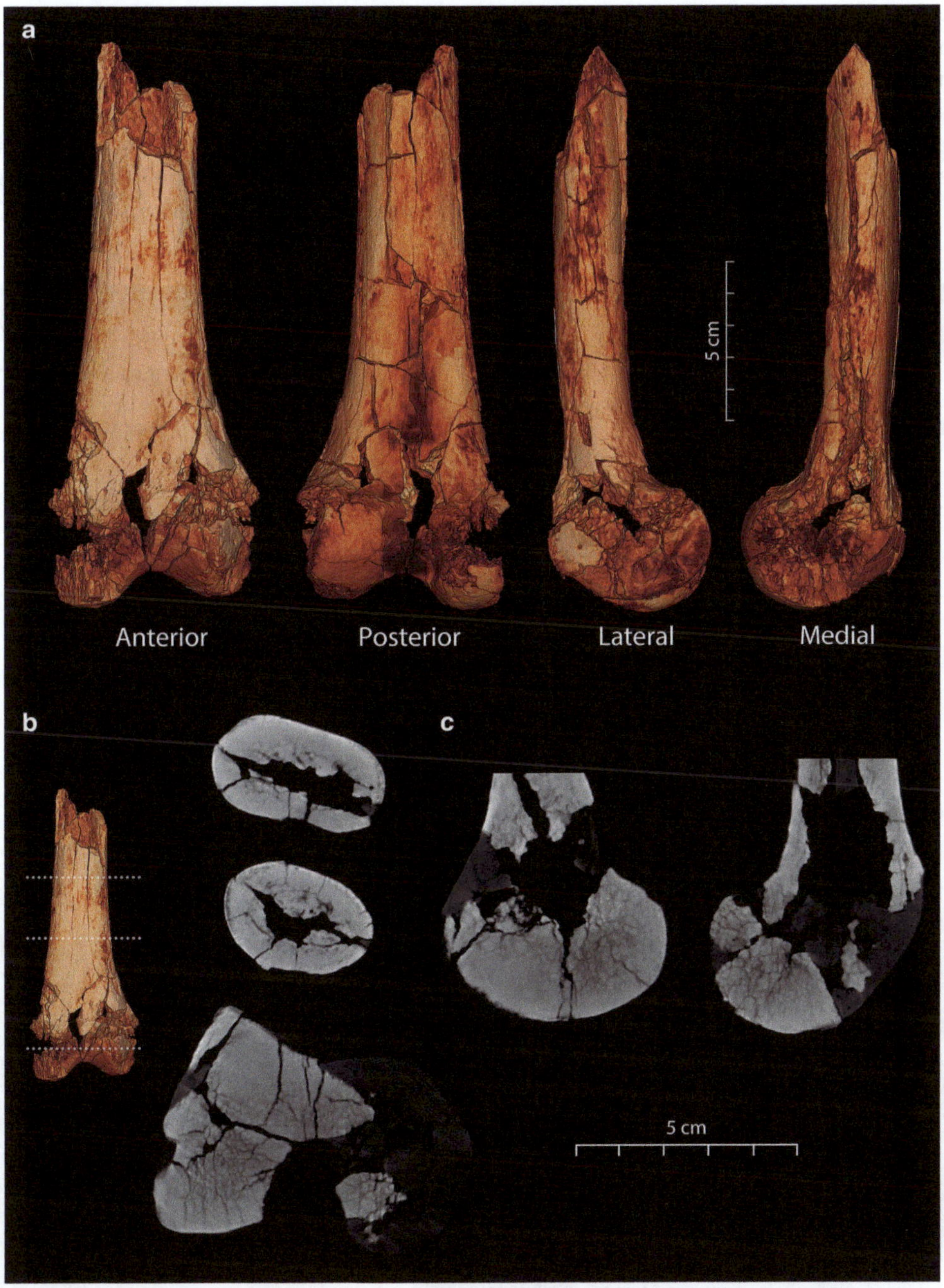

Fig. 4.10 Three-dimensional volume renderings and 2D HRCT sections of KSD-VP-1/1c, left femur. **a** Anterior, posterior, lateral, and medial views. The distortion of the diaphysis is evident in the 3D renderings. **b** Three transverse HRCT sections are taken from the diaphyseal, metaphyseal, and condylar regions of the distal femur specimen. The 2D sections show the distortion of the diaphysis as well as the extent of the missing portions of the medial condyle. The endosteal border of diaphyseal cortical bone is not well preserved in this specimen. **c** Two parasagittal sections of the distal condylar region reveal internal trabecular bone structure that is preserved as low-density regions. The dark region in the cross sections is the material added to stabilize the specimen

patellar lip and a deep patellar groove are features associated with the habitual valgus position of the knee during bipedal gait.

The specimen was crushed anteroposteriorly during preservation, so it consists of a large number of fragmentary pieces. The displacement results in a flattened anterior portion of the distal femur and creates an artificial depression directly proximal to the femoral condyles on the posterior aspect of the femur (Fig. 4.10). As a result, the dimensions of the specimen are clearly altered from its original condition, rendering measurements and descriptions of external morphology and measurements of internal cross-sectional geometric properties difficult. The damage appears to have increased the mediolateral breadth of the specimen and potentially obscured an accurate measure of the bicondylar angle. For this specimen, the volume rendering reveals that many of the fragments share matching borders, but the damage displaced them to create an uneven and fragmented surface.

Proximal to the patellar surface, the anterior portion of the distal femur consists of three main fragments, separated by vertical cracks. These thin, longitudinal slivers of cortical bone appear to be aligned with one another, but a transverse cross section reveals that they are slightly staggered (Fig. 4.11). The deformation of these fragments creates a flattened anterior profile in cross section. Comparisons with

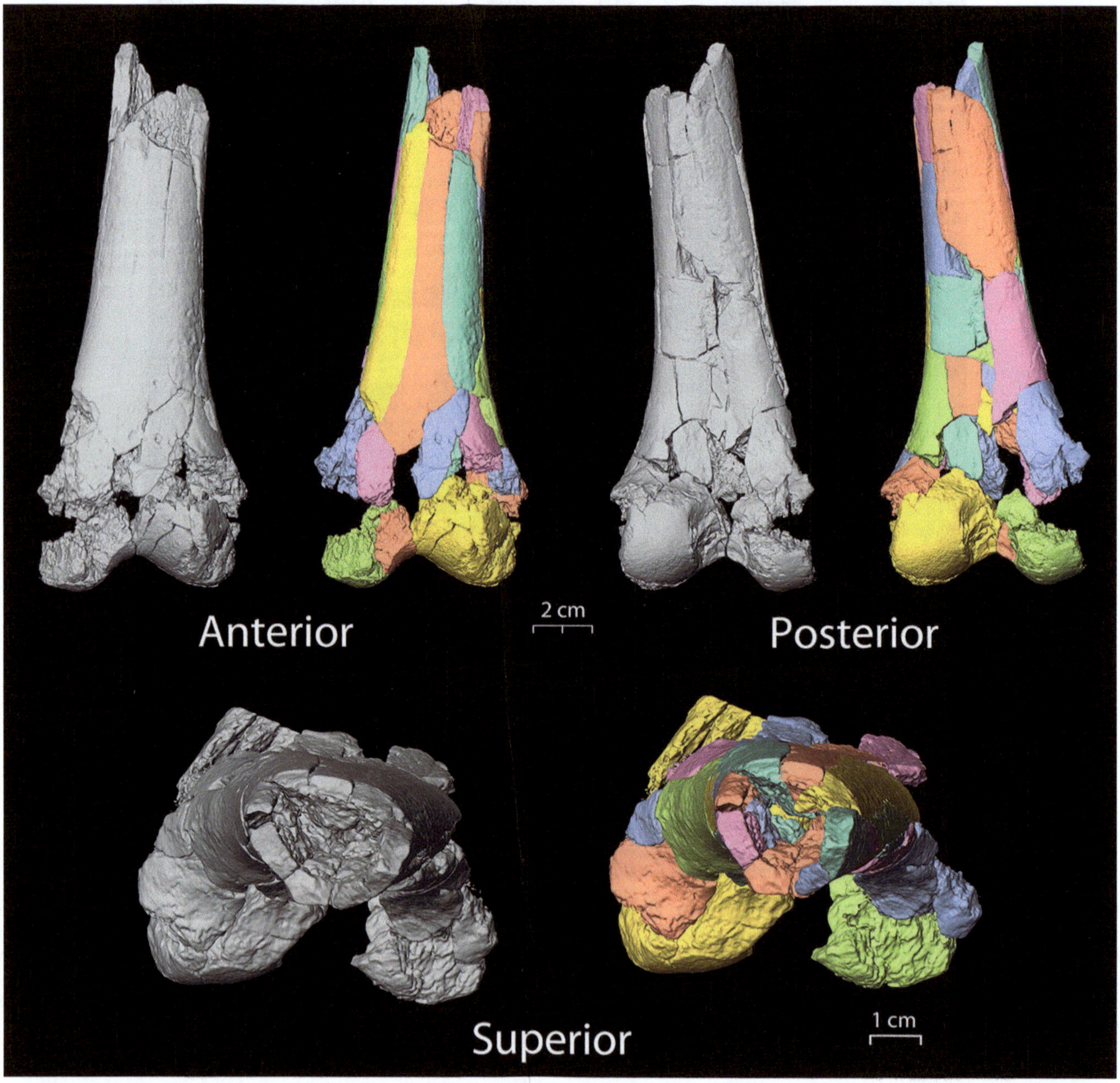

Fig. 4.11 Three-dimensional isosurface reconstruction of the distal femur shown alongside original fossil (gray surface) in anterior, posterior and superior views. For the reconstruction, each fragment was isolated and individually translated and rotated to produce a less distorted reconstruction

modern human, *Pan,* and another *Au. afarensis* distal femur (A.L. 129-1A), as well as inference from the positioning of the fragments, suggest that the flattened profile is artificially created from post-depositional taphonomic processes. The matching contours of neighboring fragments indicate that minor rigid transformation in the form of translation and rotation of the fragments may provide insight into the original shape of this anterior surface.

Medial and lateral views of the distal femur provide further evidence of the damage that flattened the anterior aspect of the femur. In medial view, the femur has a substantial gap between the anterior and posterior portions as well as a large section missing from the medial condyle. In comparison, the lateral side of the femur has a more rounded profile with no gap and a greater anteroposterior breadth. These differences suggest that the flattening, or crushing, differentially affected the two sides of the distal femur with the medial side expanding anteroposteriorly. However, it appears that the edges of the anterior and posterior fragments on the medial side have matching contours which suggests that the medial fragments can be translated using the shape and fit of the border fragments as a guide for reconstruction.

The posterior view of KSD-VP-1/1c reveals the most significant damage to the original structure. The distal portion of the popliteal surface is depressed, creating a central concavity. The popliteal surface is broken up by multiple fractures, creating several small and staggered fragments. Distally, most of the intercondylar fossa is preserved, but there is a large gap directly proximal to the intercondylar fossa. The distal aspect of the lateral femoral condyle is also well-preserved compared to the medial condyle, which has sustained considerable damage and is missing a large portion of its posterior surface.

Several cracks are visible on the surface, although not all of them extend very far into the cortical bone. The HRCT data reveals subtle differences in surface height and spacing between fragments, but some fragments are difficult to separate due to under-resolved fractures or incomplete separation. Reconstruction is best achieved if the fragmentary pieces are divided by recognizable fracture lines with a reasonable degree of certainty. Matching contours of these posterior fragments in combination with the findings from the other aspects of the distal femur suggest that reconstruction is a viable approach for KSD-VP-1/1c.

The 3D isosurface reconstruction of KSD-VP-1/1c was carefully divided into 30 individual surface objects, each representing a fragment of bone that was visibly distinct from its neighboring fragments. If a distinction could not be made, two or more fragments were grouped together into a single surface object. The fragments of the articular surface appear to be in approximately the correct anatomical position relative to one another, so the reconstruction was initiated from this location. Bone fragments bordering the articular surface were carefully adjusted to line up with the articular surface, although these adjustments were somewhat limited due to missing bone structure in this region. Realignment of the long fragments on the anterior surface created a smoother external surface and also resulted in the rotation of the medial and lateral fragments inward toward the midline of the shaft.

The collapsed posterior surfaces were expanded posteriorly to create a smoother and more continuous popliteal surface. Once the popliteal surface was expanded outward, the remaining fragments were translated and rotated into position using the shared edges for guidance. The rotation of the medial and lateral fragments resulted in the narrowing of the artificially broad distal metaphysis. In addition, the transverse cross section of the proximal-most portion of the specimen became more circular, with a moderate linea aspera defining the posterior outline. The comparisons between the original fossil and the reconstruction are shown in Fig. 4.11, with the reconstruction in multiple colors and the original in gray.

The reconstruction of the KSD-VP-1/1c distal femur allows for the measurement of cross-sectional geometric properties at specific locations along the preserved section of the diaphysis. Because no distinct medullary cavity is evident from the HRCT scan data, only the external contour of the femur, representing total subperiosteal area, was extracted for analysis of the cross-sectional geometric properties. Several cross sections were extracted using the estimates for minimum and maximum femoral length of 418 mm and 438 mm from Haile-Selassie et al. (2010) as a guide. For the total length of 418 mm, three sections were extracted representing 20, 30, and 35% of total estimated femoral length. For the estimated length of 438 mm, cross sections at 20 and 30% were extracted. Even though these cross sections lack information on internal medullary area and shape, solid subperiosteal cross sections have been shown to accurately reflect cross-sectional geometric properties (Macintosh et al. 2013).

These "solid" cross sections were used to calculate common geometric cross-sectional properties such as total area, maximum and minimum second moments of area (I_{max}, I_{min}), the polar second moment of area (J), and the polar section modulus (Z_{pol}). The same values were calculated from HRCT data for a sample of modern *H. sapiens* from the Norris Farms #36 site (Illinois, USA) dating to 1300 AD (Shaw and

Table 4.5 Cross-sectional geometric properties at three locations along the femoral shaft, based on the reconstructed femur. Section locations indicate the percent of total length from distal end of bone

Taxon	TA/BM	I_{min}/(BM*FL)	I_{max}/(BM*FL)	I_{max}/I_{min}	Z_{pol}/(BM*FL)	*J*/(BM*FL)
20% section						
Homo sapiens	9.79	0.99	1.21	1.23	0.14	2.21
	(1.02)	(0.19)	(0.21)	(0.15)	(0.02)	(0.39)
Pan troglodytes	16.02	1.80	2.85	1.60	0.32	4.65
	(4.11)	(0.49)	(0.63)	(0.17)	(0.09)	(1.09)
KSD-VP-1/1 418*	15.01	1.20	3.55	2.96	0.24	4.75
KSD-VP-1/1 438*	15.13	1.21	3.32	2.74	0.23	4.52
30% section						
Homo sapiens	9.00	0.82	1.06	1.29	0.12	1.88
	(0.93)	(0.14)	(0.22)	(0.18)	(0.02)	(0.35)
Pan troglodytes	14.42	1.61	2.11	1.33	0.28	3.71
	(3.68)	(0.48)	(0.48)	(0.11)	(0.08)	(0.95)
KSD-VP-1/1 418*	12.30	1.12	1.70	1.51	0.17	2.82
KSD-VP-1/1 438*	12.18	1.03	1.62	1.58	0.16	2.64
35% section						
Homo sapiens	8.75	0.77	1.02	1.33	0.12	1.79
	(0.90)	(0.13)	(0.22)	(0.18)	(0.01)	(0.33)
Pan troglodytes	14.23	1.60	2.03	1.29	0.27	3.63
	(3.76)	(0.55)	(0.50)	(0.12)	(0.08)	(1.04)
KSD-VP-1/1 418*	12.19	1.11	1.65	1.48	0.17	2.76

TA Total area, *BM* body mass, *FL* femur length
*Section locations were based on length estimates, 418 and 438 mm, which are used to correct CSG properties for bone length. For the estimate of 438 mm, the 35% section was not complete and therefore not included in the analysis

Ryan 2012; Macintosh et al. 2013) and for a sample of *P. troglodytes* using medical CT data from the Digital Morphology Museum, Kyoto University Primate Research Institute. Cross-sectional geometric properties were standardized relative to body mass and femoral length, following standard procedures. Total subperiosteal area was standardized by dividing it by body mass. The other calculated values, I_{max}, I_{min}, J, and Z_{pol}, were standardized by dividing their values by the product of body mass and femoral length.

Body mass was estimated in the *H. sapiens* sample using equations from Ruff et al. (1991). The body masses for the chimpanzees were associated with the scan data. Body mass and femoral length estimates for KSD-VP-1/1 are more problematic. Body mass was estimated using multiple regression equations based on human femoral head size, following the methods outlined by Ruff (2010). The estimated femoral head diameter of 38 mm based on acetabular dimensions reported by Haile-Selassie et al. (2010) was used to estimate body mass. The average of four body mass estimates derived using equations from Ruff (1991), McHenry (1992) and Grine et al. (1995) was calculated as 47.69 kg. The two femoral length estimates of 418 mm and 438 mm were used to standardize the KSD-VP-1/1 cross-sectional geometric properties.

The results of the analyses of cross-sectional properties are listed in Table 4.5 and shown in Fig. 4.12. At all three section locations, there are clear differences between humans and chimpanzees in most of the cross-sectional properties. At all three locations, *Pan* has higher values for standardized TA, Z_{pol}, and J than modern humans, indicating significantly more robust distal femora buttressed against torsional and bending loads. The I_{max}/I_{min} cross-sectional shape index is generally more similar between the two extant species at all three section locations with the distal-most 20% section having the only significant difference between chimps and humans. Chimpanzees have significantly more asymmetrical femoral cross sections at this location (Fig. 4.12).

Cross-sectional properties for the reconstructed KSD-VP-1/1c femur generally suggest a moderately robust femur with structural characteristics within the *Pan* range, or intermediate between chimpanzees and humans. Standardized total subperiosteal area at all three section locations falls squarely within the chimpanzee range. By contrast, Z_{pol} and J at all three locations lie at the extreme lower end of the *Pan* or intermediate between *Pan* and *Homo*. The only exception to this is the result for J at the 20% section which is more robust, falling in the chimpanzee range. The observed I_{max}/I_{min} cross-sectional shape index is possibly affected by the

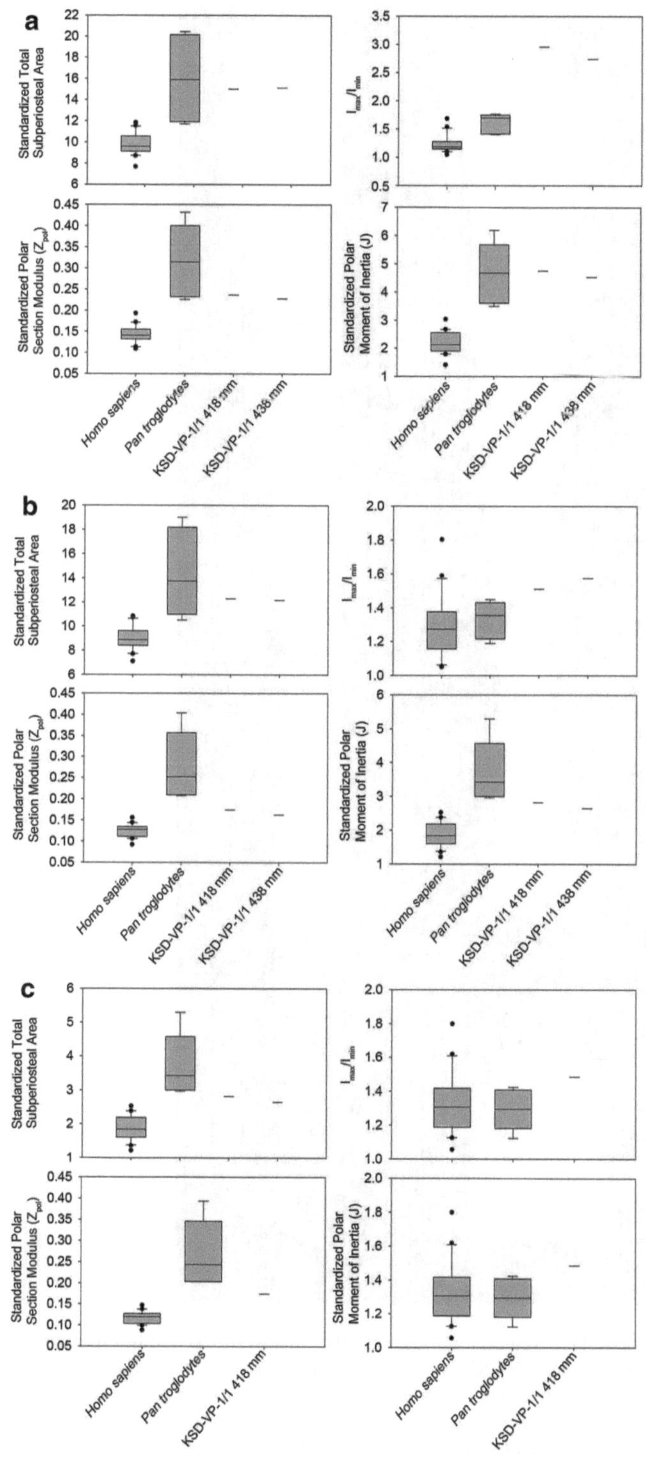

Fig. 4.12 Results of analysis of cross-sectional geometric properties of the distal femur. **a** 20% section. **b** 30% section. **c** 35% section. The two values for KSD-VP-1/1 for the 20 and 30% sections are derived from two separate estimates of total femoral length

reconstruction, which may still retain some displacement from the original form, but the results for I_{max}/I_{min} indicate a highly asymmetrical distal femoral shaft with significantly higher mediolateral breadth at the 20% section. The femoral cross section is somewhat more circular in the two more proximal section locations, but still displays higher asymmetry than either humans or chimpanzees, which do not differ appreciably in I_{max}/I_{min}.

While the high J and Z_{pol} values for KSD-VP-1/1 suggest a robust femoral shaft buttressed against torsional and bending loads, it is possible that the significant postmortem deformation of the femur is influencing these results. The process of repositioning the various femoral fragments certainly brought the cross-sectional shape more in line with what was likely the original form, but the high values for I_{max}/I_{min}, especially at the 20% location, suggests that the cross-sectional geometric properties may not be entirely reflective of the true bone structure.

Left Tibia (KSD-VP-1/1e)

KSD-VP-1/1e is a mostly complete left tibia preserving nearly the entire length of the bone from just above the tibial tuberosity to the distal plafond, including most of the distal articular surface (Fig. 4.13, also see Fig. 8.10 in Lovejoy et al. 2016). The proximal plateau is almost entirely absent with condylar articular surfaces missing, and a considerable portion of the lateral metaphysis missing as well. The cortical bone of the diaphysis is heavily eroded along nearly the entire length, especially between the 50 and 35% sections (Fig. 4.13). A volume rendering reveals that most of the periosteal cortical bone surface of the diaphysis has been exfoliated, exposing eroded internal cortical bone and, in some locations, the matrix-filled medullary cavity.

Transverse cross sections extracted from regular intervals along the tibial diaphysis reveal poor preservation internally as well. The endosteal cortical bone surface is not clearly defined along much of the shaft, especially in the distal half. Cortical bone surfaces are discernable in the proximal-most sections along the posterior aspect of the bone (Fig. 4.13), but the cross sections are either distorted due to apparent mediolateral deformation or not complete. This apparent plastic deformation in the tibia cannot be resolved through simple rigid transformations as performed with the femur. The infilling of the medullary cavity with dense matrix and the erosion of the periosteal cortical surface renders analysis

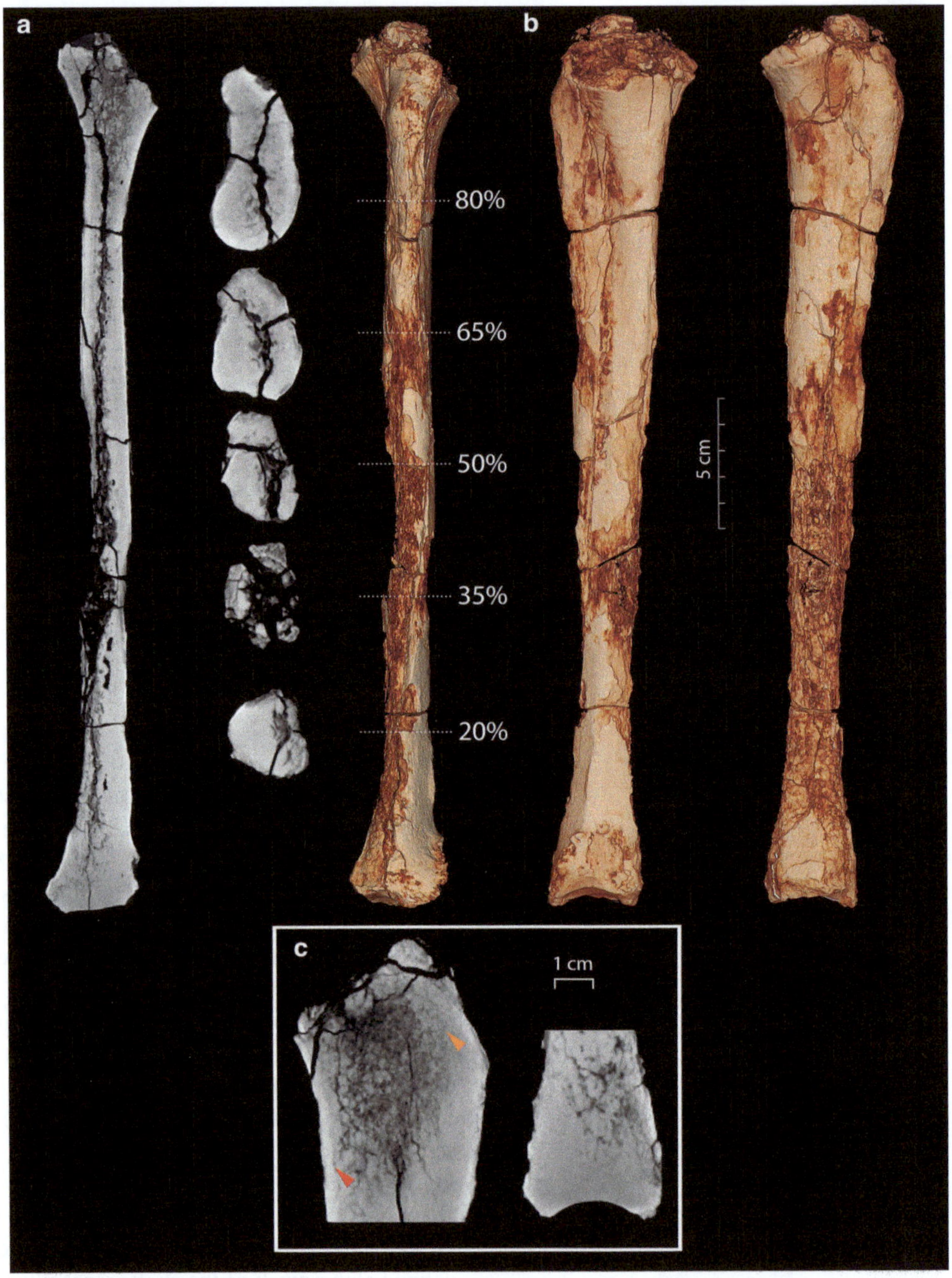

Fig. 4.13 **a** A single coronal CT section at the far left demonstrates the significant erosion of the cortical bone along the medial side of the KSD-VP-1/1 tibia. HRCT slices are extracted along the diaphysis at standard locations indicated on the three-dimensional volume rendering of the tibia in anterior view, and the cross-sectional shapes are presented to the left, revealing significant damage to the cortical and medullary regions throughout the diaphysis. **b** Additional volume renderings of KSD-VP-1/1e left tibia shown in lateral and medial views, respectively. **c** A parasagittal section of the proximal tibia and distal tibia show evidence of cortical bone (red arrow) and possible preservation of trabecular structure in the proximal end of the tibia (orange arrow)

of sectional geometric properties impossible. Sagittal slices in the proximal and distal ends of the tibia reveal preservation of the posterior cortical shell proximally, as well as relatively distinct trabecular bone structures (see inset Fig. 4.13). The trabecular bone in the tibia is preserved as low-density regions within a higher density surrounding matrix with little contrast, making quantitative structural analyses untenable.

The exception to this poor internal and external preservation is the relatively well-preserved distal articular region. The distal plafond is missing the medial malleolus, which has been cleanly sheared off post-mortem, but this region is

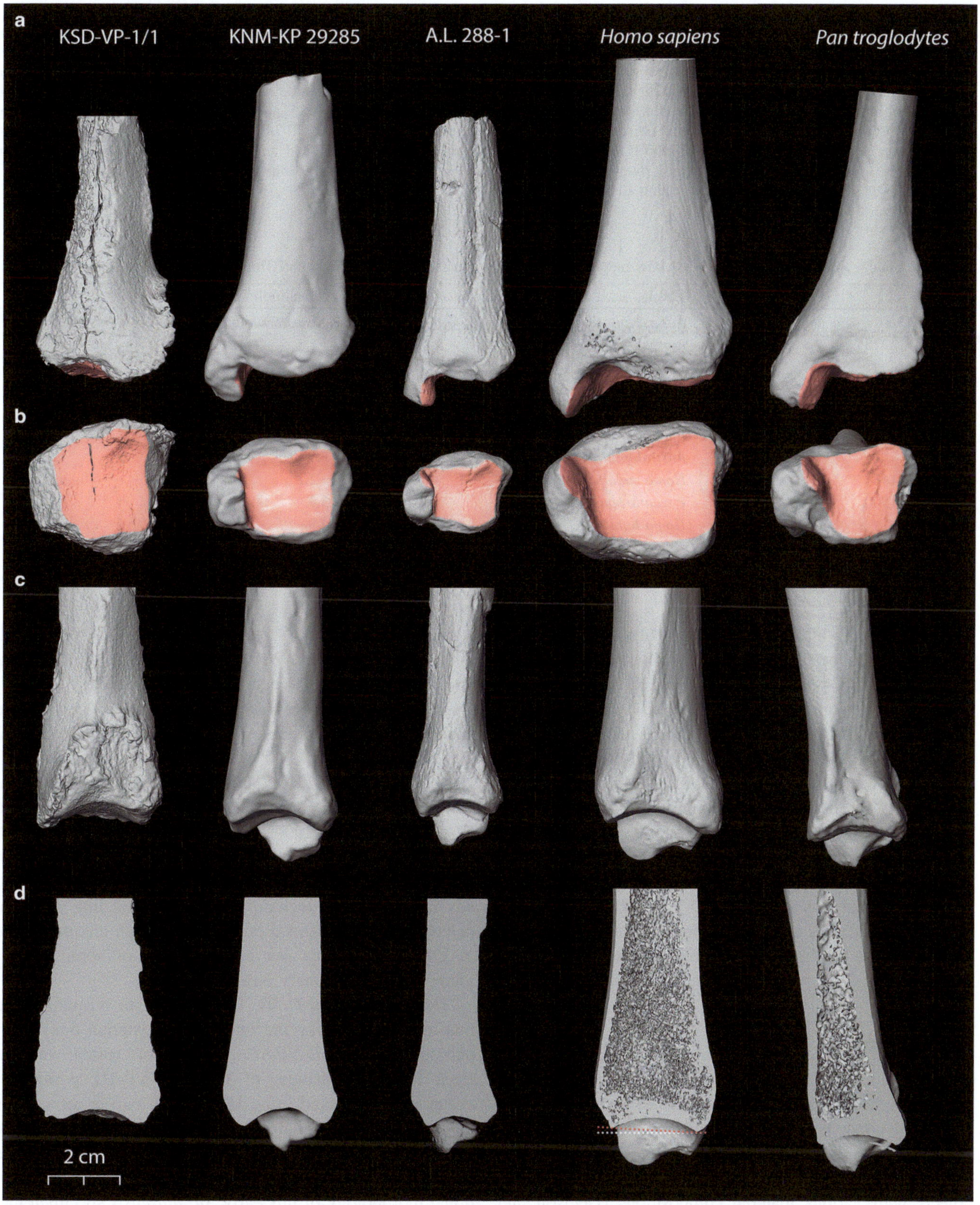
a
KSD-VP-1/1
KNM-KP 29285
A.L. 288-1
Homo sapiens
Pan troglodytes
b
c
d
2 cm

Fig. 4.14 Three-dimensional isosurface reconstructions of the distal tibia of KSD-VP-1/1e, KNM-KP 29285 (*Au. anamensis*), A.L. 288-1 (*Au. afarensis*), *H. sapiens*, and *P. troglodytes* showing articular surface shape and orientation. **a** Anterior view of distal tibia with articular surface for each specimen highlighted in red. **b** Inferior view of tibial plafond to show the highlighted articular surface shapes. The preserved portion of the KSD-VP-1/1 distal tibial articular surface reveals a square outline similar to that of KNM-KP 29285, while A.L. 288-1 and *H. sapiens* exhibit a mediolaterally elongated rectangular shape. *Pan troglodytes* is distinguished by having a trapezoidal shape, with a shorter posterior border of the articular surface. **c** Lateral views of distal tibia showing tibial arch angle. Arch angle measured after Desilva and Throckmorton (2010) and is indicated on the *H. sapiens* distal tibia. Anterior is to the left, posterior to the right. **d** Parasagittal sections through the middle of the tibiotalar articular surface. For all images, KNM-KP 29285 and A.L. 288-1 were flipped. Human, chimpanzee, and KSD-VP-1/1 surface data are from the CT scans. Surface data for A.L. 288-1 and KNM-KP 29285 were collected from fossil casts using a Next Engine 3D laser surface scanner

Table 4.6 Tibial arch angles from KSD-VP-1/1e and several other hominins taken from Desilva and Throckmorton (2010)

Specimen	Taxon	Tibial arch angle (degrees)
KSD-VP-1/1e	*Australopithecus afarensis*	3.0
KNM-KP 29285	*Australopithecus anamensis*	−1.8
A.L. 333-6	*Australopithecus afarensis*	2.9
A.L. 333-7	*Australopithecus afarensis*	5.5
A.L. 288-1	*Australopithecus afarensis*	−5.0
Stw 358	*Australopithecus africanus*	4.2
Stw 389	*Australopithecus africanus*	3.7
KNM-ER 1481	*Homo rudolfensis*	−2.1
KNM-ER 1500	*Paranthropus boisei*	3.7
O.H. 35	*Homo habilis*	4.8
Stw 567	*Homo*	−3.0
KNM-WT 15000	*Homo erectus*	1.8

otherwise mostly intact. The tibiotalar articular surface of KSD-VP-1/1c appears relatively square in shape, similar to KNM-KP 29285, while A.L. 288-1 and *H. sapiens* appear mediolaterally elongated. The *P. troglodytes* articular surface shape is distinct from the hominins, appearing more trapezoidal (Fig. 4.14). The relatively well-preserved distal plafond allows for the calculation of the tibial arch angle as defined by DeSilva and Throckmorton (2010). For this measure, KSD-VP-1/1c has an angle of 3.0, which falls well within the range of most other fossil hominins (Table 4.6). Compared to results reported in DeSilva and Throckmorton (2010), the tibial arch angle in KSD-VP-1/1c falls above the range of extant apes and well within the range of modern humans. These results suggest a developed rear-foot arch in KSD-VP-1/1.

Discussion

High-resolution computed tomography has been applied to a variety of contexts within the fields of biological anthropology and paleoanthropology. Imaging technology provides non-destructive access to the internal structure of objects with various material compositions (Denison and Carlson 1997; Ketcham and Carlson 2001) and has been used successfully for anatomical analyses of fossil materials (Thompson and Illerhaus 1998; Zollikofer et al. 2005; Alemseged et al. 2006; Lovejoy et al. 2009a, b; Suwa et al. 2009; Carlson et al. 2011; Leakey et al. 2012). This computer-assisted approach to paleoanthropology uses 3D reconstructions of fossil material to conduct 3D quantitative analyses, such as geometric morphometrics (Zollikofer 2002; Zollikofer et al. 2005; Gunz et al. 2009; Balzeau et al. 2010; Grine et al. 2010), and to investigate the structural variation of cortical and trabecular bone across the skeleton (Ryan and Krovitz 2006; Fajardo et al. 2007; Cotter et al. 2011; Ryan and Shaw 2012; Shaw and Ryan 2012; Su et al. 2013). The application of high-resolution imaging and the advanced analytic techniques to understanding hominin paleobiology represents a powerful approach that is capable of greatly expanding our understanding of hominin phylogeny and evolutionary morphology.

This chapter used 3D HRCT data to visualize and non-invasively survey the internal and external bony morphology of the KSD-VP-1/1 *Au. afarensis* partial skeleton. Each fossilized element of the KSD-VP-1/1 postcranial skeleton was scanned using HRCT with two primary goals in mind. The first goal was to evaluate and analyze the internal cortical and trabecular bone anatomy of the postcranial elements in order to gain insight into the functional morphology of this early *Au. afarensis* specimen that would not be accessible through traditional methods of observing and describing external morphology alone. The second aim was to produce 3D visualizations of the skeletal elements, and where possible, to generate 3D reconstructions of fragmentary bones through mirror imaging and rigid transformation in virtual anatomical space.

The HRCT scan data of the KSD-VP-1/1 skeleton reveal a heavily mineralized set of fossils with only minimal preservation of internal cortical and trabecular bone structure. As noted by Haile-Selassie et al. (2010), much of the external morphology of the skeleton is poorly preserved with significant post-depositional fracturing and exfoliation of the periosteal cortical bone surfaces. The HRCT data show similarly poor preservation internally with little to no observable differentiation between fossilized bone and matrix. The endosteal cortical bone surfaces in most bones are not easily discernable from the HRCT data. In all bones, except portions of the cervical vertebral bodies, the trabecular bone architecture manifests as low density, discontinuous features that are not well-delineated from surrounding matrix. The presence of higher density matrix infilling the intertrabecular spaces and abutting the endosteal cortical surface makes effective segmentation of bone from matrix nearly impossible.

The data also reveal significant variation in preservation within and between skeletal elements. The variation in the quality of preservation underscores the fact that preservation of internal morphology in fossils must be evaluated on a case by case basis. The most important consideration is not age, but rather the particular taphonomic and diagenetic conditions at work within specific sites. The diagenetic processes at work on these 3.6 Ma fossils clearly produced very dense fossils with poor mineralogical contrast between bone and matrix. Much older fossils have been effectively imaged and analyzed (Ryan and Ketcham 2002; Ryan et al. 2012) using HRCT.

Even without the ability to effectively segment bone and matrix for the purpose of quantitative analyses of internal bone morphology, these data allow robust scientific visualizations of anatomical structure. These reconstructions and visualizations allowed for the collection of unique metric data not available from the original fossils. Analyses of cortical cross-sectional geometric properties of the distal femur reveal a robust femoral diaphysis relative to modern humans, with robusticity values similar to those of chimpanzees. Future analyses may be able to utilize 3D geometric morphometric techniques to produce more insight into overall morphometric variation in the early *Au. afarensis* skeleton.

Acknowledgments We thank the editors, Y. Haile-Selassie and D. Su, for inviting us to contribute to this volume. Thanks also to J. Kappelman and K. Carlson, who provided constructive comments on an earlier version of this chapter. Grant sponsorship: National Science Foundation BCS-1124713 (T.M.R.).

References

Alemseged, Z., Spoor, F., Kimbel, W. H., Bobe, R., Geraads, D., Reed, D., & Wynn, J. G. (2006). A juvenile early hominin skeleton from Dikika, Ethiopia. *Nature, 443*, 296–301.

Balzeau, A., Crevecoeur, I., Rougier, H., Froment, A., Gilissen, E., Grimaud-Hervé, D., et al. (2010). Applications of imaging methodologies to paleoanthropology: Beneficial results relating to the preservation, management and development of collections. *Comptes Rendus Palevol, 9*, 265–275.

Carlson, K. J., Stout, D., Jashashvili, T., de Ruiter, D. J., Tafforeau, P., Carlson, K., et al. (2011). The endocast of MH1, *Australopithecus sediba. Science, 333*, 1402–1407.

Churchill, S. E., Holliday, T. W., Carlson, K. J., Jashashvili, T., Macias, M. E., Mathews, S., et al. (2013). The upper limb of *Australopithecus sediba. Science, 340*, 1233477.

Churchill, S. E., & Trinkaus, E. (1990). Neandertal scapular glenoid morphology. *American Journal of Physical Anthropology, 83*, 147–160.

Cotter, M. M., Loomis, D. A., Simpson, S. W., Latimer, B., & Hernandez, C. J. (2011). Human evolution and osteoporosis-related spinal fractures. *PLoS ONE, 6*, e26658.

Denison, C., & Carlson, W. D. (1997). Three-dimensional quantitative textural analysis of metamorphic rocks using high-resolution computed X-ray technology: Part I. Methods and techniques. *Journal of Metamorphic Geology, 15*, 29–44.

DeSilva, J. M., & Throckmorton, Z. J. (2010). Lucy's flat feet: The relationship between the ankle and rearfoot arching in early hominins. *PLoS ONE, 5*, e14432.

Dobson, S. D. (2005). Are the differences between Stw 431 (*Australopithecus africanus*) and A.L. 288-1 (*Au. afarensis*) significant? *Journal of Human Evolution, 49*, 143–154.

Drapeau, M. S. (2004). Functional anatomy of the olecranon process in hominoids and plio-pleistocene hominins. *American Journal of Physical Anthropology, 124*, 297–314.

Drapeau, M. S., Ward, C. V., Kimbel, W. H., Johanson, D. C., & Rak, Y. (2005). Associated cranial and forelimb remains attributed to *Australopithecus afarensis* from Hadar, Ethiopia. *Journal of Human Evolution, 48*, 593–642.

Fajardo, R. J., Muller, R., Ketcham, R. A., & Colbert, M. (2007). Nonhuman anthropoid primate femoral neck trabecular architecture and its relationship to locomotor mode. *Anatomical Record, 290*, 422–436.

Green, D. J., & Alemseged, Z. (2012). *Australopithecus afarensis* scapular ontogeny, function, and the role of climbing in human evolution. *Science, 338*, 514–517.

Grine, F. E., Jungers, W. L., Tobias, P. V., & Pearson, O. M. (1995). Fossil *Homo* femur from Berg Aukas, northern Namibia. *American Journal of Physical Anthropology, 97*, 151–185.

Grine, F. E., Gunz, P., Betti-Nash, L., Neubauer, S., & Morris, A. G. (2010). Reconstruction of the late Pleistocene human skull from Hofmeyr, South Africa. *Journal of Human Evolution, 59*, 1–15.

Gunz, P., Mitteroecker, P., Neubauer, S., Weber, G. W., & Bookstein, F. L. (2009). Principles for the virtual reconstruction of hominin crania. *Journal of Human Evolution, 57*, 48–62.

Haile-Selassie, Y., Latimer, B. M., Alene, M., Deino, A. L., Gibert, L., Melillo, S. M., et al. (2010). An early *Australopithecus afarensis* postcranium from Woranso-Mille, Ethiopia. *Proceedings of the National Academy of Sciences of the United States of America, 107*, 12121–12126.

Johanson, D. C., Lovejoy, C. O., Kimbel, W. H., White, T. D., Ward, S. C., Bush, M. E., et al. (1982). Morphology of the Pliocene partial hominid skeleton (A.L. 288-1) from the Hadar Formation. *Ethiopia. American Journal of Physical Anthropology, 57*, 403–451.

Ketcham, R. A., & Carlson, W. D. (2001). Acquisition, optimization and interpretation of X-ray computed tomographic imagery: applications to the geosciences. *Computers and Geosciences, 27*, 381–400.

Larson, S. G. (2013). Shoulder morphology in early hominin evolution. In K. E. Reed, J. G. Fleagle, & R. E. Leakey (Eds.), *The paleobiology of Australopithecus* (pp. 247–261). New York: Springer.

Larson, S. G., Jungers, W. L., Morwood, M. J., Sutikna, T., Jatmiko Saptomo, E. W., et al. (2007). *Homo floresiensis* and the evolution of the hominin shoulder. *Journal of Human Evolution, 53*, 718–731.

Latimer, B. M., Lovejoy, C. O., Spurlock, L., & Haile-Selassie, Y. (2016). The thoracic cage of KSD-VP-1/1. In Y. Haile-Selassie & D.F. Su (Eds.), *The postcranial anatomy of* Australopithecus afarensis *: new insights from KSD-VP-1/1* (pp. 143–153). Dordrecht: Springer.

Leakey, M. G., Spoor, F., Dean, M. C., Feibel, C. S., Antón, S. C., Kiarie, C., & Leakey, L. N. (2012). New fossils from Koobi Fora in northern Kenya confirm taxonomic diversity in early *Homo*. *Nature, 488*, 201–204.

Lorensen, W. E., & Cline, H. E. (1987). Marching cubes: A high resolution 3D surface construction algorithm. *Computer Graphics, 21*, 163–169.

Lovejoy, C. O., & Heiple, K. G. (1970). A reconstruction of the femur of *Australopithecus africanus*. *American Journal of Physical Anthropology, 32*, 33–40.

Lovejoy, C. O., Latimer, B., Suwa, G., Asfaw, B., & White, T. D. (2009a). Combining Prehension and Propulsion: The Foot of *Ardipithecus ramidus*. *Science, 326*, 72e1–72e8.

Lovejoy, C. O., Suwa, G., Spurlock, L., Asfaw, B., & White, T. D. (2009b). The Pelvis and Femur of *Ardipithecus ramidus*: The emergence of upright walking. *Science, 326*, 71e1–71e6.

Lovejoy, C. O., Latimer, B. M., Spurlock, L., & Haile-Selassie, Y. (2016). The pelvic girdle and limb bones of KSD-VP-1/1. In Y. Haile-Selassie & D.F. Su (Eds.), *The postcranial anatomy of* Australopithecus afarensis *: New insights from KSD-VP-1/1* (pp. 155–178). Dordrecht: Springer.

Macintosh, A. A., Davies, T. G., Ryan, T. M., Shaw, C. N., & Stock, J. T. (2013). Periosteal versus true cross-sectional geometry: a comparison along humeral, femoral, and tibial diaphyses. *American Journal of Physical Anthropology, 150*, 442–452.

McHenry, H. M. (1992). Body size and proportions in early hominids. *American Journal of Physical Anthropology, 87*, 407–431.

Melillo, S. (2016). The shoulder girdle of KSD-VP-1/1. In Y. Haile-Selassie & D.F. Su (Eds.), *The postcranial anatomy of* Australopithecus afarensis*: New insights from KSD-VP-1/1* (pp. 113–141). Dordrecht: Springer.

Meyer, M. R. (2016). The cervical vertebrae of KSD-VP-1/1. In Y. Haile-Selassie & D. F. Su (Eds.), *The postcranial anatomy of* Australopithecus afarensis*: New insights from KSD-VP-1/1* (pp. 63–111). Dordrecht: Springer.

Ruff, C. (2010). Body size and body shape in early hominins—implications of the Gona pelvis. *Journal of Human Evolution, 58*, 166–178.

Ruff, C. B., Scott, W. W., & Liu, A. Y. C. (1991). Articular and diaphyseal remodeling of the proximal femur with changes in body mass in adults. *American Journal of Physical Anthropology, 86*, 397–413.

Ryan, T. M., & Ketcham, R. A. (2002). Femoral head trabecular bone structure in two omomyid primates. *Journal of Human Evolution, 43*, 241–263.

Ryan, T. M., & Krovitz, G. E. (2006). Trabecular bone ontogeny in the human proximal femur. *Journal of Human Evolution, 51*, 591–602.

Ryan, T. M., & Shaw, C. N. (2012). Unique suites of trabecular bone features characterize locomotor behavior in human and non-human anthropoid primates. *PLoS ONE, 7*, e41037.

Ryan, T. M., Silcox, M. T., Walker, A., Mao, X., Begun, D. R., Benefit, B. R., et al. (2012). Evolution of locomotion in Anthropoidea: the semicircular canal evidence. *Proceedings of the Royal Society of London B, 279*, 3467–3475.

Schroeder, W., Martin, K., & Lorensen, B. (1998). *The visualization toolkit* (Vol. 2). Upper Saddle River, NJ: Prentice Hall PTR.

Shaw, C. N., & Ryan, T. M. (2012). Does skeletal anatomy reflect adaptation to locomotor patterns? Cortical and trabecular architecture in human and nonhuman anthropoids. *American Journal of Physical Anthropology, 147*, 187–200.

Su, A., Wallace, I. J., & Nakatsukasa, M. (2013). Trabecular bone anisotropy and orientation in an Early Pleistocene hominin talus from East Turkana, Kenya. *Journal of Human Evolution, 64*, 667–677.

Suwa, G., Asfaw, B., Kono, R. T., Kubo, D., Lovejoy, C. O., & White, T. D. (2009). The *Ardipithecus ramidus* skull and its implications for hominid origins. *Science, 326*, 68e1–68e7.

Thompson, J. L., & Illerhaus, B. (1998). A new reconstruction of the Le Moustier 1 skull and investigation of internal structures using 3-D-μCT data. *Journal of Human Evolution, 35*, 647–665.

Voisin, J. L. (2006). Clavicle, a neglected bone: Morphology and relation to arm movements and shoulder architecture in primates. *Anatomical Record. Part A, Discoveries in Molecular, Cellular and Evolutionary Biology, 288*, 944–953.

Voisin, J. L. (2008). The Omo I hominin clavicle: Archaic or modern? *Journal of Human Evolution, 55*, 438–443.

Vrba, E. S. (1979). A new study of the scapula of *Australopithecus africanus* from Sterkfontein. *American Journal of Physical Anthropology, 51*, 117–129.

Zollikofer, C. P. E. (2002). A computational approach to paleoanthropology. *Evolutionary Anthropology, 11*, 64–67.

Zollikofer, C. P. E., & Ponce de Leon, M. S. (2005). *Virtual reconstruction: A primer in computer-assisted Paleontology and Biomedicine*. Hoboken: Wiley.

Zollikofer, C. P. E., de Leon, M. S. P., Lieberman, D. E., Guy, F., Pilbeam, D., Likius, A., et al. (2005). Virtual cranial reconstruction of *Sahelanthropus tchadensis*. *Nature, 434*, 755–759.

Chapter 5
The Cervical Vertebrae of KSD-VP-1/1

Marc R. Meyer

Abstract A series of six partial cervical vertebrae were recovered in association with the KSD-VP-1/1 postcranial remains from the C2 axis to the C7 vertebral level, representing the oldest adult cervical column known in the hominin fossil record. The vertebrae of this large male australopith are more derived than those of its smaller female counterparts, and as a whole, present a biomechanical and enthesopathological signature typical of the dynamic vertical loading regime of orthograde humans. Differences between KSD-VP-1/1 and humans observed in the most cranial cervical levels appear to have insignificant functional implications, and are likely developmental reciprocates of australopith cranial morphogenesis. Despite their antiquity, the KSD-VP-1/1 vertebrae produce a surprisingly human-like kinematic signal, with a highly mobile neck, a head carriage consistent with habitual upright posture and bipedalism, and spinal cord dimensions most similar to that of modern humans.

Keywords *Australopithecus* • Cervical vertebrae • Functional anatomy • Lordosis • Posture • Spinal cord

Introduction

Six cervical vertebrae from the C2 axis to the C7 vertebral level were recovered in association with the KSD-VP-1/1 *Australopithecus afarensis* postcranial skeleton (Fig. 5.1). Although the KSD-VP-1/1 vertebrae are fragmentary, they represent the oldest series of hominin cervical vertebrae known. Because they are so fragile, hominin vertebrae are especially prone to biolytic and diagenetic degradation, and only a small number of cervical elements have survived to make their way into the fossil record. Serial patterns across the vertebral column are more informative than isolated elements, but other than the vertebral column of KSD-VP-1/1 very few series of cervical vertebrae are known for early hominins. The 3.3 Ma *Au. afarensis* from Dikika, Ethiopia (Alemseged et al. 2006) and at least two *Au. sediba* individuals from the 1.95 Ma Malapa site in South Africa (Berger et al. 2010) have been recovered with serial cervical vertebrae, but details on these remains are presently forthcoming. These fossils will eventually enrich our understanding of ontogeny and axial evolution, but at present the KSD vertebrae provide the first opportunity to examine serial cervical functional anatomy in an adult *Au. afarensis*.

KSD Cervical Vertebrae Descriptions

C2 Vertebra (Axis)

KSD-VP-1/1h is the axis, or second cervical vertebra (Fig. 5.2) represented by a partial centrum contiguous with the left pedicle, left lamina, and left inferior articular process. An isolated right inferior articular process with a small lamina fragment was also recovered for this element. Missing from this fossil vertebra is the superior aspect of the centrum and odontoid process, and the superior articular facets except for a small posterolateral margin of the left superior articular facet.

The articular surface shape of the inferior centrum is somewhat pentagonal, with a transverse width of 16 mm and dorsoventral height of 17.1 mm. Similar to that of humans, the centrum is dorsally eccentric, with the widest aspect of the inferior centrum distributed well toward the dorsal margin, in contrast to the ventrally eccentric centrum in apes. In lateral profile, the inferior centrum exhibits dorsoventral flexion at an angle of 142° and the inferior articular facet is angled at 123°, sweeping posteroinferiorly from the transverse plane, two features that signal a kinematic

M.R. Meyer (✉)
Department of Anthropology, Chaffey College, 5885 Haven Ave, Rancho Cucamonga, CA 91739, USA
e-mail: marc.meyer@chaffey.edu

Yohannes Haile-Selassie and Denise F. Su (eds.), *The Postcranial Anatomy of Australopithecus afarensis: New Insights from KSD-VP-1/1*, Vertebrate Paleobiology and Paleoanthropology, DOI 10.1007/978-94-017-7429-1_5

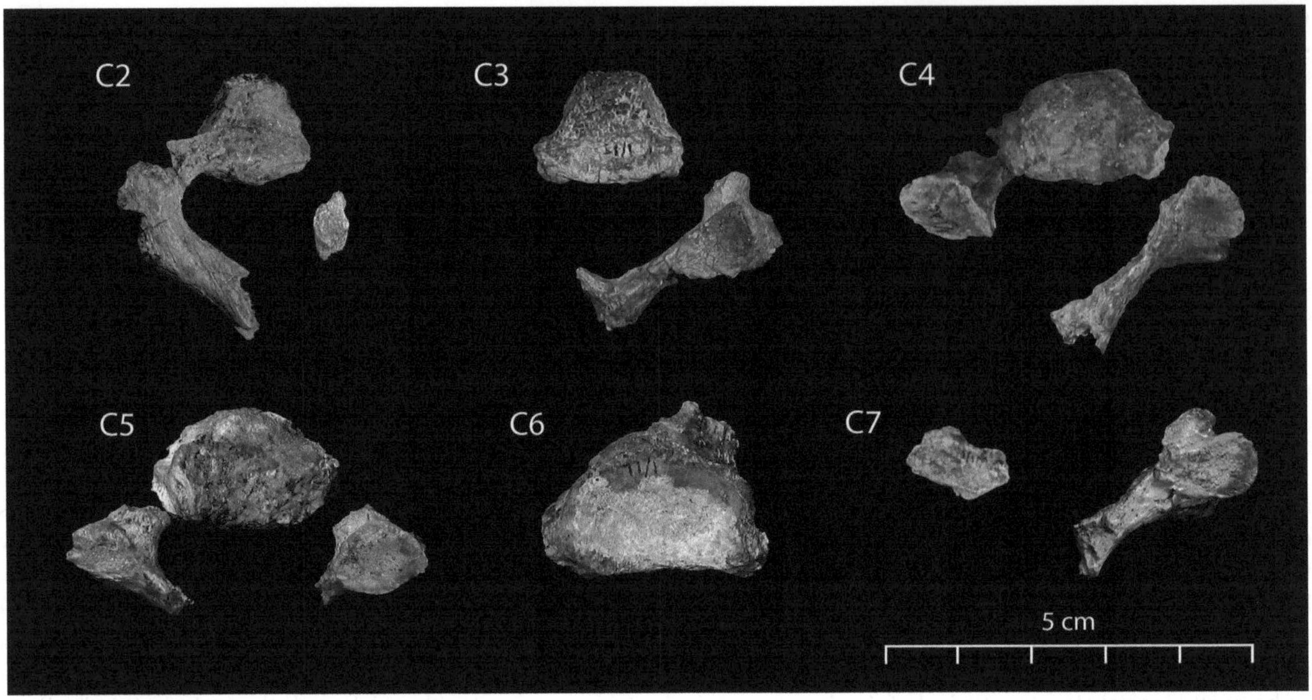

Fig. 5.1 KSD-VP-1/1 cervical vertebrae. All vertebral elements are in the superior view with the exception of the C2 right articular process, which is inverted inferiorly

predisposition for a lordotic articulation with the subjacent C3 vertebra. A moderately defined inferior median ventral trigon is preserved on the centrum, which is flanked laterally by shallow fossae for the tendinous insertion of the *longus colli*. The ring apophysis of the inferior centrum surface is fused, as is seen in adult humans (Byrd and Comiskey 2007), signaling the mature status of the vertebral remains.

The preserved left lamina is thick and rugose, with the superior aspect exhibiting a rounded superior margin. The inferior and medial aspects of the lamina are slightly abraded. The lamina sweeps dorsomedially for approximately 2 cm but is broken just short of the midsagittal axis anterior to the laminar dorsal curve and the root of the spinous process. The minimum width of the lamina at its midpoint is 6.7 mm. Together the centrum and preserved aspect of the lamina form the left half of a heart-shaped neural canal, with anterolateral shoulders scooping ventral to the midsagittal dorsal margin of the centrum. Based on doubling a measure of the preserved left neural canal to the midline, the transverse width of the neural canal is about 20 mm. Based on a mirror reconstruction of the vertebra, the dorsoventral neural canal height is estimated to be 12 mm and reconstructed neural canal cross-sectional area is estimated to be 2.1 cm^2 using NIH ImageJ software. The spinous process is missing except for the root of its inferior margin, which seems to indicate a somewhat superior angle of inflection.

The inferior articular facet is pear shaped and flat on the ventral articular surface, measuring 11 mm on the superoinferior major axis and 9.1 mm on the transverse minor axis. The reconstructed minimum width between the facets of the inferior articular processes is 15 mm and the maximum width between facets is 37 mm, resulting in an inferior interarticular facet width of 26 mm. The left pedicle is slightly abraded laterally, but its minimum transverse width and superior–inferior height may be estimated at 5.4 and 7.5 mm, respectively. The vertebral notch (inferior aspect of the pedicle representing the dorsoventral dimension of the intervertebral foramen) measures 6.8 mm. The anterior and posterior tubercles (cf. transverse processes) are also absent, but a portion of the left pedicle preserves the medial half of an obliquely oriented foramen transversarium.

The small posterolateral margin of the left superior articular facet inclines 55° inferolaterally, an orientation significantly steeper than hominins or medium- to large-bodied primates. Because the small remnant of the superior articular facet represented may not accurately represent the whole facet orientation, this anomaly is difficult to assess. It may simply reflect postmortem distortion, as the left inferior articular process appears to be slightly skewed inferomedially.

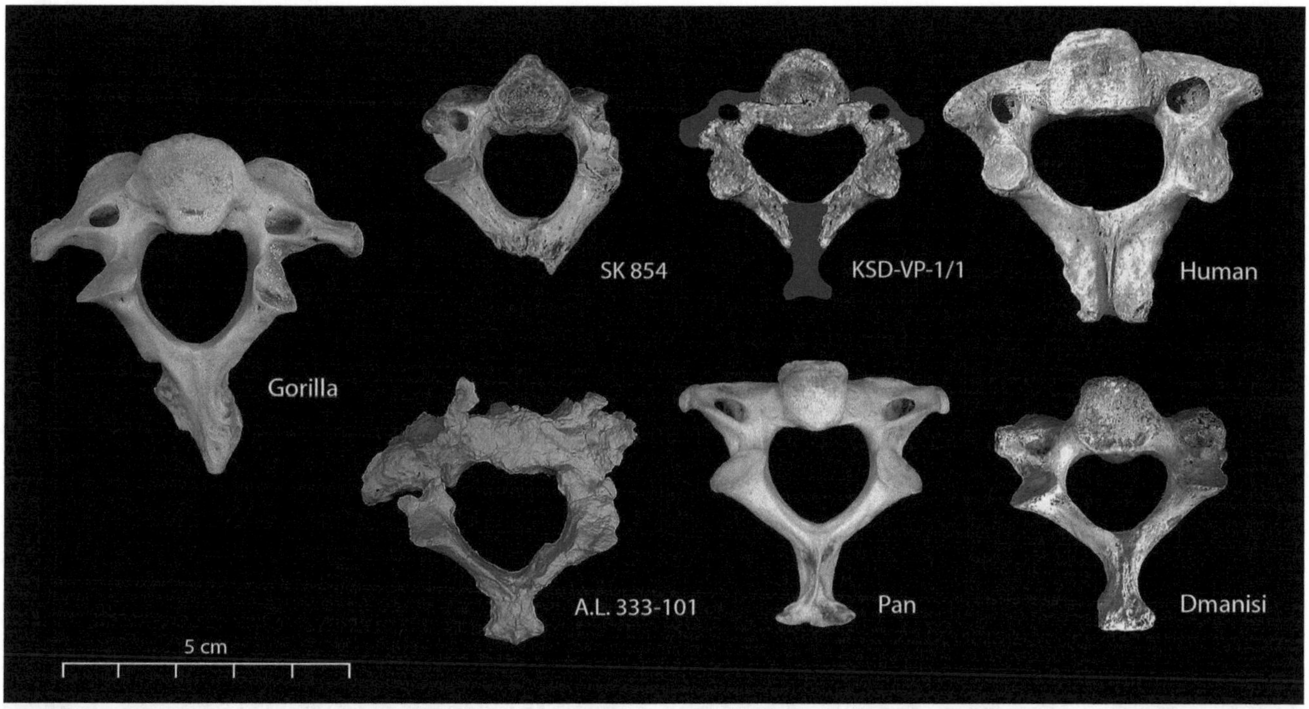

Fig. 5.2 Comparison of C2 vertebrae (inferior view) with KSD-VP-1/1h mirror reconstruction. KSD-VP-1/1 and A.L. 333-101 represent *Australopithecus afarensis*; SK 854 represents *Paranthropus robustus*; Dmanisi represents the D2673 *Homo erectus* specimen. Note the anteroinferior tubercle of SK 854, which is the inferior terminus of the median ventral keel. Inferior views are shown here as the odontoid processes of the KSD-VP-1/1 and SK 854 specimens are broken rendering the superior views less informative

The C3 Vertebra

The C3 vertebra (Fig. 5.3) is represented by a mostly complete centrum (KSD-VP-1/1i) and the right half of some of the posterior vertebral components (KSD-VP-1/1x), including the right superior and inferior articular processes, complete right half of the lamina and remnant of the left lamina marking the dorsal apex of the neural arch. The posterior aspect of the right pedicle preserved at its junction with the ventral pars interarticularis, and although the spinous process is broken and not preserved, the confluence of the right and left laminae begin a curved dorsomedial trajectory signaling the transition from the lamina to root of the spinous process.

The centrum features an abraded right superior centrum surface and abraded left inferior surface. This damage exaggerates the degree of ventral constriction of the vertebra, which renders a somewhat ape-like appearance. The resultant superior centrum surface is somewhat pentagonal, with a transverse dimension of 20.5 mm and dorsoventral height of 14.7 mm. The inferior centrum surface is less angular and ventrally constricted than the superior surface. Both the superior and inferior centrum surfaces are dorsally eccentric, with the widest aspect closest to the posterior margin. The uncinate processes are dorsally positioned, shallow, and feature lateral margins that slightly inflect superolaterally. The superior ventral margin is slightly abraded but apparently reveals a rounded margin, as is the case in many non-human primates. Due to the abrasion damage, it is challenging to discern whether the round sloping margin is the product of postmortem damage or the true anatomy of the vertebra. However, like *Gorilla* C3 vertebrae with the rounded margins, the superior mid-centrum surface plane is broken by a very slight bulge that serves as a preamble to the round sloping ventral margin. The KSD ventral surface descends and narrows slightly to a moderate transverse waist and gradually transitions inferiorly to the transversely wider inferior centrum ventral margin.

The superior articular facet is extremely tall superoinferiorly (14.9 mm) and slightly concave, angled approximately 45° to the transverse plane (based on the orientation of the superior lamina). The posteroinferior margin of the facet merges smoothly into the superior aspect of the lamina. There is a small osteophyte on the right lateral margin of the superior articular facet at its midpoint that would appear to be inconsequential. A superior interarticular facet width of 25.7 mm is calculated by mirror imaging the preserved half, and the facets are obliquely oriented to the dorsal sagittal midline (line of the spinous process). The pedicle is relatively large, with a transverse width and superior–inferior

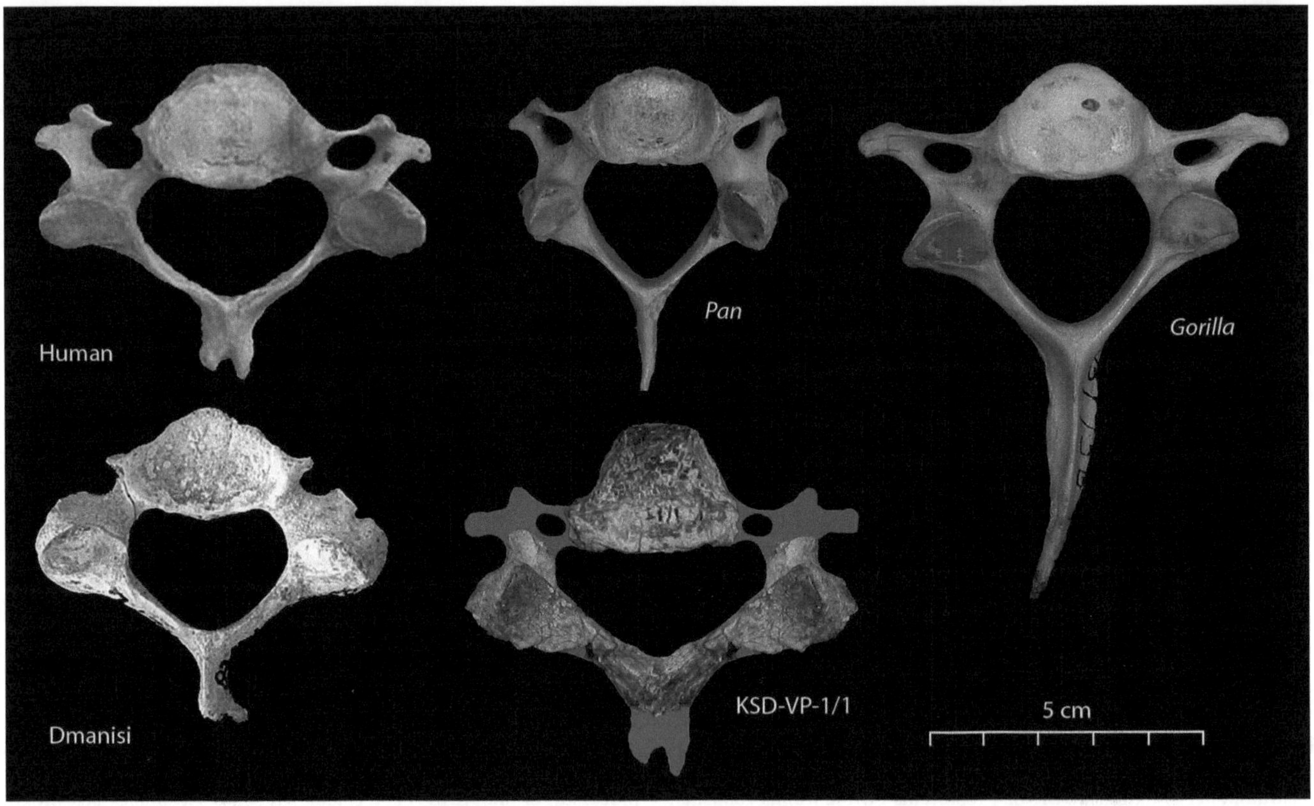

Fig. 5.3 Comparison of C3 vertebrae (superior view) with KSD-VP-1/1i mirror reconstruction. Presently, the Dmanisi *Homo erectus* D2674 specimen represents the only other early hominin fossil at this cervical level. Note the transversely expanded neural arch of the KSD-VP-1/1 fossil relative to the apes

height measuring 6.5 and 7 mm, respectively. The preserved right lamina superior margin is damaged immediately posterior to the superior articular facet, but demonstrates a relatively thick structure with an intact rounded inferior margin. The right lamina sweeps dorsomedially where it intersects at the dorsal apex of the neural arch with a remnant of the left lamina. Together the right and left laminae converge to form the root of the spinous process. The width of the lamina at its midpoint is 5 mm with a height of 12 mm. Preservation of the lamina allows for the reconstruction of a transversely expanded neural arch measuring 24 mm wide with an estimated dorsoventral length of 12 mm. A reconstructed neural canal cross-sectional area measures 2.28 cm^2, nearly at the mean for anatomically modern humans of 2.33 cm^2.

The C4 Vertebra

The C4 vertebra (Fig. 5.4) consists of a mostly complete centrum (KSD-VP-1/1j) and left pedicle and articular process (KSD-VP-1/1ac) that fit together neatly with the centrum, as well as an associated right articular process, right lamina, and spinous process root (KSD-VP-1/1z). The superior centrum shape is oval measuring 21 mm in transverse width and 15.4 in dorsoventral length. Because of damage to the superior ventral margin on the midline, it is difficult to assess whether the C4 vertebra shares the round, inferior slope of the superior ventral margin with the superjacent C3. However, the flat consistency of the preserved superior surface and geometry of the lateral portions of the superior surface and orientation of the ventral centrum suggest that unlike the C3, the ventral superior margin of this vertebra dropped vertically. The inferior centrum surface is shaped similar to the superior surface, but has a small degree of osteophytosis on the periphery of the right ventral margin. There is damage to the left side of the inferior centrum that has sheared off a portion of the ventrolateral surface, resulting in a diagonal truncation extending from the ventral margin to the lateral left side.

The fused annular ring is preserved, and especially pronounced at the dorsal margin of the inferior centrum. Although human C4 vertebrae tend to have a more kidney-shaped centrum at this level, the centrum is dorsally eccentric as in humans, with the majority of articular centrum surfaces dorsal to the widest transverse dimension. The uncinate processes are dorsally positioned as well, with a relatively shallow profile. Although the superior aspect of the

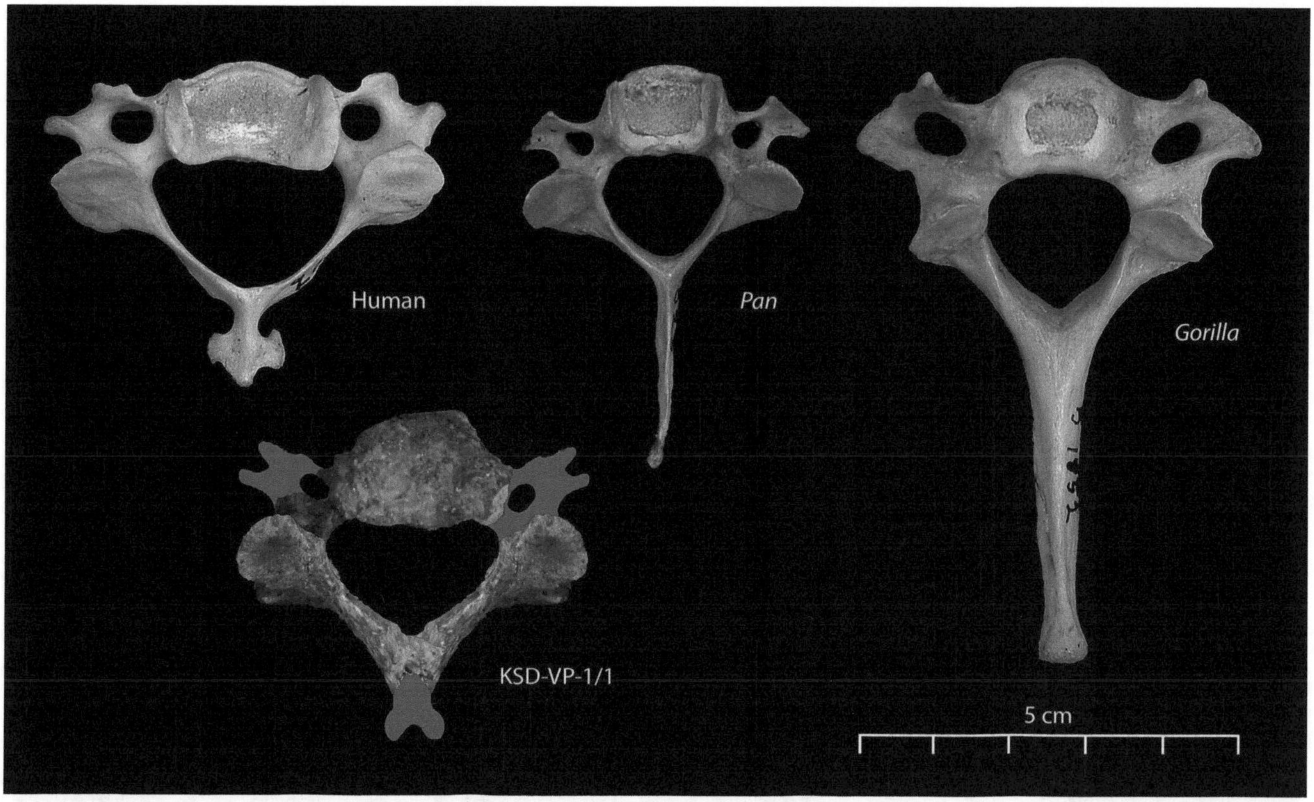

Fig. 5.4 Comparison of C4 vertebrae (superior view) with KSD-VP-1/1j centrum, KSD-VP-1/1ac left pedicle, and KSD-VP-1/1z right articular process and neural arch mirror reconstruction. The KSD-VP-1/1 C4 vertebra is presently the sole representative of this cervical element for early hominins older than 3 Ma

right uncinate process is abraded, both appear to have lateral margins that smoothly slope superolaterally, unlike the sharply demarcated and vertically oriented uncinate processes of *Pan*. The dorsal aspect of the centrum is notable for the presence of a median dorsal keel, an hourglass-shaped buttress of bone reinforcing the dorsal centrum often seen in humans at the middle cervical levels. Two small vascular foramina enter the dorsal centrum on the lateral sides of the median keel at the midpoint of constriction.

The left pedicle is tall, with slight damage to the inferior margin, but can be reliably measured at 8.9 mm superoinferiorly. Greater damage to the ventral aspect of the pedicle precludes measurement of its transverse dimensions. Anterior to the left pedicle is the posterior half of the foramen transversarium with a diameter of 3.5 mm. The right articular processes are represented by a complete superior articular facet measuring 14.9 mm superoinferiorly and 11.8 mm transversely, the pars interarticularis (lateral mass), and an inferior articular facet broken in half in an inferoblique direction. Like the left superior articular facet, the inferior facet is flat, but more oblique and transverse in its orientation to the sagittal midline axis. The preserved portion of the right inferior articular process presents a slightly cuplike or convex facet. The dorsal aspect of the right inferior articular process features a small but pronounced lateral tubercle for the multifidus spinae, a small muscle that attaches from the dorsal aspect of the articular processes in C4–C7 vertebrae to a superjacent spinous process for joint stabilization and proprioception, as well as assisting in extension.

The articular process complex is broken anteriorly at the root of the pedicle, but is continuous with the right lamina posteriorly which preserves half of the neural arch. The superior margin of the right lamina is somewhat sharp immediately posterior to the superior articular process but thickens dorsally as it continues to form the dorsal apex of the neural arch at the confluence with the left lamina. The spinous process is absent, broken at its root just past the point of laminar curvature. Preservation of portions of both the right and left neural arch and articular processes allow for a superior interarticular facet width measurement of 35.5 mm and a neural canal area of 2.23 cm^2.

The C5 Vertebra

The C5 vertebra (Fig. 5.5) is represented by a mostly complete centrum (KSD-VP-1/1k), the left articular process, left

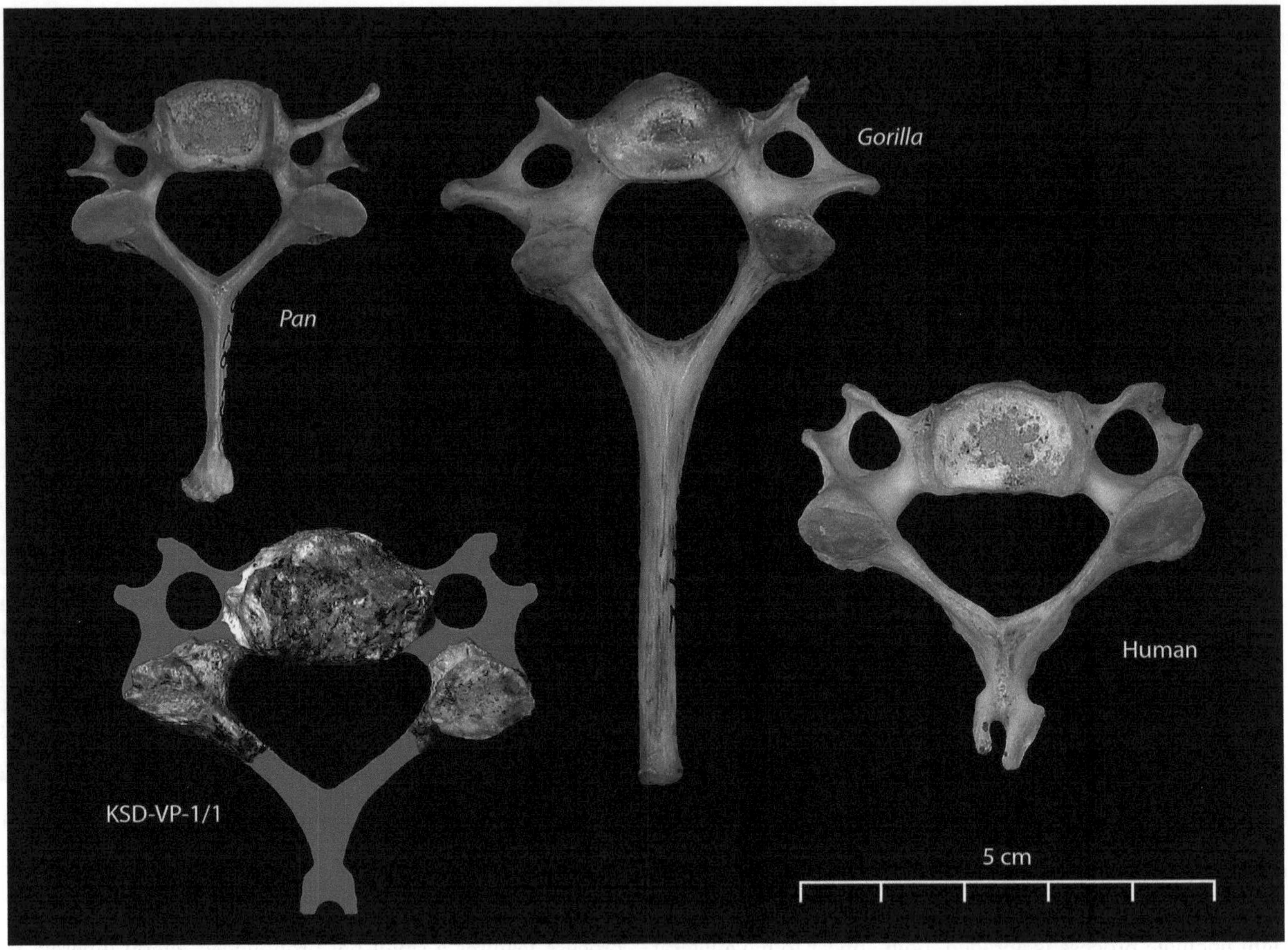

Fig. 5.5 Comparison of C5 vertebrae (superior view) with KSD-VP-1/1 k mirror image reconstruction. Presently, the KSD-VP-1/1 C5 vertebra is the sole representative of this cervical element for early hominins

pedicle and small extension of lamina (KSD-VP-1/1aa), and right articular process (KSD-VP-1/1y). The superior centrum shape is damaged on the right dorsolateral side but renders a clear oval in shape that is dorsally eccentric, with a transverse width of 24 and 15.6 mm dorsoventrally. A very shallow left uncinate process is preserved with low margins sloping superolaterally. The uncinate process is dorsally distributed with its superior apex distributed toward the dorsal centrum margin. The right uncinate is damaged but its morphology suggests bilateral symmetry with the left process. The inferior centrum surface is abraded with some of the posterior bone missing. The inferior centrum surface ventral margin is rounded, countered by a more angular dorsal surface. A thin osteophytic lip runs along the periphery of the inferior ventral centrum forming a small demi-ring of bone projecting anteroinferiorly approximately 1.5 mm at its maximum along the right margin. This lipping is more pronounced than the minimal osteophyte expression on the superjacent C3 vertebra.

This vertebra features both a bilateral concavity on the ventral surface and the faint indication of a median keel on the dorsal surface. The thickened bone along the midline between the concavities is often seen in humans and is reminiscent of the buttressing or median ventral keel, but differs in being transversely expanded relative to the thinner and more everted buttress in non-human primates. This bilateral concavity is most noticeable in the C5 level, and to a much lesser extent on the other KSD cervical centra. The median dorsal keel in this vertebra is not as pronounced as in the C3, apparently due to postmortem erosion of the dorsal surface.

The left pedicle is separated from the centrum by a break that obliterates its superior and ventral aspect, but preserves a curved medial border making up the left neural canal's transverse maximal. A small remnant of the right pedicle curves anteromedially from the articular process providing the right neural canal lateral border. The inferior articular facets are damaged; however, both superior articular facets preserve very tall facet surfaces measuring 14.4 mm superoinferiorly and 11.8 mm transversely, with a reconstructed interarticular facet width measurement of 36.8 mm. Both superior articular facets are relatively flat and orient obliquely to the dorsal sagittal midline, blending smoothly into the left lamina posteriorly. The lamina tapers narrowly dorsally and breaks roughly halfway along its formation of the left neural arch. The right lamina is broken and absent immediately posterior to the superior articular process, but its dorsomedial curved trajectory provides the root of the neural arch. A reconstructed neural canal transverse dimension of 27.5 mm and estimated anteroposterior length of 16 mm produce a heart-shaped neural canal with an area of 2.23 cm^2.

The C6 Vertebra

The C6 vertebra (Fig. 5.6) is represented only by a large centrum (KSD-VP-1/1l). Its superior shape is reniform and dorsally eccentric, measuring 26 mm transversely and 16.5 mm dorsoventrally. There is considerable breakage on the right lateral margin, but the right ventral margin presents a small degree of osteophytic lipping matching that of the superjacent C5 inferior surface. The left superior surface is nearly intact, preserving a low uncinate process with very shallow superolaterally sloping margins. The right uncinate process margins are poorly represented due to damage, but the root of the uncinate matches the smooth slope of the complete left side, signaling a very shallow uncovertebral joint morphology as in humans.

The inferior surface is marked by two pathological indications: osteophytic lipping (Spondylosis deformans) and the apparent presence of a Schmorl's node, indicative of rupture of the intervertebral disc and invagination through the plane of the centrum endplate. The Schmorl's node is relatively shallow, irregularly shaped, and localized slightly dorsal to the midpoint of the inferior centrum surface. There is moderately severe osteophytic lipping that maximally projects 5 mm ventrolaterally from the right inferior centrum margin. The left side of the inferior centrum is damaged and missing aspects of the posterolateral surface. The dorsal centrum surface is perforated by two vascular foramina, and presents the ventralmost root of the right pedicle. The anterior centrum is marked by invasive damage, but offers a transversely concave profile sharply delineated by the superior and inferior margins.

The C7 Vertebra

The C7 vertebra (Fig. 5.7) is represented by the right articular processes, articular pillar, root of the posterior tubercle of the transverse process, and right lamina (KSD-VP-1/1ad,

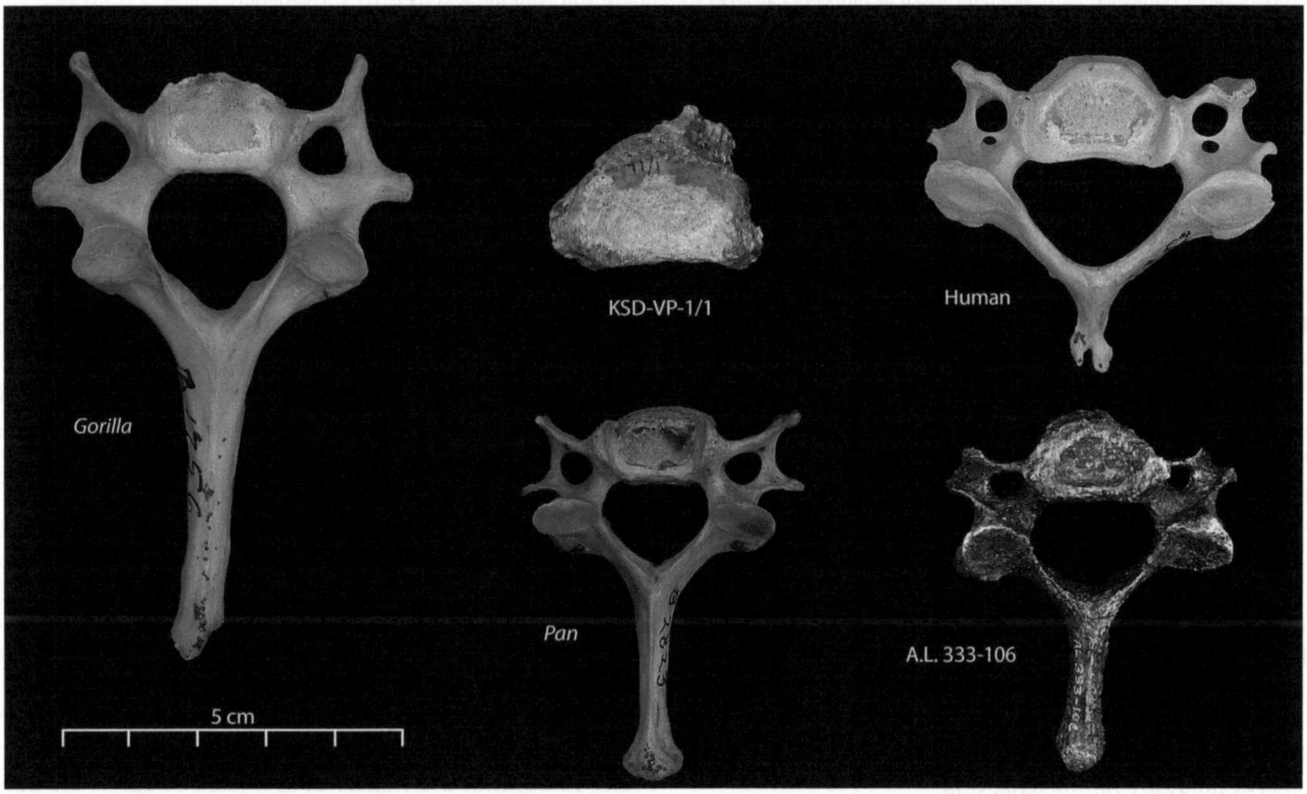

Fig. 5.6 Comparison of C6 vertebrae (superior view). Presently, A.L. 333-106 (*Au. afarensis*) represents the only penecontemporaneous early hominin fossil at this cervical level for comparison with the KSD-VP-1/1 C6 vertebra

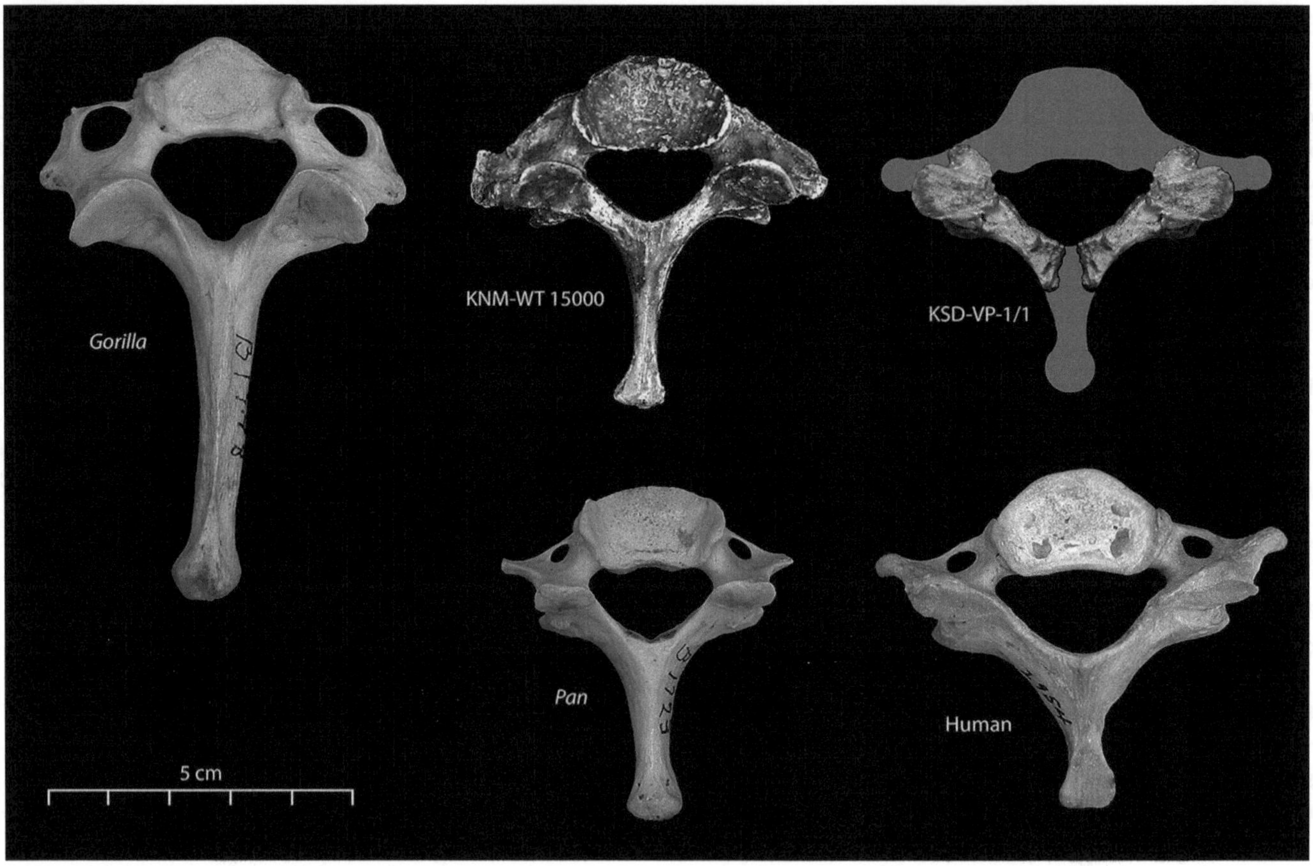

Fig. 5.7 Comparison of C7 vertebrae (superior view) with KSD-VP-1/1 mirror reconstruction. Presently, KNM-WT 15000 (*H. erectus*) represents the only other published early hominin fossil at this cervical level for comparison with the KSD-VP-1/1 C7 vertebra

KSD-VP-1ae, KSD-VP-1/1af and KSD-VP-1/1ag), and by an isolated left articular process and articular pillar (KSD-VP-1/1ab). The right superior articular facet is obliquely oriented to the dorsal sagittal midline, with a pear-shaped flat surface. There is damage perforating the facet's inferior aspect at the junction with the lamina. The superior articular facet is tall with a superoinferior height of 14.8 mm and transverse width of 11.8 mm, and a reconstructed interarticular facet width of 39.5 mm.

Abrasion marks the superior margin of the right lamina, which extends posteriorly to preserve most of the right half of neural arch. As is the pattern in the superjacent vertebrae, there is no spinous process, as the lamina is broken at the point where the dorsomedial trajectory of the neural arch makes its transitional curve to the sagittal axis. The pars interarticularis is broken at the superior limit of the inferior articular process. The right lamina is similar in robusticity to modern humans at 14.9 mm in superoinferior height and 5.2 mm thick. There is no laminar fossa behind the superior articular facet, which can indicate the presence of lordosis in the human lumbar region as it signals imbrication of the superjacent inferior articular process inferiorly during extension.

The right inferior articular facet resembles its superior counterpart in shape but exhibits more of a cuplike concavity. The superior lateral shoulder of the dorsal inferior articular process has a pronounced tubercle for the multifidus spinae muscle. The two facets are separated by an anterorlaterally concave articular pillar, from which the root of the transverse process emerges. The articular pillar is similar to humans and much thicker than that of *Pan.* Although the right transverse process is broken just proximal to the posterior tubercle, its caliber and trajectory indicate a relatively gracile morphology with an inferolateral orientation.

Materials and Methods

The extant sample selected for comparison is composed of 520 cervical vertebrae from anatomically modern humans ($N = 59$ individuals), *Pan* ($N = 26$), and *Gorilla* ($N = 21$). The African ape vertebrae are all from wild-shot adults with complete secondary synostosis. Human vertebrae are also fully fused from individuals 17–50 years of age at the time of death (N = 59), the majority of which come from the Hamann-Todd Osteological Collection housed at the Cleveland Museum of Natural History. The human sample also includes indigenous San and Bantu individuals (N = 6) from the Transvaal Museum and Witwatersrand University, indigenous Kikuyu from the National Museums of Kenya (N = 11), and Natufians from El Wad and Kebara curated at the Peabody Museum at Harvard University (N = 5). Because aspects of human vertebral morphology are related to somatotype, ancestry, climate, and lifeways (Abbie 1957; Grave et al. 1999; Huggare 1992; Snodgrass and Galloway 2001), this study avoided the use of a Eurocentric comparative population (8% of the sample). The sex ratio across the sample group was nearly even for each taxon, although this varies slightly in some analyses where vertebral elements are broken or missing and/or measurements were not collected. None of the vertebrae in the extant comparative sample exhibit pathological, developmental, traumatic, or degenerative malformations.

Because of the fragile nature of cervical vertebrae, the comparative fossil sample is very limited, especially for early hominins. *Australopithecus afarensis* is solely represented by the A.L. 333-101 C2 vertebra and by the A.L. 333-106 C6 vertebra dated to 3.3 Ma (Lovejoy et al. 1982). *Paranthropus robustus* is represented by SK 854, a C2 vertebra dated to 1.8 Ma (Broom and Robinson 1949; Napier 1959; Robinson 1972). *Homo erectus* is represented by KNM-WT 15000R, a C7 vertebra dated to 1.6 Ma (Brown et al. 1985; Walker and Leakey 1993), and D2673 and D2674, a C2 and C3 vertebra, respectively, from Dmanisi (Meyer 2005; Lordkipanidze et al. 2007). Not considered in this analysis was the C7 vertebra KNM-ER 164 from Koobi Fora originally attributed to *Homo erectus* (Day and Leakey 1974), which was later identified as modern human (Meyer 2003) from an intrusional deposit (Alan Walker, *personal communication*). Although many Neandertal cervical vertebrae are known (i.e., Shanidar 2, La Chapelle-aux-Saints), only casts of cervical vertebrae from Kebara and La Ferrassie from the American Museum of Natural History in New York were examined for this analysis. Limited data from *Homo heidelbergensis* were also collected from scaled images and published measurements from Gomez-Olivencia et al. (2007).

Linear measurements (Table 5.1) are in accordance with Cook et al. (1983), Ward (1991), Latimer and Ward (1993), and Meyer (2005). Angles and areas were calculated from the ImageJ software platform (National Institute of Health). Estimation of area using the traditional mathematical method (PI × long axis × short axis)/4) is incumbent with error in assuming perfect circularity or ovality in a feature, which is not the case in vertebral elements which are often kidney shaped, pentagonal, triangular, or heart shaped. Several analyses here employ the geometric mean, which has been shown to be effective in combining multiple measurements into a single proxy of size (Mosimann 1970; Jungers et al. 1995; Gordon et al. 2008). This statistic is useful for isolated vertebrae where dividing variables by the geometric mean serves to control for size in the absence of associated skeletal remains (cf. Darroch and Mosimann 1985; Lleonart et al. 2000) and captures size variation more consistently than single linear variables (Jungers et al. 1995). Simple ratios of paired variables were utilized in some cases to examine shape as per Thompson (1961), which effectively standardizes individuals to the same size, but may not remove undesired size effects because they maintain their size-dependent shape due to allometry (Lleonart et al. 2000) and are meaningful only if growth is isometric.

One way to eliminate size effects with allometric growth is to perform principal component analysis (PCA) and discard the first component (a method known as "shearing") and interpret subsequent axes as shape data (Blackith and Reyment 1971; Palmqvist et al. 1999), as the first principal component typically correlates positively with size (Jolicoeur and Mosimann 1960; Shea 1985). Although shearing did not radically change the results of analyses, both standard PCA and sheared PCA results are presented here for comparison. For the comparison of two groups, p values were calculated by two-tailed unpaired student's t test. Multiple comparisons were performed by the Tukey–Kramer HSD method. Tukey–Kramer HSD is an extremely conservative statistical test for multiple comparisons, which corrects for experiment-wise error rate.

Vertebral Centrum

Centrum Size

The sizes and shapes of cervical centra (vertebral bodies) are important to understand axial scaling and overall function, as the pattern of load transmission in the thoracic and lumbar column of modern *Homo* differs from that of all other extant primate taxa. Superoinferiorly short and transversely expanded centra in modern humans transmit loads vertically, whereas the long and dorsoventrally expanded centra in African great apes effectively act as a strut-like beam interposed between the hindlimbs and forelimbs better suited to support ventral viscera and resist flexion in the sagittal plane (Cartmill et al. 1987; Boszczyk et al. 2001).

In terms of overall size, cervical vertebrae of KSD-VP-1/1 are large, especially those inferior to the C3 level (Fig. 5.8, Table 5.2). Australopithecine lumbar and thoracic centra have been described as relatively small relative to humans (Robinson 1972; Sanders 1996, 1998), yet the sizes of the KSD-VP-1/1 cervical vertebrae caudal to the C3 level are larger than the human mean and closer to gorilla values.

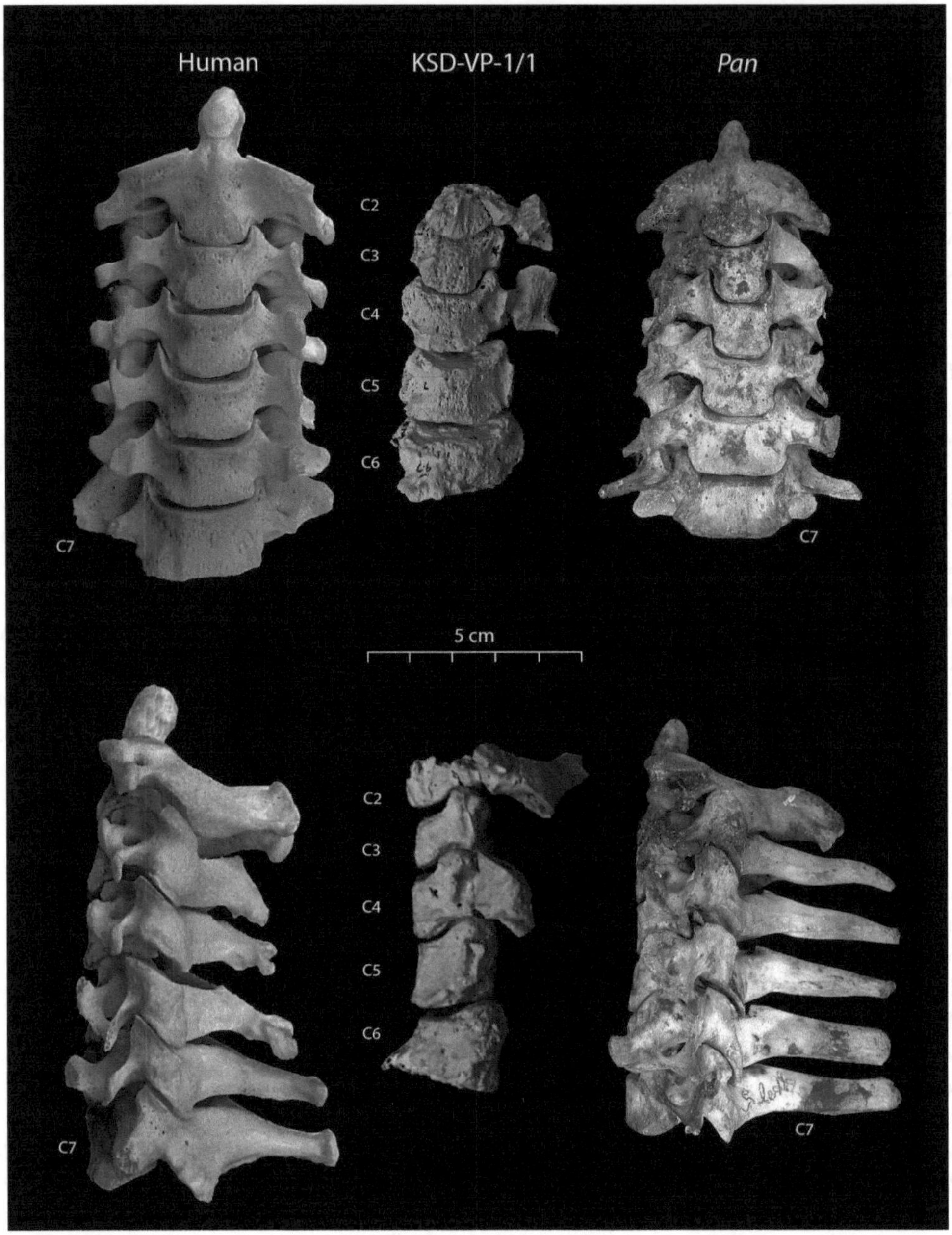

Fig. 5.8 Anterior view (top) and lateral view (bottom) of anatomically modern human, KSD-VP-1/1, and *Pan* cervical vertebrae from the C2 to C7 level

Table 5.1 KSD-VP-1/1 selected vertebral measurements (mm) by element

Vertebral level	Neural canal transverse width	Neural canal length	Superior centrum transverse width	Superior centrum length	Inferior centrum transverse width	Inferior centrum length	Dorsal centrum height	Ventral centrum height	Superior interfacet width
C2	20	12	–	–	18	17.1	–	–	–
C3	24	12	20.5	14.7	16	14.5	12.9	9.9	32.3
C4	23	13.5	21	15.4	20.5	16.2	13.7	13.5	35.5
C5	27.5	16	24	15.6	21	17	14.5	13	36.8
C6	–	–	26	16.5	24	19	17.8	12.1	–
C7	26	14.5	–	–	–	–	–	–	39.5

Table 5.2 Centrum geometric mean (*N*th root of 7 centrum dimensions) by element

Group	*N*	C2	*N*	C3	*N*	C4	*N*	C5	*N*	C6	*N*	C7	Mean C2–C6
Gorilla	18	14.2* (1.9)	19	15.6* (1.9)	10	15.6* (1.7)	10	15.9* (1.5)	10	17.4* (2.0)	10	18.6* (2.0)	15.7
Pan	18	10.9* (0.8)	19	11.7* (0.8)	10	11.6* (0.4)	10	11.9* (0.4)	10	12.9* (0.5)	10	14.1* (0.7)	11.8
AMH	51	13.3h (1.0)	50	14.4h	22	14.0h (1.1)	20	14.0h (1.0)	21	14.7h (1.5)	21	15.9h (1.4)	14.1
KSD	1	12.0	1	12.0	1	15.8	1	16.7	1	17.3		–	14.8
A.L. 333		–		–		–		–	1	13.1		–	–
SK 854	1	10.9		–		–		–		–		–	–
WT-15K		–		–		–		–		–	1	13.2	–
Dmanisi	1	12.2	1	10.7		–		–		–		–	–
ATA		–	2	12.2 (1.0)		–		–		–		–	–
Neand		–	2	–		–	1	13.2		–	1	14.2	–

Despite the small upper cervical elements, the mean size of the vertebral bodies in the whole KSD-VP-1/1 column is slightly larger than the mean for anatomically modern humans. Because the KSD-VP-1/1 C2 vertebra is missing aspects of the centrum, a similar but separate geometric mean was calculated for this element for reference. AMH = anatomically modern humans, KSD = KSD-VP-1/1, A.L. 333 = *Au. afarensis,* SK 854 = *P. robustus,* WT 15 K = KNM-WT 15000 *H. erectus,* Dmanisi = *H. erectus,* ATA = *H. heidelbergensis* (Atapuerca), Neand = Neandertal. * = different from humans $p < 0.05$; h = different from apes $p < 0.05$. Standard deviations appear in parentheses

Vertebrae can be used as an effective tool for sex assessment (Wescott 2000; Ostrofsky and Churchill 2013), and like the femoral head, centrum geometric mean values in the present sample are sexually dimorphic in *Gorilla* ($p < 0.0001$) and humans ($p < 0.0002$), although not in *Pan*. At the C6 level, for example, the centra of anatomically modern human males are 10% larger (CV = 9.8), 22% larger in male *Gorilla* (CV = 11.7), but only 3% larger in *Pan* males (CV = 4.1). The KSD-VP-1/1 C6 centrum is 32% greater than its putative female counterpart A.L. 333-106, supporting studies that note a degree of sexual dimorphism in *Au. afarensis* comparable to *Gorilla* (McHenry 1992; Lockwood et al. 1996; Gordon et al. 2008; Kimbel and Delezene 2009).

The centrum geometric mean for each vertebra was calculated as the 7th root of the following seven linear dimensions: superior and inferior centrum transverse widths, superior and inferior centrum dorsoventral lengths, and ventral, lateral, and dorsal centrum superoinferior heights (Meyer 2005). The advantage of using the centrum geometric mean is that it offers a size control for isolated vertebral elements without associated postcranial materials and avoids the confounding effect of differing locomotor regimes on femoral head size. All centrum geometric means significantly correlate with femoral head measures (e.g., C3 level, $r^2 = 0.80$, $p > 0.001$, $N = 89$), highlighting the utility of centrum geometric mean as basic size measure across species.

As in humans and apes, the centra of KSD-VP-1/1 exhibit a caudal size increase, but the increase is dramatically greater than that of the other sample groups (Fig. 5.9). The cervical spine in humans is typically pyramidal, with centrum sizes increasing in the caudal direction (Pope 2001; van Roy et al. 2001). In the lumbar column, centrum size increases may reflect the cumulative load borne by each element (Nordin and Weiner 2001) and be especially large in cursorial mammals due to amplified load magnitudes in running (Cartmill and Brown 2013). However, because non-cursorial apes share caudally progressive increases in cervical centrum size, the caudal size increases almost certainly relate to lordosis. In the hominin lumbar spine, pyramidal increases in centrum widths are correlated with articular facet widths enabling contiguous facets to imbricate

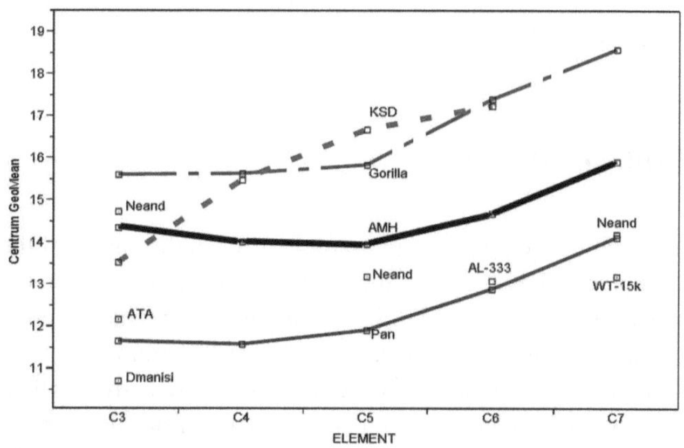

Fig. 5.9 Centrum geometric mean in C3–C7 vertebral levels. In all groups, centra sizes increase caudally. Note that A.L. 333-106 falls on the *Pan* mean and, by contrast, below the C3 level, the KSD-VP-1/1 centra (dotted line) are larger than the mean for anatomically modern humans (AMH). KSD = KSD-VP-1/1, A.L. 333 = *Au. afarensis,* SK 854 = *P. robustus,* WT-15k = KNM-WT 15000 *H. erectus,* Dmanisi = *H. erectus,* ATA = *H. heidelbergensis* (Atapuerca), Neand = Neandertal

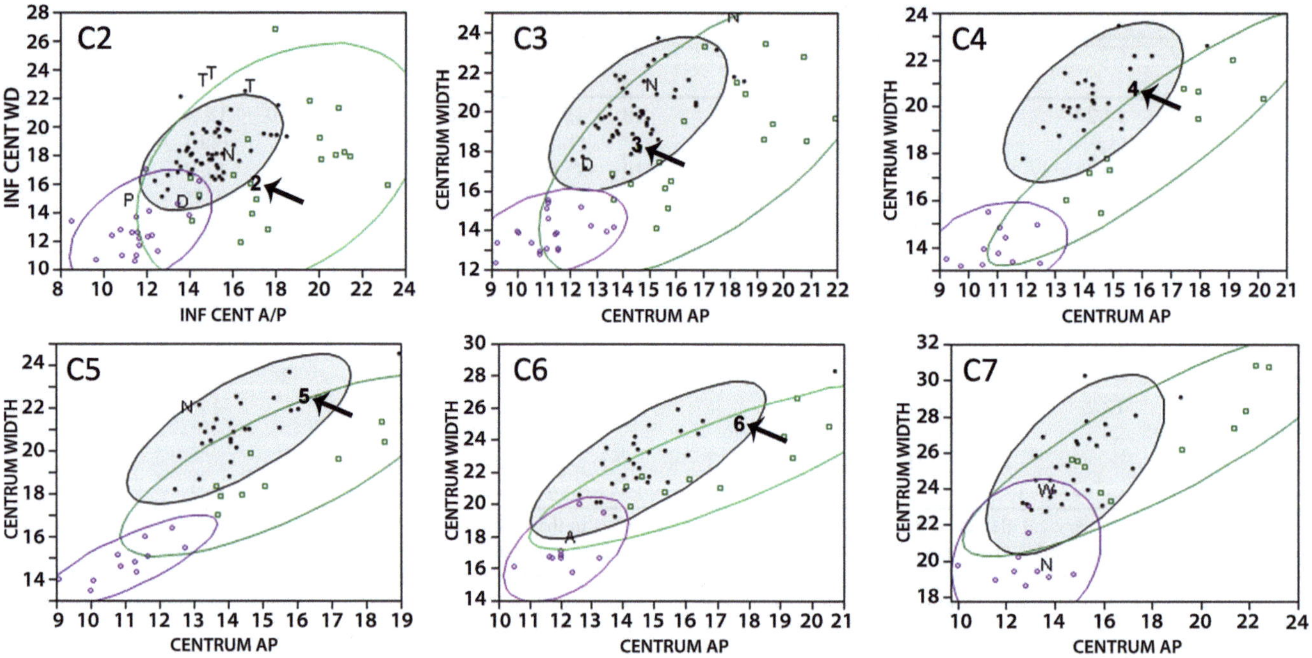

Fig. 5.10 Centrum shape index in C2–C7 vertebral levels (centrum transverse width/centrum anteroposterior length). Except for the C2 vertebra, the KSD-VP-1/1 specimens (indicated by arrows) fall within the 95% distribution ellipse for modern humans (shaded). Note that the more diminutive hominin fossils fall closer to (or within) the distribution for *Pan* than do the KSD-VP-1/1 fossils. The C2 bivariate plot represents only inferior centrum dimensions as the odontoid and centrum dimensions are not analogous. KSD = numerals indicating vertebral level, modern humans = solid points, *Pan* = open circles, *Gorilla* = open squares, A = A.L. 333-106 *Au. afarensis,* D = Dmanisi *H. erectus*, W = WT 15000 *H. erectus*, T = *H. heidelbergensis* (Atapuerca), *N* = Neandertals, *P* = SK-854 *P. robustus*. Bivariate normal ellipses *P* = 0.95%. All measurements are in millimeter

in extension, which facilitates lumbar lordosis (Latimer and Ward 1993). Similarly, the shared pattern of increasing centrum size in the cervical column may reflect the fact that humans and apes all exhibit cumulative lordosis in the cervical vertebral column (see analyses below) as all sample groups in the present study exhibit an increase in centrum size in the caudal direction (Fig. 5.10).

Centrum Width

As in all of the comparative sample groups, the shape of the vertebrae of KSD-VP-1/1 increases in transverse width in the caudal direction (Table 5.3). Although the shape index (centrum transverse width/dorsoventral length) in the KSD-VP-1/1 vertebrae indicates centra that are narrower than the modern human mean and closest to the mean shape index for *Pan,* they fall within the range of modern humans. In response to the subtle lateral shifts offsetting higher vertical load vectors during bipedal locomotion, modern humans feature especially enlarged transverse thoracic and lumbar centrum dimensions relative to quadrupeds (Meyer 2008). By contrast, the typical primate vertebra is transversely narrower and dorsoventrally expanded to optimally resist flexion in the sagittal plane (Boszczyk et al. 2001). Shape data here are consistent with this pattern, as centra at each cervical level in humans tend to be relatively wider than those other taxa, although shape index values broadly overlap across taxa, and may partially be a product of scale rather than indicate postural differences (Fig. 5.3).

Transverse Centrum Distribution

Like modern humans and other bipeds, the widest transverse dimensions of the KSD-VP-1/1 centra are skewed toward the dorsal centrum margins (dorsally eccentric), whereas at each vertebral level, the African great apes have centra surfaces that are wider toward the ventral margins (ventrally eccentric) (Fig. 5.11; Table 5.4). As a result, cervical centra in apes are structured to better resist flexion in the sagittal plane and ventral shear that accompanies pronograde posture and locomotion. Dorsal eccentricity of the centrum surfaces in bipedal hominins results in centrum better apportioned to resist higher magnitudes of loading toward the dorsal aspect of the centrum associated with a lordotic cervical column in upright posture. Accordingly, the dorsally eccentric centrum in modern humans corresponds to a nucleus pulposus that is

Table 5.3 Centrum shape index by element (centrum transverse width/centrum dorsoventral length)

Group	*N*	C2	*N*	C3	*N*	C4	*N*	C5	*N*	C6	*N*	C7	Mean C2–C6
Gorilla	19	0.96* (0.2)	19	1.09* (0.1)	10	1.14* (0.1)	10	1.25* (0.1)	10	1.33* (0.1)	10	1.47* (0.2)	1.15
Pan	20	1.10* (0.2)	20	1.25* (0.1)	11	1.28* (0.1)	11	1.34* (0.1)	10	1.42* (0.1)	11	1.60 (0.2)	1.28
AMH	55	1.22^{h} (0.1)	56	1.39^{h} (0.1)	28	1.42^{h} (0.1)	26	1.48^{h} (0.1)	28	1.55^{h} (0.1)	28	1.72 (0.1)	1.41
KSD	1	1.05	1	1.25	1	1.32	1	1.38	1	1.41	–	–	1.28
A.L. 333	–	–	–	–	–	–	–	–	1	1.45	–	–	–
SK 854	1	1.33	–	–	–	–	–	–	–	–	–	–	–
WT-15K	–	–	–	–	–	–	–	–	–	–	1	1.74	–
Dmanisi	1	1.08	1	1.36	–	–	–	–	–	–	–	–	–
ATA	3	1.5^{*h} (0.1)	2	1.61^{h} (0.4)	–	–	–	–	–	–	–	–	–
Neand	1	1.14	2	1.39 (0.0)	0	–	1	1.71^{h}	–	–	1	1.44	–

Ratio values were logarithmically transformed for statistical testing. AMH = anatomically modern humans, KSD = KSD-VP-1/1, A.L. 333 = *Au. afarensis*, SK 854 = *P. robustus*, WT 15 K = KNM-WT 15000 *H. erectus*, Dmanisi = *H. erectus*, ATA = *H. heidelbergensis* (Atapuerca), Neand = Neandertal. * = different from humans $p < 0.05$; h = different from apes $p < 0.05$. Standard deviations in parentheses

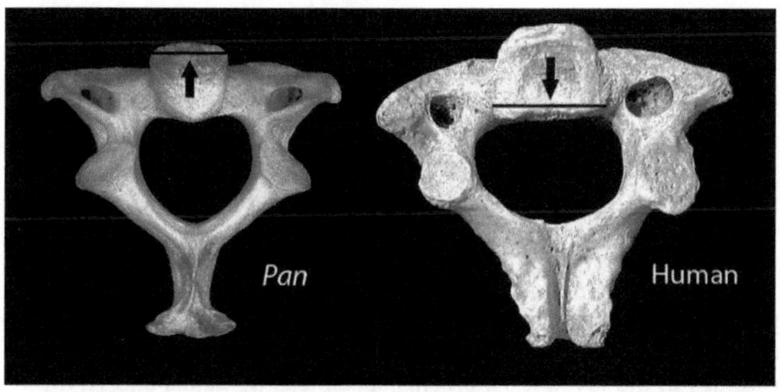

Fig. 5.11 Transversely widest point of inferior centrum in *Pan* (left) and anatomically modern humans (right). At each vertebral level, apes have ventrally eccentric centra, whereas hominin centra are dorsally eccentric

positioned slightly dorsal to the center of the intervertebral disc and functions to distribute vertical compressive forces (Humzah and Soames 1988). Dorsally eccentric centra in the KSD-VP-1/1 vertebrae signal that the nucleus pulposus in each vertebra was skewed toward the dorsal centrum surface and provides evidence that load mitigation in its cervical spine was comparable to modern humans and essentially dissimilar from the African great apes.

Centrum Heights

Because the KSD-VP-1/1 vertebral series possesses a number of consecutive elements, the length of the cervical column may be compared to other taxa, and by extension the length of the physiological neck segment. Although australopiths are typically described as smaller than humans in most postcranial features, the total physiological length of the KSD-VP-1/1 cervical column from C3 to C6 is 51.8 mm, which is greater than the modern human mean of 47 mm. This is despite a superoinferior height in its C3 vertebra that is shorter than the human mean and similar to that of *Pan* (Table 5.5). As the putatively female A.L. 333-106 is at the mean value of modern humans, it appears that neck length in *Au. afarensis* was equal to or greater in length than that of modern humans. Centrum midsagittal heights are thought to be genetically determined and unaffected by mechanical loading (Hinck et al. 1966; Ogden and Ganey 2003), rendering absolute centrum heights functionally uninformative, as there is broad distributional overlap among taxa of differing locomotor regimes in the sample. Centrum heights were calculated as the average of each element's ventral and dorsal craniocaudal midsagittal heights (Ward 1991).

By contrast, relative superoinferior centrum heights produce size-corrected values that are morphologically and biomechanically more informative as they neatly partition most bipeds from the knuckle-walking African great apes. Despite their large absolute heights, relative centrum heights (centrum superoinferior height/centrum transverse width) of the KSD-VP-1/1 fossils match the modern human pattern of craniocaudally short cervical centra (Table 5.6). As in the lumbar column, cervical centra superoinferior heights identify general postural adaptation among primates, as more orthograde primate taxa exhibit reduced centrum heights (Nalley 2013). Thus, superoinferiorly short centra in the KSD-VP-1/1 skeleton are entirely expected for an obligate biped; yet, the centrum height of the C6 vertebra of A.L.333-106 falls on the mean for *Gorilla*, which, like *Pan*, exhibits significantly narrower and taller cervical vertebrae than hominins

Table 5.4 Percentage of centrum posterior to widest transverse dimension by element

Group	*N*	C2	*N*	C3	*N*	C4	*N*	C5	*N*	C6	*N*	C7	Mean C2–C6
Gorilla	20	66* (16)	18	67* (13)	12	57* (16)	12	62* (13)	14	54 (11)	13	51 (5)	61
Pan	17	66* (13)	18	68* (18)	12	70* (17)	11	73* (15)	12	63* (15)	13	52 (10)	68
AMH	43	28h (10)	42	25h (9)	28	26h (8)	23	29h (13)	22	43 (15)	19	41 (15)	30
KSD	1	20h	1	30	1	34	1	42	1	30	–	–	31
A.L. 333	–	–	–	–	–	–	–	–	1	25	–	–	–
S.K. 854	1	37	–	–	–	–	–	–	–	–	–	–	–
WT-15000	–	–	–	–	–	–	–	–	–	–	1	45	–
Dmanisi	1	38	1	44	–	–	–	–	–	–	–	–	–
ATA	–	–	2	40 (1)	–	–	–	–	–	–	–	–	–
Neand	–	–	–	–	–	–	–	–	–	–	–	–	–

KSD-VP-1/1 is very near the mean for anatomically modern humans (AMH). Ratio values were logarithmically transformed for statistical testing. AMH = anatomically modern humans, KSD = KSD-VP-1/1, A.L. 333 = *Au. afarensis*, SK 854 = *P. robustus*, WT 15 K = KNM-WT 15000 *H. erectus*, Dmanisi = *H. erectus*, ATA = *H. heidelbergensis* (Atapuerca), Neand = Neandertal. * = different from humans $p < 0.05$; h = different from apes $p < 0.05$. Standard deviations in parentheses

Table 5.5 Superoinferior centrum height (mm) by element

Group	*N*	C3	*N*	C4	*N*	C5	*N*	C6	*N*	C7	Sum C3–C6
Gorilla	19	14.2* (2.0)	10	14.2* (1.7)	10	14.8* (1.8)	10	15.4* (2.0)	10	15.2* (1.2)	58.6
Pan	20	11.1 (1.6)	11	10.8 (1.7)	11	11.0 (1.6)	11	11.1 (1.5)	11	11.3* (1.7)	44.1
AMH	56	12.2 (1.5)	27	11.8 (1.4)	24	11.5 (1.3)	26	11.6 (1.4)	26	12.9h (1.5)	47.0
KSD	1	11.4	1	13.6	1	13.8	1	13.1	–	–	51.8
A.L. 333	–	–	–	–	–	–	1	11.5	–	–	–
WT-15K	–	–	–	–	–	–	–	–	1	8.7	–
Dmanisi	1	6.9*	–	–	–	–	–	–	–	–	–
ATA	2	8.0* (1.2)	–	–	–	–	–	–	–	–	–
Neand	2	10.9 (1.2)	1	11.6	2	10.5 (1.0)	2	10.7 (0.5)	1	11.5	43.6

The mid-cervical vertebrae in KSD-VP-1/1 are taller than the mean for anatomically modern humans, whereas *Au. afarensis* C6 (A.L. 333-106) matches human heights. Low values in Neandertal (Neand) and *H. heidelbergensis* (ATA) accord with the observations of Gomez-Olivencia et al. (2007) of craniocaudally low vertebrae compared to modern humans. Low values in the juvenile *H. erectus* specimens (Dmanisi and KNM-WT 15000) are due to lack of secondary ossification of end-plate epiphyses forming the ring apophysis and are not comparable with fused adult centra. AMH = anatomically modern humans, KSD = KSD-VP-1/1, A.L. 333 = *Au. afarensis*, WT-15 k = KNM-WT 15000 *H. erectus*, Dmanisi = *H. erectus*, ATA = *H. heidelbergensis* (Atapuerca), Neand = Neandertal. * = different from humans $p < 0.05$; h = different from apes $p < 0.05$. Standard deviations in parentheses

superior to the C7 level. Not surprisingly, C7 vertebrae across the sample taxa converge toward a common value, as the cervicothoracic joint experiences the lowest range of functional variability in the cervical column in primates and serves primarily as a fixed haft between the neck and trunk (Graf et al. 1995).

Centrum Morphology

Human cervical centra differ from those of quadrupedal primates in that the ventral surfaces are less constricted transversely and lack the narrow ventral "waist" interposed between the superior and inferior surfaces typical of the great apes. With the exception of the C7, transverse ventral constriction (waisting) is prominent in the apes, and because it is an expression of shape, not size, absolute transverse values are uninformative and require size correction for analysis. Relative centrum waist values (minimum ventral transverse width/ventral centrum superoinferior height) capture these shape variances and produces significant differences between the transversely expanded human centra and the narrow waisted centra of the African great apes (Table 5.7). Although accurate measurements of the ventral centrum waist on the KSD-VP-1/1 C5 and C6 centra are not reliable due to postmortem damage of the lateral margins of the centra, the well-preserved C3 and C4 vertebrae are transitional in this dimension, with values intermediate to modern humans and *Gorilla*.

The functional implications of ventral centrum waisting are presently unexplored, but the reduction of ventral waisting probably relates to the general pattern of craniocaudal height reduction and transverse expansion of the hominin cervical column. The centrum waist, or pronounced transverse constriction of the ventral centrum, is a feature

Table 5.6 Relative centrum height (centrum superoinferior height/centrum transverse breadth) by element

Group	*N*	C3	*N*	C4	*N*	C5	*N*	C6	*N*	C7	Mean C3–C6
Gorilla	19	0.88* (0.1)	10	0.87* (0.1)	10	0.88* (0.1)	10	0.76* (0.0)	10	0.67 (0.1)	0.85
Pan	20	0.97* (0.1)	11	0.92* (0.1)	11	0.87* (0.1)	11	0.75* (0.1)	11	0.68 (0.1)	0.88
AMH	56	0.73^{h} (0.1)	27	0.70^{h} (0.1)	24	0.66^{h} (0.1)	26	0.62^{h} (0.1)	26	0.65 (0.1)	0.68
KSD	1	0.75	1	0.74	1	0.72	1	0.61	–	–	0.71
A.L. 333	–	–	–	–	–	–	1	0.76	–	–	–
WT-15K	–	–	–	–	–	–	–	–	1	0.47	–
Dmanisi	1	0.48^{h}	–	–	–	–	–	–	–	–	–
ATA	2	0.47^{*h} (0.1)	–	–	–	–	–	–	–	–	–
Neand	2	0.60^{h} (0.0)	1	0.63	2	0.58 (0.0)	2	0.50^{h} (0.0)	1	0.68	0.58

Despite their large absolute sizes, the KSD-VP-1/1 vertebrae share a short and broad vertebral structure with modern humans. Ratio values were logarithmically transformed for statistical testing. AMH = anatomically modern humans, KSD = KSD-VP-1/1, A.L. 333 = *Au. afarensis*, WT-15 k = KNM-WT 15000 *H. erectus*, Dmanisi = *H. erectus*, ATA = *H. heidelbergensis* (Atapuerca), Neand = Neandertal. * = different from humans $p < 0.05$; h = different from apes $p < 0.05$. Standard deviations in parentheses

Table 5.7 Centrum ventral waist (minimum ventral transverse width/ventral centrum superoinferior height) by element

Group	*N*	C3	*N*	C4	*N*	C5	*N*	C6	*N*	C7	Mean C3–C4
Gorilla	18	1.11* (0.2)	8	1.19* (0.1)	9	1.28* (0.1)	9	1.55* (0.1)	9	2.05 (0.2)	1.15
Pan	17	1.05* (0.2)	8	1.10* (0.1)	8	1.29* (0.2)	8	1.55* (0.3)	8	2.15 (0.2)	1.08
AMH	41	1.61^{h} (0.2)	16	1.73^{h} (0.3)	14	1.76^{h} (0.3)	14	1.97^{h} (0.3)	14	2.07 (0.3)	1.67
KSD	1	1.48	1	1.32	–	–	–	–	–	–	1.40
A.L. 333	–	–	–	–	–	–	1	1.52	–	–	–
ATA	–	–	–	–	–	–	–	–	–	–	–
Neand	1	1.78^{h}	–	–	–	–	–	–	–	–	–

With the exception of the C7, African great ape ventral centra are significantly more constricted transversely than those of KSD-VP-1/1, modern humans (AMH) and Neandertals (Neand). Note that the putatively female A.L. 333-106 (C6 vertebra) exhibits the primitive ape-like morphology. Ratio values were logarithmically transformed for statistical testing. AMH = anatomically modern humans, KSD = KSD-VP-1/1, A.L. 333 = *Au. afarensis*, WT-15 k = KNM-WT 15000 *H. erectus*, Dmanisi = *H. erectus*, ATA = *H. heidelbergensis* (Atapuerca), Neand = Neandertal. * = different from humans $P < 0.05$; h = different from apes $p < 0.05$. Standard deviations in parentheses

that is often accompanied by a median ventral keel, a feature which serves to buttress ventral loading (Kikuchi et al. 2012) and appears variably in the African great apes. None of the KSD-VP-1/1 cervical vertebrae exhibit median ventral keels; instead, the C4 vertebra possesses a median dorsal keel, a structural buttressing often seen in modern humans (Fig. 5.12). The dorsal keel of KSD-VP-1/1 complements its dorsally eccentric superior and inferior centrum endplate surfaces, signaling a shift from the ventral axial loading regime typical of primates to the dorsally eccentric loading regime of orthograde bipedalism. These findings are inconsistent with the polemical "aquatic ape" hypothesis (Westenhöfer 1942; Hardy 1960; Morgan 2002) as a strong median ventral keel is universal among medium-bodied aquatic mammal vertebrae (author's unpublished data).

In contrast to some of the derived aspects of vertebral anatomy in the KSD-VP-1/1 vertebrae, the morphology of the superior ventral margin of the C3 centrum is rounded and slopes inferiorly, as is the case in many primates (e.g., *Papio*, *Gorilla* and often in *Pan*) but not in humans (Fig. 5.13). Rounded, inferiorly sloping margins in primate centra are associated with enhanced joint excursion (Deane et al. 2012) relative to modern human C3 vertebrae with sharply defined ventral margins that would allow for an increased range of ventral flexion on the sagittal plane. Some

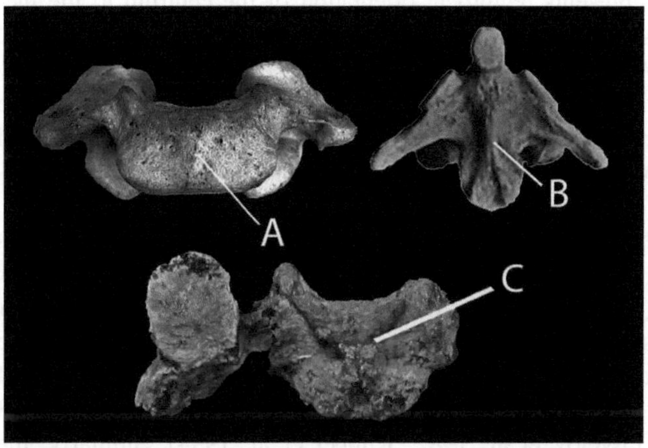

Fig. 5.12 Median anterior keel versus posterior keel (images not to scale); *A* median anterior keel on *Gorilla* C7 vertebra, *B* median anterior keel on *Ateles* C2 vertebra, *C* median posterior keel on the KSD-VP-1/1 C4 vertebra

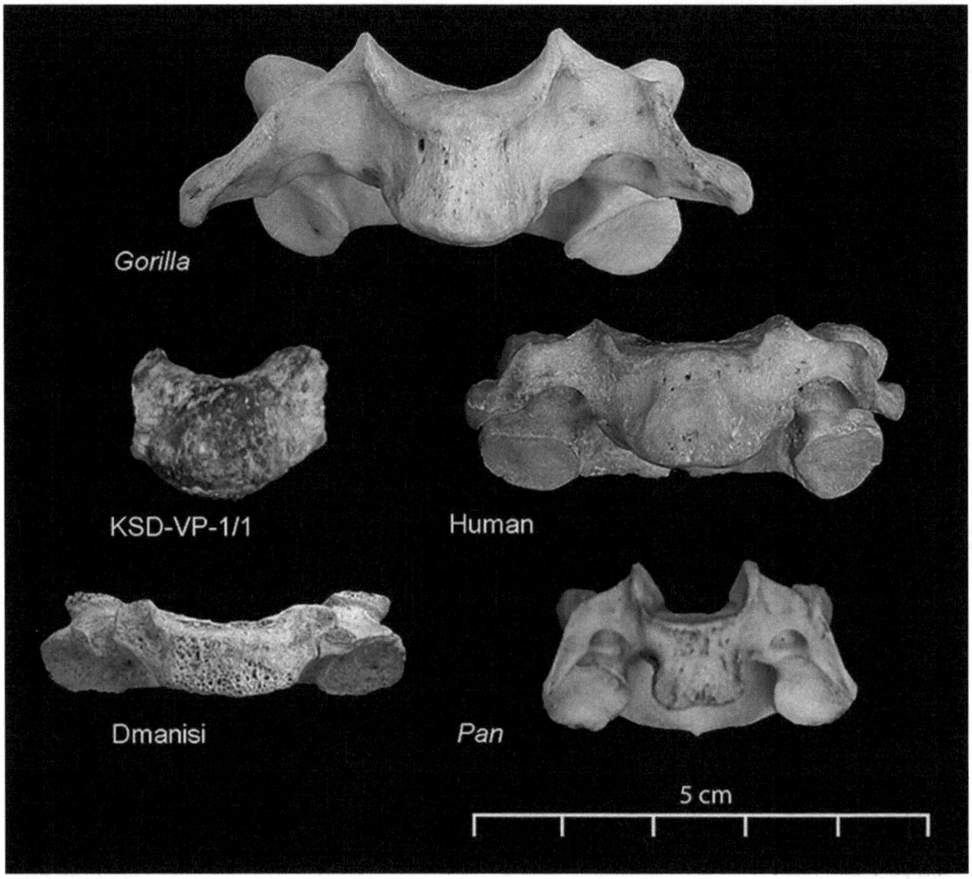

Fig. 5.13 Comparison of C3 vertebrae in anterior view. Unlike modern humans and Dmanisi *H. erectus*, the superior ventral margin of the KSD-VP-1/1 C3 centrum is rounded and slopes inferiorly. As in many primates (e.g., *Papio*, *Gorilla* and often in *Pan*), this morphology in the KSD-VP-1/1 specimen may represent reciprocal orientation of the centrum margin to the plesiomorphic orientation of the occiput and foramen magnum

of the morphology may reflect postmortem distortion, although it is very possible that the ventrally sloping C3 represents reciprocal orientation of the centrum to the plesiomorphic orientation of the occiput in *Au. afarensis.* Although the foramen magnum of *Au. afarensis* is relatively anterior (Kimbel et al. 1984) and closer in position to modern humans than many later hominins (Nevell and Wood 2008), the basicrania of *Au. afarensis* feature a posteroinferiorly oriented foramen magnum closest to *Pan*, unlike the anteroinferior orientation of modern humans (Kimbel and Rak 2010). Accordingly, Kimbel and Rak (2010) note the strongly angled atlanto-occipital articular surfaces on the A.L. 333-45 and A.L. 822-1 crania and the possibility of a reciprocal angulation in the articular facets of the fragmented australopith C1 vertebra of A.L. 333-83 (Lovejoy et al. 1982; Kimbel et al. 1994). Similarly, the rounded slope morphology of the superior surface of the KSD-VP-1/1 C3 centrum would have the effect of orienting the superior centrum margin and intervening C2–C3 intervertebral disc toward loads emanating from the basicranium.

Uncinate Processes

The uncinate processes and the Luschka joints between the uncinate processes and the superjacent vertebral body (uncovertebral joints) develop during childhood. They impart the typical saddle shape of the lower five cervical vertebrae and play an important biomechanical role in stabilizing the cervical spine during coupled motions, such as lateral bending with rotation. Finite-element models indicate that the uncovertebral joints generally serve to reduce motion, especially in axial rotation and lateral bending (Iai et al. 1993; Moskovich 2001; Storvik and Stemper 2011). In the African great apes, the uncinate processes are relatively tall, creating a vertically oriented and deep joint extending approximately halfway up the superoinferior length of the superjacent centrum (see Figs. 5.8 and 5.13). Tall uncinate processes in the apes contribute toward their stiff-necked appearance, whereas reduced or absent uncinate processes across species increase the range of cervical motion (Clausen et al. 1997; Breit and Kunzel 2002). Human uncinate processes are much lower, creating a shallower and more flexible joint, with slight superolateral inflections, especially at the most caudal levels allowing for an unprecedented range of head and neck movement relative to the torso movement among hominoids. The derived human morphology may be part of a suite of traits associating a functionally decoupled head from the torso (Meyer 2005). Like the cervical vertebrae of modern humans and early *Homo*, the KSD-VP-1/1 vertebrae share the low, sloping uncinate margins, suggesting an increased range of cervical mobility and signal the potential for a decoupled head from the torso. A more cogent signal of the decoupled head and torso would also include human-like spinous process and articular facet kinematics, as well as marked reduction of the nuchal musculature, which is difficult to assess in the absence of cranial remains for the individual. Relative to *Pan* and *Gorilla*, the crania of male *Au. afarensis* seem to exhibit reduced nuchal musculature (Kimbel and Rak 2010), implying a degree of cranial independence from the torso in KSD-VP-1/1 *at least* intermediate to humans and the apes.

Centrum Wedging Angles

Cervical lordosis is one of the four curves that the adult human spinal column expresses on the sagittal plane. Thoracic and sacral kyphosis are primary ventral curvatures present from birth, whereas lordosis in the lumbar and cervical column are developmental secondary curvatures thought to relate to axial loading. Although intervertebral discs contribute toward mid-cervical lordosis (Johnson 1998), assessment of vertebral wedging angles assists in recognizing spinal curvatures in skeletal remains.

Subaxial centra across the sample are similarly wedged ventrally, until the hominin divergence at the C7 level where the pattern transitions from increasing kyphosis toward lordosis. The C7 is the only cervical level where the wedging angle significantly differs between humans and the apes. At 11.4°, the ventral wedging angle of the KSD-VP-1/1 C3 is outside the sample range, a divergence explained by the difficulty in determining the superior ventral margin of the centrum, and as its rounded, inferiorly sloping morphology is not amenable to angular measurement and may be a function of postmortem distortion. Wedging angles of the subjacent KSD-VP-1/1 C5 and C6 vertebrae fall within the sample range but fall closest to the mean wedging angles for *Gorilla*. Because its C7 is absent, it is not possible to assess whether it exhibited the diagnostic anticlinal pattern of bipeds or continued the pattern of increasing kyphosis in the caudal direction as in apes.

Cervical lordosis has been explained as the development of dorsal centrum wedging as a response to holding the head upright (Aiello and Dean 1990); however, below the C2 level none of the groups exhibit a mean pattern of dorsal wedging (Table 5.8). Instead, the sample data demonstrate ventral wedging (kyphosis) in the subaxial cervical vertebrae of all sample groups (including humans). Yet, high dorsal wedging angles (lordosis) of the C2 vertebra overcome cumulative subaxial kyphosis and are responsible for overall skeletal lordosis in the cervical column across sample groups (Table 5.9). A similar pattern is seen in the lumbar lordosis of bipedal hominins where ventral wedging in the upper lumbar levels (kyphosis) is negated by higher dorsal wedging values, resulting in an aggregate lumbar column that is lordotic (Shi et al. 1995, 1999; Meyer 2005). Because there is no superior centrum on the C2 axis vertebra, wedging angles for this level were calculated as a measure of the inferior centrum on the transverse plane perpendicular to the most superior point of the ventral odontoid process, and are not precise analogs of subaxial wedging angles. As human and great ape values are very similar, the value for KSD-VP-1/1 was reconstructed using a bipedal (human) model (Figs. 5.14 and 5.15).

Table 5.8 Subaxial centrum wedging angle (degrees) by element

Group	*N*	C3	*N*	C4	*N*	C5	*N*	C6	*N*	C7	Mean C4–C6	Sum C4–C6
Gorilla	19	0.3 (3.4)	10	1.6 (3.5)	10	1.5 (4.3)	10	4.7 (3.7)	10	5.9* (3.5)	2.58	7.75
Pan	20	0.6 (4.1)	11	−0.2 (3.6)	11	1.6 (5.5)	10	3.7 (4.0)	11	5.7* (2.7)	1.71	5.14
AMH	55	0.8 (3.1)	27	0.6 (3.2)	25	3.0 (3.7)	26	3.5 (2.9)	26	2.1[h] (2.2)	2.37	7.10
KSD	1	11.4*	1	0.6	1	5.5	1	6.9	–	–	4.32	12.96
A.L. 333	–	–	–	–	–	–	1	3.2	–	–	–	–
WT-15K	–	–	–	–	–	–	–	–	1	0.7	–	–
Dmanisi	1	1.0	–	–	–	–	–	–	–	–	–	–
ATA	2	1.8 (1.9)	–	–	–	–	–	–	–	–	–	–
Neand	2	0.5 (0.7)	–	–	1	1.6	1	2.2	1	−4.5[h]	1.88	3.76

Positive values reflect ventral wedging (kyphosis) and negative values reflect dorsal wedging (lordosis). The formula for calculating wedging angles is from Digiovanni et al. (1989) and Latimer and Ward (1993): 2 TAN^{-1} {[(centrum dorsal height – centrum ventral height)/2]/centrum anteroposterior diameter}. AMH = anatomically modern humans, KSD = KSD-VP-1/1, A.L. 333 = *Au. afarensis*, WT-15 k = KNM-WT 15000 *H. erectus*, Dmanisi = *H. erectus*, ATA = *H. heidelbergensis* (Atapuerca), Neand = Neandertal. * = different from humans $p < 0.05$; h = different from apes $p < 0.05$. Standard deviations in parentheses

Table 5.9 C2 wedging angle (in degrees) by group and cumulative angle sums

Group	*N*	C2	SD	Sum degrees C3–C7	Sum degrees C2–C7
Gorilla	17	−22	7.2	14.0 Kyphosis	−8 Lordosis
Pan	20	−17	8.2	11.4 Kyphosis	−6 Lordosis
AMH	55	−21	5.7	9.9 Kyphosis	−11 Lordosis
KSD	1	−17	–	–	–
Dmanisi	1	−19	–	–	–

Note that the C2 level is exclusively responsible for sum lordosis in the cervical spine. The value for KSD-VP-1/1 was reconstructed. Positive values reflect ventral wedging (kyphosis) and negative values reflect dorsal wedging (lordosis). The Dmanisi *H. erectus* C2 vertebra is the oldest hominin C2 with a preserved odontoid process. AMH = anatomically modern humans, KSD = KSD-VP-1/1, Dmanisi = *H. erectus*

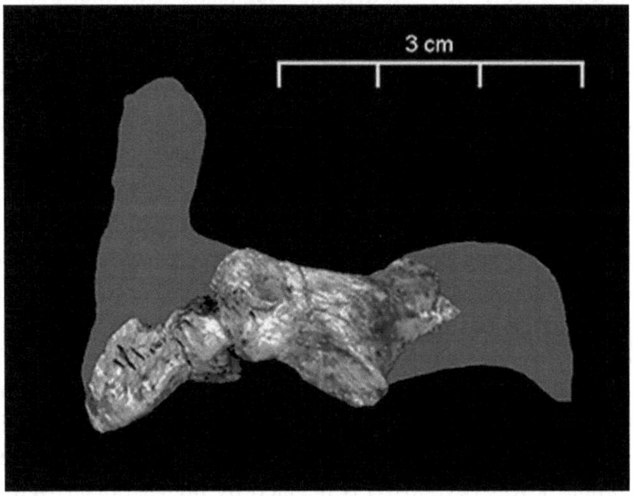

Fig. 5.14 Reconstruction of the lateral profile of the KSD-VP-1/1 C2 vertebra (based on a human analog). The inferior centrum surface projects inferiorly with a predicted a ventral length approximately 15% taller in superoinferior height than the dorsal length, resulting in an effective dorsal wedging angle of 17°

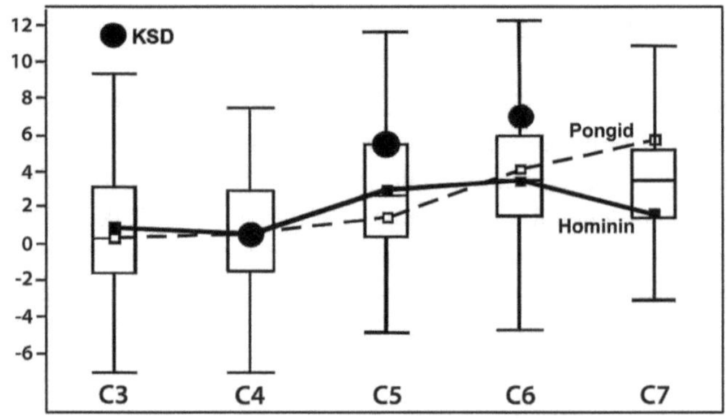

Fig. 5.15 Boxplots of combined subaxial cervical wedging angles (in degrees). Wedging angles in the KSD-VP-1/1 vertebrae are indicated by large solid points. There is considerable similarity between the combined ape (gray line) and combined hominin means (black line) at each level except at *C7*. Positive values represent kyphosis, negative values represent lordosis. Boxplots indicate 25, 50, and 75th percentiles, and whiskers indicate sample ranges excluding outliers

The human mean value of −11° in the present study of total cervical column lordosis is much less than the −34° value reported for humans from radiographic analyses (Harrison et al. 1996, 2004). This is largely because intervertebral discs, which do not typically preserve, are a large factor in the physiological angulation of the axial spine (Johnson 1998), whereas lumbar intervertebral discs (except for the lumbosacral joint) do not significantly influence wedging angles (Gogen et al. 2007). However, it does illustrate the lordotic nature of the KSD-VP-1/1 C2 despite substantial subaxial kyphosis. The observed sweeping inferoposterior angle of the inferior articular process is also a morphological trait consistent with lordosis, as seen in the inferior facets of lordotic lumbar columns (Latimer and Ward 1993; Meyer 2005). The combination of these features implies that the aggregate cervical spine of KSD-VP-1/1 was lordotic, as is the case across the sample and primates in general. Despite wedging values in KSD-VP-1/1 exceeding the mean for anatomically modern humans and approaching those of *Gorilla*, there is considerable overlap among sample groups.

Based on wedging angles of the KSD mid-cervical vertebrae, it is tempting to infer that head carriage of KSD-VP-1/1 was flexed more ventrally than in modern humans. This inference is supported by a strong correlation between more flexed head posture in humans and large facial dimensions combined with a short posterior cranial base (Solow and Tallgren 1976; Huggare and Raustia 1992). But because the intervertebral discs (which engender a considerable portion of cervical lordosis) and the diagnostic C7 centrum are absent, this conclusion is speculative. Moreover, because the KSD vertebrae share dorsally eccentric centra with humans that mirror the dorsally skewed distribution of lordotic intervertebral discs, an equally reasonable interpretation would be that the KSD-VP-1/1 cervical column exhibited lordosis similar to that of modern humans.

Centra Summary

The upper cervical levels in KSD-VP-1/1 are smaller than those of anatomically modern humans, but caudal to the C3 level, they are larger than those of humans and closer to *Gorilla* values. The size of its C6 vertebra signals a degree of sexual dimorphism comparable to *Gorilla* and a neck length equal to or greater in length than that of modern humans. The KSD-VP-1/1 superior and inferior centrum surface shapes are somewhat ape-like in that they are relatively narrower than those of modern humans and are intermediate in the degree of ventral waisting. In addition, its C3 vertebra resembles *Gorilla* in the rounded, inferiorly sloping form of the ventral superior centrum margin. The KSD-VP-1/1 vertebrae are also closer to *Gorilla* in the degree of ventral wedging, which could imply that it had a more flexed, forward tilt of the neck than humans, but the dorsal wedging angle of the C2 vertebra and apparent dorsal distribution of the intervertebral discs make it likely that the overall cervical column in KSD-VP-1/1 possessed a degree of lordosis more similar to modern humans than the African great apes.

Despite some similarities with *Gorilla*, the KSD-VP-1/1 centra are relatively shorter craniocaudally than the apes, lack a ventral keel, have low sloping uncinate processes, and exhibit a shift from the ventral axial loading regime characteristic of primates to the high dorsal vertical peak force loading regime of orthograde bipedalism. This suite of features is not observed in the African great apes where a habitually flexed hip and knee posture produces lower vertical peak forces (Schmitt 2003). Enlarged centra areas in KSD-VP-1/1 agree with studies linking large articular surface areas across joints to bipedal locomotion (Rose 1984; Jungers 1988), as joint enlargement mitigates load force at heel strike (Bramble and Lieberman 2004). While there are small differences between KSD-VP-1/1 and anatomically modern humans, there are no features of the centra indicative of anything other than upright posture and orthograde bipedality in KSD-VP-1/1.

Interarticular Facet Widths

In addition to possessing the largest transverse centrum dimensions, the human vertebral column possesses the widest distances between the articular facets in response to the high degree of axial load and torsion in bipedalism (Boszczyk et al. 2001). In the lumbar spine, the pattern of caudally increasing interfacet widths in bipedal hominins makes it possible for adjacent vertebrae to imbricate and extend posteriorly in concert with dorsally wedged centra, which allows for lumbar lordosis (Latimer and Ward 1993). Accordingly, increasing interfacet widths may serve as an indication of lordosis, whereas progressively decreasing or static widths may signal kyphosis. Across the sample groups, interarticular facet widths gradually increase caudally, with a peak interarticular dimension at the C5 and C6 levels for African great apes and humans, respectively. As a result, a broad, pyramidal configuration of support forms at the base of the cervical column, which serves as a mechanism for lordosis and stable balance of the head. Despite vertebral wedging of the centra, vertebrae across taxa, including KSD-VP-1/1, share a kinematic signature consistent with cervical lordosis.

The KSD-VP-1/1 superior interarticular facet widths are very close to the mean for anatomically modern humans in the mid-cervical levels, but are larger than the mean at the C7 level, which implies a peak in the KSD-VP-1/1 cervical region caudal to the C5 level. This distinguishes KSD-VP-1/1 from both the ape and human patterns of decreasing interarticular facet widths at the C7 level, but the nearly identical mean value for interarticular widths in modern humans and KSD-VP-1/1 indicates similar overall kinematic structure. Interarticular facet width is measured as the average of the minimum and maximum interarticular superior facet widths. Measures for incomplete KSD-VP-1/1 vertebrae are computed by doubling preserved unilateral articular facet distances from the sagittal midline.

The large interfacet distance of A.L. 333-101, an *Au. afarensis* C2, may reflect postmortem distortion, as this element is highly fragmented. Alternatively, this may reflect a considerable degree of variation in the australopith sample, or sexual dimorphism, as the trajectory of the male KSD-VP-1/1 interfacet distances forecast C6 values well above the putatively female A.L. 333-106 C6 value, which lies on the mean for *Pan* (as does SK 854, a *P. robustus* C2 specimen).

Although absolute interarticular facet widths are informative, the pattern observed in relative interarticular facet widths (superior interarticular facet width/superior centrum transverse width) reveals more cogent differences between anatomically modern humans and other groups (Fig. 5.16). The African great apes have a marked increase in relative interarticular facet widths in the mid-cervical region, whereas both KSD-VP-1/1 and modern humans exhibit decreases caudal to the C4 level. The flatter human curve in the data indicates that proportional load transfer between the centrum and facets is more uniform throughout the bipedal cervical column than that of apes, as there is only a 17% difference between maximum and minimum interarticular facet widths in humans, compared to a 28% and 29% increase in *Gorilla* and *Pan*, respectively. Ratios of the difference between C3 levels to peak values display the same pattern, with those of KSD-VP-1/1 essentially matching the human increase of 8% as opposed to the 14% increase in the African great apes.

KSD-VP-1/1 mirrors the human pattern of a reduced and more cranial peak in interarticular facet widths, implying that

Table 5.10 Superior interarticular facet width (mm) by element

Group	*N*	C2	*N*	C3	*N*	C4	*N*	C5	*N*	C6	*N*	C7	Mean C3–C5, C7
Gorilla	18	30.0* (2.6)	20	31.9* (3.7)	10	34 (4.1)	10	35.4 (5)	10	35.2* (4.9)	10	34.2 (4.3)	33.9
Pan	21	22.1* (1.6)	21	25.7* (1.7)	12	28.3* (1.6)	12	28.9* (1.6)	12	28.6 (1.7)	12	28.4* (1.7)	27.8
AMH	52	28.0^{h} (1.9)	52	33.8^{h} (1.9)	24	35.8 (2)	23	37.0 (2.1)	23	37.3 (3.1)	23	36.8 (2.7)	35.9
KSD	–	–	1	32.3	1	35.5	1	36.8	–	–	1	39.5	36.0
A.L. 333	1	33.4	–	–	–	–	–	–	1	28.2	–	–	–
SK 854	1	23.3	–	–	–	–	–	–	–	–	–	–	–
WT-15k	–	–	–	–	–	–	–	–	–	–	1	33.7	–
Dmanisi	1	29.2	1	29.8	–	–	–	–	–	–	–	–	–
Neand	1	34.5*	1	41.3*	2	39.5 (1.8)	2	41.2 (0.7)	2	44.8* (3.3)		–	40.7

The Neandertal divergence in the pattern is not informative and is likely due to sample error, as the C2 and C3 represent only one individual each of different sizes (the smaller Kebara C2, and larger La Ferrassie C3), while the C4–C6 levels represent the mean of both individuals. When this size discrepancy is taken into account, the Neandertal peak at the C3 level would diminish and it is probable that the Neandertals would follow the typical hominoid pattern. AMH = anatomically modern humans, KSD = KSD-VP-1/1, A.L. 333 = *Au. afarensis,* SK 854 = *P. robustus,* WT-15k = KNM-WT 15000 *H. erectus,* Dmanisi = *H. erectus,* Neand = Neandertal. * = different from humans $p < 0.05$, h = different from apes $P < 0.05$. Standard deviations in parentheses

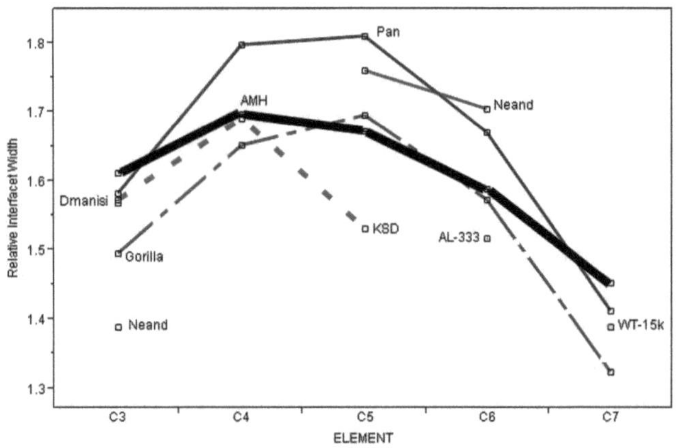

Fig. 5.16 Relative interarticular facet width by element (superior interarticular facet width/superior centrum transverse width). The mid-cervical peak of the African great apes demonstrates a shift of proportionally greater axial loads to the corresponding articular facets, whereas humans exhibit a more uniform load profile. AMH = anatomically modern humans, A.L. 333 = *Au. afarensis,* WT-15k = KNM-WT 15000 *H. erectus,* Dmanisi = *H. erectus,* ATA = *H. heidelbergensis* (Atapuerca), Neand = Neandertal

like humans, KSD-VP-1/1 mid-cervical centra channeled proportionately greater load magnitudes than apes that shift proportionally greater axial loads to the corresponding articular facets (Table 5.10). The Neandertal distribution does not provide enough data for full comparison; however, the slope of the decrease between the C5 and C6 levels matches the slope in humans and lacks the steep descent of the ape distribution, suggesting the possibility of a more cranial peak in relative facet distribution as in modern humans.

Articular Process Height

Large dorsal cervicothoracic musculature in the African great apes inhibits head drop and flexion of the upper cervical spine from gravitational forces acting on a head cantilevered on the neck. Ventral forces are furthered amplified by loading from powerful ventral muscles at work during mastication, increasing dorsoventral translation amplitude away from the physiological axis. It is the articular facets that serve as the initial site of bone-to-bone contact in response to these forces and serve as the primary fulcrum for sagittal movement (Hoffmann 2006). Accordingly, cantilevered, prognathic heads and large nuchal musculature in the apes mandate relatively taller articular processes and differentiate them from modern humans (Table 5.11). Relative articular process height (total superoinferior height of both superior and inferior articular facets/superior interarticular facet width) expresses the total combined superoinferior height of the articular processes. In this dimension, KSD-VP-1/1 vertebrae are closest to the human mean at every level except C3, where it approaches the mean for *Pan*. This morphology might be expected considering the increased centrum wedging at this level, as high wedging angles intensify loads on the articular processes (Cailliet 2004) and are associated with tall articular processes for resisting ventroflexion and shear. Likewise, A.L. 333-106 retains the more primitive morphology, falling between *Pan* and *Gorilla* means for facet height.

Table 5.11 Relative articular process height (superoinferior height of superior and inferior articular facets/superior interarticular facet width)

Group	*N*	C3	*N*	C4	*N*	C5	*N*	C6	*N*	C7	Mean C3–C5, C7
Gorilla	20	0.75* (0.1)	10	0.73 * (0.1)	10	0.76* (0.1)	10	0.85* (0.1)	10	0.89* (0.1)	0.78
Pan	21	0.68* (0.1)	12	0.67* (0.1)	12	0.67* (0.1)	12	0.74* (0.1)	12	0.75* (0.1)	0.69
AMH	52	0.6^{h} (0.0)	24	0.59^{h} (0.1)	23	0.56^{h} (0)	23	0.58^{h} (0.1)	23	0.68^{h} (0.1)	0.61
KSD	1	0.67	1	0.59	1	0.59	–	–	1	0.67	0.63
A.L. 333	–	–	–	–	–	–	1	0.77*	–	–	–
WT-15K	–	–	–	–	–	–	–	–	1	0.59	–
Dmanisi	1	0.48^{h}	–	–	–	–	–	–	–	–	–
Neand	1	0.5^{h}	2	0.51^{h} (0)	2	0.47^{h} (0.1)	2	0.5^{h} (0)	–	–	–

AMH = anatomically modern humans, KSD = KSD-VP-1/1, A.L. 333 = *Au. afarensis,* WT-15 k = KNM-WT 15000 *H. erectus,* Dmanisi = *H. erectus,* Neand = Neandertal. * = different from humans $p < 0.05$, h = different from apes $p < 0.05$. Standard deviations in parentheses. Ratio values were logarithmically transformed for statistical testing

Articular Facet Orientation

The amount and type of movement that occurs in the cervical column is determined by the discs, uncinate processes, ligaments, and most significantly, the articular processes. Throughout the cervical spine, lateral flexion of a given segment is strictly coupled to rotation of that segment, and the degree of coupling is determined by the orientation of the articular facet joints (zygapophyseal joints) (Penning and Wilmink 1987; Yamamoto et al. 1989; Maurel et al. 1997; Onan et al. 1998). In human cervical vertebrae, the line of gravity runs just anterior to the articular facet joints, whereas the line of gravity runs close to, or through, the centra of other vertebrae (Standring 2008). As a result, articular facets are oriented at an angle of about 45° to the coronal plane (Shapiro and Frankel 1989; Moskovich 2001), and at a slightly higher angle in the apes (Meyer 2005), instead of the near vertical superoinferior orientation of the subjacent thoracic and lumbar articular facets. As a result of their orientation, cervical articular facets facilitate considerably more yaw (rotational movement) than the lumbar and thoracic vertebrae, but bear more weight than their thoracic or lumbar counterparts (Aiello and Dean 1990; Kirpalani and Mitra 2008). Although humans and apes do not differ in the transverse orientation of their articular facets, the articular facets in modern humans differ from those of the African great apes in that their major axis is obliquely oriented relative to the midsagittal line of the spinous process (Fig. 5.17). The more acutely oriented articular facets serve to limit yaw in the ape cervical spine and, along with superoinferiorly taller uncinate processes, contribute to the stiff coupling of the head and trunk in the African great apes.

Although A.L. 333-106 exhibits the acute articular facet major axis orientation of the apes, the articular facets in KSD-VP-1/1 are obliquely oriented as in modern humans (Table 5.12). This implies that the cervical spine of KSD-VP-1/1 had evolved essentially human-like articular process morphology, which allows for greater rotation between motion segments in the cervical spine relative to the African great apes and provides another example of skeletal kinematics that may be more derived in male specimens of *Au. afarensis* than in its female counterpart. It is curious that A.L. 333-106 and the C7 of KNM-WT 15000 are the only vertebrae among the hominins to display an ape-like orientation of the articular processes. The juvenile status of the KNM-WT 15000 *H. erectus* specimen probably does not explain its primitive morphology, as there is no correlation between age and the superior articular facet angle in humans ($p > 0.8$), or vertebral size in apes ($p > 0.89$) or humans ($p > 0.49$). These two hominins, however, do exhibit distinctly long spinous processes, which correlate with articular facet angles in *Pan*

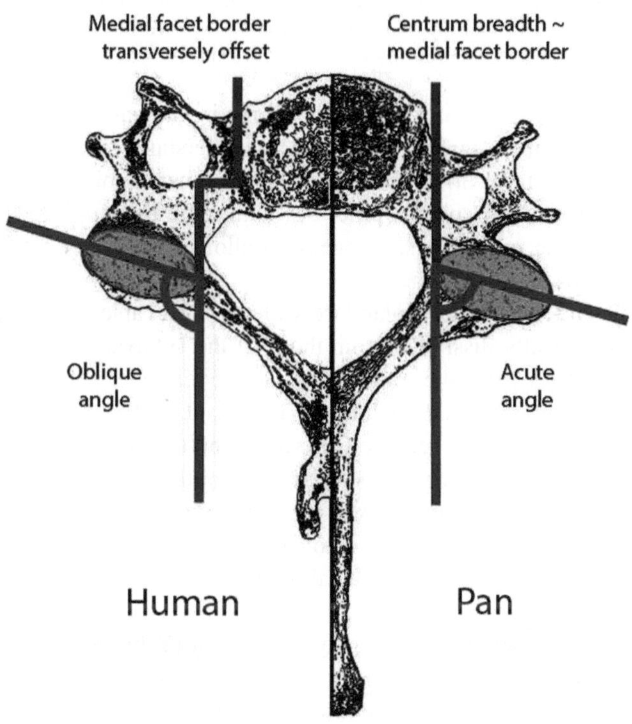

Fig. 5.17 Typical subaxial cervical facet orientation comparison in modern humans (left half) and *Pan* (right half)

Table 5.12 Superior articular facet angle (degrees) by element

Group	*N*	C3	*N*	C4	*N*	C5	*N*	C6	*N*	C7	Mean C3–C5, C7
Gorilla	19	74.6* (14)	15	66.2* (9)	14	72.8* (11)	14	66.7* (11)	15	71.3* (6.5)	71.2
Pan	15	60.6* (8.8)	13	69.2* (9.4)	13	73.8* (11)	13	77.2* (7.7)	13	81.3* (7.8)	71.2
AMH	22	103h (9.1)	23	105h (9)	24	106h (5.9)	23	108h (5.5)	24	104h (3.7)	104.5
KSD	1	121h	1	125h	1	127h	–	–	1	99h	118
A.L. 333	–	–	–	–	–	–	1	81*	–	–	–
WT-15K	–	–	–	–	–	–	–	–	1	80*	–
Dmanisi	1	98	–	–	–	–	–	–	–	–	–
Neand	1	103	2	87.8 (3.9)	2	94h (1.4)	2	98h (8.5)	2	92.8 (4.6)	94.4

The male *Au. afarensis* KSD-VP-1/1 fossils exhibit the derived oblique morphology, whereas the female A.L. 333-106 C6 vertebra exhibits the primitive acute morphology. AMH = anatomically modern humans, KSD = KSD-VP-1/1, A.L. 333 = *Au. afarensis,* WT-15 k = KNM-WT 15000 *H. erectus,* Dmanisi = *H. erectus,* Neand = Neandertal. * = different from humans $p < 0.05$, h = different from apes $P < 0.05$. Standard deviations in parentheses

Table 5.13 Mean C2 superior articular facet angle (degrees) across primate taxa

Group	*N*	Mean	SD
Pan	16	100	9.3
Pongo	2	115	8.9
Papio	3	96	6.1
Colobus	3	86	3.9
Sypmhalangus	3	85	7.3
Ateles	3	75	4.4
AMH	68	133	7.7
KSD	1	55	–
Neandertal	2	134	9.2
Dmanisi	1	110	–
SK 854	1	120	–
A.L. 333-101	1	107	–

The anomalous KSD value is likely a product of postmortem deformation. AMH = anatomically modern humans, KSD = KSD-VP-1/1, A.L. 333 = *Au. afarensis*, SK 854 = *P. robustus*, Dmanisi = *H. erectus*

($r^2 = 0.43$, $N = 53$, $p < 0.0001$), but not in humans or *Gorilla*. Further investigation is needed to understand variation in articular facet orientation, but because other hominins have similarly long spinous processes with more human-like orientations, epigenetics, ontogeny, or allometry cannot explain *Pan*-like values in these cases.

The superior articular facets of the C2 (Table 5.13) are functionally dissimilar from those of the subaxial vertebrae as C2 provides the majority of yaw, pitch, and roll movements of the head in humans (Iai et al. 1993; Chapman 2008). However, because C2 facets simultaneously function as surrogate vertebral bodies in transmitting vertical loads, they are larger than subaxial articular facets and are oriented obliquely relative to the coronal plane in large-bodied primates to counter craniocaudal loads.

With an inclination of 55°, the KSD-VP-1/1 C2 superior articular facet angle is highly anomalous (Table 5.14). It is possible that the small remnant represented may not accurately represent the whole facet orientation, or more probable, may reflect postmortem distortion, as the left inferior articular process seems to be somewhat skewed inferomedially

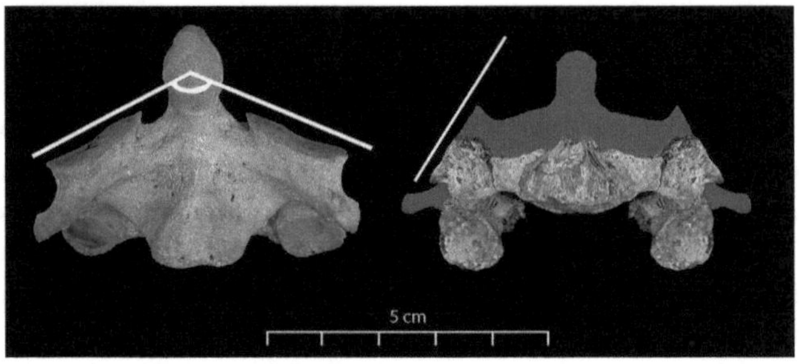

Fig. 5.18 The C2 superior articular facet angle of a human (left) and KSD-VP-1/1 (right). The KSD-VP-1/1 angle is highly anomalous and likely due to postmortem distortion

(Fig. 5.18), especially when considering the more human-like value of 120° in SK 854, a *Paranthropus* specimen. Small-brained, non-human primates generally present C2 superior articular processes that are oriented more inferolaterally on the coronal plane (as in KSD-VP-1/1) compared with the more horizontally oriented weight-bearing platform configuration in larger brained primates (Delattre 1924; Gommery 2000), making it unlikely that an australopith would exhibit such a low value. It should be noted that age or stature has no bearing on this angle as subadult human values match those of adults (Meyer 2005).

Inferior Articular Facet Orientation

The angles of the inferior articular facets relative to the transverse plane throughout the spine are associated with the incidence of the fundamental curvatures, where the spine is kyphotic, and the inferior articular facet orientation tends to be essentially vertical and sweep inferoposteriorly in the lordotic spine (Meyer 2005). The KSD-VP-1/1 C2 and all other vertebrae at each cervical level for the sample group share mean values above 90°, indicating the kinematic predisposition for cervical lordosis across taxa. The only

Table 5.14 Inferior articular facet angle (degrees) by element

Group	*N*	C2	*N*	C3	*N*	C4	*N*	C5	*N*	C6	*N*	C7	Mean C2–C7
Gorilla	17	129* (5.8)	16	128 (11)	10	132 (7.2)	10	127 (13)	10	115* (7.7)	9	113 (7.9)	124
Pan	15	121 (8.7)	15	125 (6.5)	9	119 (7.5)	8	116 (7.4)	9	112* (4.9)	9	110 (11)	117
AMH	23	116 (6.2)	9	122 (7.9)	23	120 (7.8)	23	122 (6.9)	24	124[h] (6.7)	23	119 (8.6)	121
KSD	1	123	–	–	–	–	–	–	–	–	–	–	–
A.L. 333	1	120	–	–	–	–	–	–	1	127	–	–	–
SK 854	1	120	–	–	–	–	–	–	–	–	–	–	–
WT-15K	–	–	–	–	–	–	–	–	–	–	1	118	–
Dmanisi	1	119	–	–	–	–	–	–	–	–	–	–	–

AMH = anatomically modern humans, KSD = KSD-VP-1/1, A.L. 333 = *Au. afarensis*, SK 854 = *P. robustus*, WT-15 k = KNM-WT 15000 *H. erectus*, Dmanisi = *H. erectus*, Neand = Neandertal. * = different from humans $p < 0.05$, h = different from apes $p < 0.05$. Standard deviations in parentheses

systematic difference between humans and African great apes in this angle is at the C6 vertebra, where humans tend to have a more oblique orientation of the inferior facets. The inferior articular facet angles of australopiths fall within the human distribution, indicating that the degree of cervical lordosis in australopiths was similar to modern humans. The inferoposterior orientation of the inferior articular facets with lordosis in humans, *Gorilla*, and *Pan* often produces laminar fossae, which are small depressions on the posterior aspect of the lamina caused by imbrication of the superjacent inferior articular facets. Dorsal convexity of the spine with lordosis engenders laminar fossae in the lumbar region as well (Latimer and Ward 1993; Meyer 2005).

Articular Facet Size

Torsional strains on the vertebral joints and joint capsules depend on the length of the lever arm (Yamamoto et al. 1989; Breit and Kunzel 2002), which in vertebrae are the spinous processes. Thus, it would be expected that articular facet size is partly related to spinous process length, percentage of cranial mass anterior to the line of gravity, and mode of posture and locomotion (see discussion below). With large proportions of cranial mass anterior to the line of gravity, both the spinous processes and articular facets of the African great apes and Neandertals are larger than those of modern humans. Large articular facets in Neandertals (especially at the upper cervical levels) may also relate to countering ventral loading incurred by their large and prognathic heads, although paramasticatation may explain why the largest facet areas are more cranial than the other groups (Trinkaus 1986). With a less centered foramen magnum (Dean and Wood 1981, 1982) and more prognathic cranium than humans, *Au. afarensis* would similarly have placed a greater proportion of axial loading anterior to the articular facets than humans, explaining why KSD-VP-1/1 superior articular facet areas are larger than the human mean and closer to the mean for *Gorilla* (Fig. 5.19). A similar morphology across primates is observed among taxa with large body mass paired with a heavy, prognathic cranium skewed anterior to the spine, as the posterior components of the vertebrae, especially the articular facets, bear the brunt of ventral axial loads (Breit 2002). As discussed above, in *Gorilla*, its large, heavy cranium in combination with its mode of posture and locomotion produces greater shear stresses on the articular facets at the most caudal levels, resulting in the largest subaxial articular processes at the C6 and C7 levels. Unlike *Gorilla*, KSD-VP-1/1 articular facets do not increase in size caudally; instead, they maintain a pattern of fairly consistent facet areas throughout the cervical column. Similar patterns in KSD-VP-1/1 and humans imply that posture, locomotor pattern, and head carriage in KSD-VP-1/1 were more like humans than *Gorilla*, despite their larger size. None of the

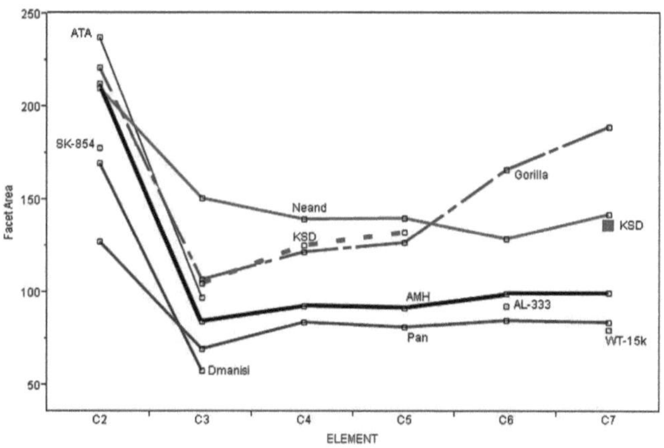

Fig. 5.19 Superior articular facet area (mm) by element. The KSD-VP-1/1 articular facets (dotted line) are *Gorilla* sized, but do not markedly increase in size at the most caudal level as in *Gorilla*. AMH = anatomically modern humans, A.L. 333 = *Au. afarensis*, SK 854 = *P. robustus*, WT-15k = KNM-WT 15000 *H. erectus*, Dmanisi = *H. erectus*, ATA = *H. heidelbergensis* (Atapuerca), Neand = Neandertal

superior articular facet areas in the KSD-VP-1/1 cervical spine differ significantly from anatomically modern humans or the African great apes.

Pars Interarticularis Size

Three of the KSD-VP-1/1 vertebrae preserve at least half of the pars interarticularis for analysis. The pars interarticularis is the vertebral region connecting the superior and inferior articular processes; in the cervical spine, it is also referred to as the lateral mass. High stresses are transmitted through this region during adjacent segment movement and the area is the frequent site of bone fractures and lytic defects in isthmic spondylolithesis, especially at the C6 level (Hession and Butt 1996). Pars interarticularis width was calculated as the transverse width between the left and right pars interarticularis at the midpoint of the superior and inferior lateral mass (Meyer 2005). In effect, this dimension is something of a misnomer as it captures the transverse dimension of the laminae using the pars interarticularis waist as an osteometric limit. Despite the fact that the articular facets in KSD-VP-1/1 are extremely tall, this expansion comes at the expense of a craniocaudally short pars interarticularis. The transverse widths of the C3 and C4 pars interarticularis are within the range of modern humans, but their shallow craniocaudal heights indicate potentially low tolerances of shear forces (Fig. 5.20). The C7 would be liable to shear stresses as both height and width of the pars interarticularis have low values; however, isthmic spondylolisthesis does not occur in the C7 in humans (Jeyapalan and Chavda 1994). Although this morphology might be an expression of the extent of individual variation, it illustrates a mid-cervical spine in KSD-VP-1/1 that is structured for countering shear magnitudes somewhat differently from humans. The other australopiths (SK 854, *P. robustus*, C2 vertebra; A.L. 333-106, *Au. afarensis*; C6 vertebra) have higher values than KSD-VP-1/1, but also fall outside of the human distribution and group with *Pan* due to their very narrow dimensions.

The Pedicles

There are three KSD-VP-1/1 vertebrae that preserve craniocaudal dimensions of the pedicles. In accordance with the results above that found a higher proportion of posterior vertebral elements in early hominins relative to modern humans, absolute heights of the subaxial pedicles in

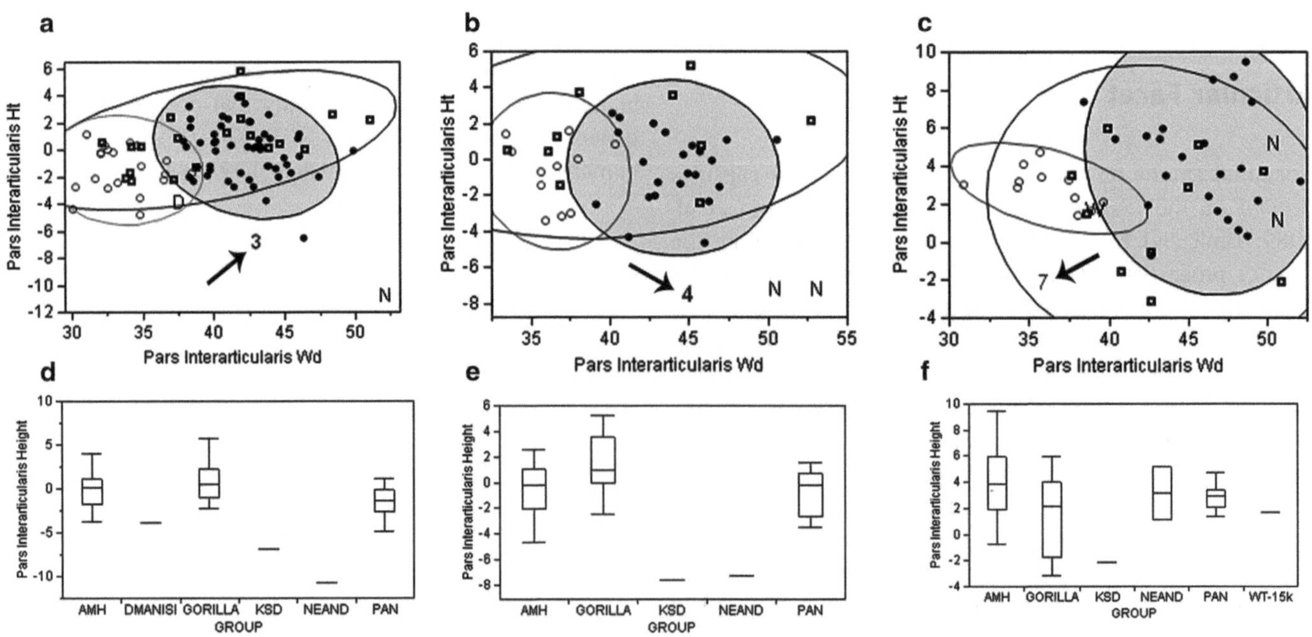

Fig. 5.20 Pars interarticularis transverse width by superoinferior height in **a** C3, **b** C4, and **c** C7 vertebrae. The KSD-VP-1/1 specimens (indicated by arrows) fall outside the 95% distribution ellipse for modern humans (shaded). KSD = numerals indicating vertebral level, modern humans = solid points, *Pan* = open circles, *Gorilla* = open squares, D = Dmanisi *H. erectus*, W = KNM-WT 15000 *H. erectus*, *N* Neandertals. Bivariate normal ellipses $p = 0.95\%$. All measurements are in mm. Boxplots of **d** C3, **e** C4, and **f** C7 pars interarticularis height by group. AMH = modern humans, DMANSI = Dmanisi *H. erectus*, KSD = KSD-VP-1/1, NEAND = Neandertals. Boxplots indicate 25, 50, and 75th percentiles, and whiskers indicate sample ranges excluding outliers

Table 5.15 Pedicle superoinferior height (mm) by element

Group	*N*	C2	*N*	C3	*N*	C4	*N*	C5	*N*	C6	*N*	C7	Mean C2–C4
Gorilla	18	9.1* (1.6)	20	8.4* (1.2)	10	8.0* (1.3)	10	7.8* (1.4)	10	7.8* (1.6)	15	7.1 (1.2)	8.5
Pan	18	7.1 (0.8)	20	6.6 (1.2)	11	7.3* (1.2)	10	6.6 (1)	11	6.1 (0.9)	13	6.4 (0.8)	7
AMH	57	7.8 (1.10)	53	6 (0.8)	27	6.3^{h} (0.9)	26	6.2 (1.2)	26	6.2 (1.2)	24	6.5 (1)	6.7
KSD	1	7.5	1	7	1	8.9	–	–	–	–	–	–	7.8
A.L. 333	1	7.8	–	–	–	–	–	–	1	8.4	–	–	–
SK 854	1	7.8	–	–	–	–	–	–	–	–	–	–	–
WT-15K	–	–	–	–	–	–	–	–	–	–	1	5.7	–
Dmanisi	1	7.4	1	5.2	–	–	–	–	–	–	–	–	–

AMH = anatomically modern humans, KSD = KSD-VP-1/1, A.L. 333 = *Au. afarensis,* SK-854 = *P. robustus,* WT-15 k = KNM-WT-15000 *H. erectus,* Dmanisi = *H. erectus,* Neand = Neandertal. * = different from humans $P < 0.05$, h = different from apes $p < 0.05$. Standard deviations in parentheses

Table 5.16 Relative pedicle size (pedicle cross-sectional area/centrum geometric mean) by element

Group	*N*	C2	*N*	C3	*N*	C4	*N*	C5	*N*	C6	*N*	C7	Mean C2–C7
Gorilla	17	3.0* (0.8)	19	3.1* (0.7)	10	3.0* (0.6)	10	2.9* (0.9)	10	2.8* (0.6)	15	2.5* (0.6)	2.9
Pan	18	2.2 (0.4)	19	2.0* (0.5)	10	2.2* (0.5)	9	1.9 (0.3)	10	2.0 (0.4)	13	2.0* (0.4)	2.1
AMH	52	2.5 (0.7)	50	1.4^{h} (0.3)	22	1.5^{h} (0.3)	21	1.6 (0.3)	20	1.6 (0.4)	24	1.6^{h} (0.3)	1.7
KSD	–	–	1	2.7	1	–	–	–	–	–	–	–	–
A.L. 333	–	–	–	–	–	–	–	–	1	2.6	–	–	–
WT-15K	–	–	–	–	–	–	–	–	–	–	1	2.2	–
Dmanisi	1	2.7	1	2.0	–	–	–	–	–	–	–	–	–
ATA	–	–	1	1.3	–	–	–	–	–	–	–	–	–
Neand	–	–	2	2.1 (0.4)	0	–	1	1.7	–	–	–	–	–

AMH = anatomically modern humans, KSD = KSD-VP-1/1, A.L. 333 = *Au. afarensis,* WT-15 k = KNM-WT-15000 *H. erectus,* Dmanisi = *H. erectus,* ATA = *H. heidelbergensis* (Atapuerca), Neand = Neandertal. * = different from humans $p < 0.05$, h = different from apes $p < 0.05$. Standard deviations in parentheses. Ratio values were logarithmically transformed for statistical testing

KSD-VP-1/1 also exceeded the mean for modern humans (Table 5.15). The pedicle heights of KSD-VP-1/1, like those of A.L. 333-106, are closest to *Gorilla* values and fall within the upper range for modern humans. The pattern of caudal change, however, appears to match *Pan* where the tallest pedicle is at C4, the only cervical level where modern humans are significantly shorter than both apes (*Gorilla*, $p = 0.0003$; *Pan*, $p = 0.027$), suggesting higher load transfer in the apes across the mid-cervical pedicles. Human pedicles are fairly consistent in height throughout the cervical spine, demonstrating the uniformity of posterior cervical loads in bipeds. Like those of the African great apes, the pedicles of KSD-VP-1/1 reveal a primitive pattern of variable load transfer between the centrum and posterior vertebral elements.

In terms of relative pedicle size (pedicle cross-sectional area/centrum geometric mean), only the C3 in KSD-VP-1/1 preserves enough morphology for comparison and resembles the other *Au. afarensis* specimen (A.L. 333-106) in its *Gorilla*-like size (Table 5.16). The African great ape mid-cervical pedicles are approximately twice as robust as humans, relative to size, as they transmit higher loads between the posterior vertebral component and the centrum. Cross-sectional area of the pedicle was calculated as follows Meyer (2005): {(pedicle height × pedicle thickness × PI)/4}.

The Lamina

The superoinferior heights of the laminae generally increase caudal to the C4 cervical level in all groups, with humans intermediate to *Gorilla* and *Pan* (Table 5.17). The laminae in KSD-VP-1/1 vertebrae are taller than the human mean and closest to that of *Gorilla*; however, lamina heights in the KSD-VP-1/1 mid-cervical vertebrae parallel the human pattern of slight height increases, unlike in *Gorilla* where mid-cervical lamina heights increase dramatically. This indicates a human-like loading profile across the KSD-VP-1/1 mid-cervical laminae. By contrast, the australopiths SK 854 and A.L. 333-106 are intermediate to *Pan* and anatomically modern humans, whereas the laminae of the two juvenile specimens of early *Homo* (Dmanisi and KNM-WT 15000) are shorter, closest to the mean for *Pan*. Dramatically low Neandertal values suggest they channeled axial loads differently from other groups across the posterior cervical spine.

There is a U-shaped distribution in lamina dorsoventral thickness in modern humans and *Pan*, with laminae thickest at the most cranial and most caudal levels and thinnest at the mid-cervical levels. Although the three KSD-VP-1/1 laminae recovered are thicker than those of humans, the pattern of decreasing mid-cervical lamina thickness in KSD-VP-1/1 C3 and C4 parallels the human pattern and is different from the pattern in *Gorilla* and Neandertals where laminae exhibit a continuous increase in thickness caudally (Table 5.18).

The lamina index (superoinferior lamina height/anteroposterior lamina thickness) combines height and width dimensions to capture their relative proportions. Both human and *Pan* display relatively taller and thinner mid-cervical lamina, while *Gorilla* and Neandertals maintain consistently short and thick lamina at each vertebral level. KSD-VP-1/1 falls below human mean values, a signal of relatively shorter and thicker lamina; however, it also displays mid-cervical reduction of lamina thickness similar to modern humans (Fig. 5.21). In humans, the lamina index is not associated with vertebral size (centrum geometric mean; $r^2 = 0.048$), body size (postcranial geometric mean; $r^2 = 0.006$), age ($r^2 = 0.005$), or sex ($p = 0.73$),

Table 5.17 Lamina superoinferior height (mm) by element

Group	*N*	C2	*N*	C3	*N*	C4	*N*	C5	*N*	C6	*N*	C7	Mean C3–C4, C7
Gorilla	19	11.9 (1.9)	20	10.9 (2.0)	10	11.6* (1.4)	10	12.5 (1.8)	10	14.1* (2.8)	10	15.2 (3.9)	12.6
Pan	19	8.8* (1.2)	20	8.1* (1.1)	11	8.3* (1.0)	11	9.3* (1.0)	10	10.6 (0.9)	11	10.6* (1.3)	9.0
AMH	48	11.3 (1.5)	53	10.1 (1.2)	24	10.2^{h} (1.3)	23	11.2 (1.7)	24	11.9 (1.6)	24	14.5 (1.6)	11.6
KSD	–	–	1	11.9	1	12.0	–	–	–	–	1	14.9	12.9
A.L. 333	1	11.6	–	–	–	–	–	–	1	11.1		–	–
SK 854	1	8.9	–	–	–	–	–	–	–	–		–	–
WT-15K	–	–	–	–	–	–	–	–	–	–	1	10.6	–
Dmanisi	1	9.6	1	7.7	–	–	–	–	–	–		–	–
Neand	1	5.9*	1	3.7*^{h}	2	3.4*^{h} (0.6)	2	4.1*^{h} (0.3)	2	5.1*^{h} (0.1)	2	6.0* (0.5)	4.4

The laminae in KSD-VP-1/1 are taller than the mean of modern humans, but parallel the human pattern of relative uniformity. AMH = anatomically modern humans, KSD = KSD-VP-1/1, A.L. 333 = *Au. afarensis,* SK 854 = *P. robustus,* WT-15 k = KNM-WT-15000 *H. erectus,* Dmanisi = *H. erectus,* Neand = Neandertal. * = different from humans $p < 0.05$, h = different from apes $p < 0.05$. Standard deviations in parentheses

Table 5.18 Lamina dorsoventral thickness (mm) by element

Group	*N*	C2	*N*	C3	*N*	C4	*N*	C5	*N*	C6	*N*	C7	Mean C3–C4, C7
Gorilla	19	5.3 (1.3)	20	4.6* (1.3)	10	5.2* (1.5)	10	6.4* (1.8)	10	7.6* (1.6)	10	8.3* (1.7)	6.0
Pan	19	4.2* (1.0)	20	2.3* (0.4)	10	2.3 (0.4)	10	2.5 (0.7)	11	3.2 (0.6)	10	4.5 (0.5)	3.0
AMH	45	5.6 (1.3)	52	3.1 (0.8)	24	2.4 (0.6)	23	2.3 (0.4)	23	3.1 (0.6)	23	4.8 (1.0)	3.4
KSD	–	–	1	5.0	1	3.9	–	–	–	–	1	5.2	4.7
A.L. 333	1	3.4	–	–	–	–	–	–	1	3.5	–	–	–
SK 854	1	5.4	–	–	–	–	–	–	–	–	–	–	–
WT-15K	–	–	–	–	–	–	–	–	–	–	1	4.7	–
Dmanisi	1	4.8	1	2.9	–	–	–	–	–	–	–	–	–
ATA	3	6.2 (1.2)	1	4.0	–	–	–	–	–	–	–	–	–
Neand	1	9.3*	1	9.3*^{h}	2	10*^{h} (0.7)	2	11*^{h} (0.5)	2	12*^{h} (0.2)	2	14*^{h} (0.4)	11.1

Although the KSD-VP-1/1 laminae are thicker than those of humans, they mirror the human pattern of decreasing thickness, unlike Neandertals and *Gorilla*. AMH = anatomically modern humans, KSD = KSD-VP-1/1, A.L. 333 = *Au. afarensis,* SK 854 = *P. robustus,* WT-15 k = KNM-WT-15000 *H. erectus,* Dmanisi = *H. erectus,* ATA = *H. heidelbergensis* (Atapuerca), Neand = Neandertal. * = different from humans $p < 0.05$, h = different from apes $p < 0.05$. Standard deviations in parentheses

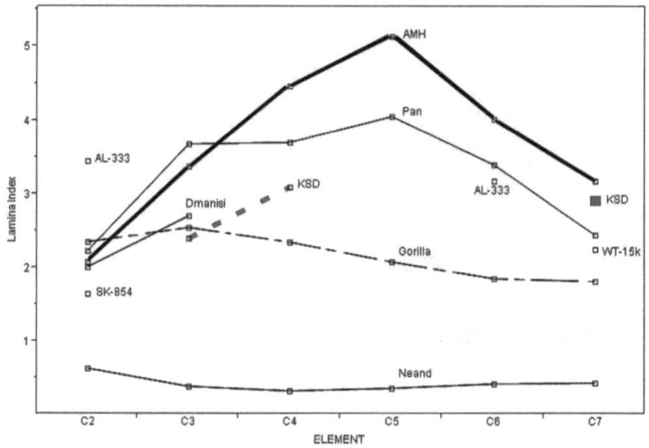

Fig. 5.21 Lamina index (superoinferior height/anteroposterior thickness) by element. Although the KSD-VP-1/1 laminae (dotted line) are shorter and thicker than modern humans they mirror the distinct human pattern of mid-cervical increase. At each cervical level, *Gorilla* and Neandertals are significantly different from modern humans, whereas *Pan* only differs at the C5 and C7 levels. AMH = anatomically modern humans, KSD = KSD-VP-1/1, A.L. 333 = *Au. afarensis*, WT-15k = KNM-WT 15000 *H. erectus*, Dmanisi = *H. erectus*, ATA = *H. heidelbergensis* (Atapuerca), Neand = Neandertal

nor does it associate in the African great apes with vertebral size ($r^2 = 0.22$), body size ($r^2 = 0.22$), or sex ($p = 0.23$). The lamina index, therefore, likely signals differing axial loading regimesacross the posterior vertebral component and highlights the functional similarity between KSD-VP-1/1 and modern humans.

Posterior Versus Anterior Load Transfer

Examinations of australopith lumbar and thoracic vertebrae have noted relatively small centra relative to the posterior features of their vertebrae, and, as a result, it has been posited that early hominins differed from modern humans whose vertebral bodies bear the bulk of compressive loads (Robinson 1972; Sanders 1996, 1998). By contrast, australopiths seem to have channeled a greater percentage of load-bearing force through dorsal columnar structures (posterior vertebral components), especially the articular facets, which are exceptionally large in australopiths (Sanders 1990, 1995).

The KSD-VP-1/1 cervical vertebrae support this observation with a high value for the posterior–anterior component ratio (posterior geometric mean/centrum geometric mean), a ratio that evaluates the relative proportions of vertebral load transfer (Fig. 5.22; Table 5.19). This ratio compares the relative size (geometric mean) of the posterior vertebral component to the size of the superior centrum (geometric mean), with high values representing greater posterior proportions. The geometric mean for the posterior component is the 9th root of the product of nine dimensions from the articular facets, pars interarticularis, lamina, and pedicle.

Because of its small centrum, the C3 vertebra of KSD-VP-1/1 falls well above the mean for anatomically modern humans, indicating a higher proportion of the posterior component. The same condition is observed in A.L. 333-106, both *H. erectus* specimens and the Neandertals. The modern human pattern exhibits steadily increasing posterior proportions in the caudal direction with no appreciable peak, whereas the African great apes demonstrate a pronounced mid-cervical emphasis on the posterior vertebral components with a peak at the C4–C5 levels. These data imply that, as in apes, the mid-cervical posterior components (i.e., articular facets, pedicle, lamina, pars interarticularis) in earlier hominins channeled proportionately greater load magnitudes than modern humans, who shift proportionally greater axial loads to the centrum. This ratio is sexually dimorphic in apes ($p = 0.004$), but not in humans ($p = 0.84$), and does not relate to body size in apes or humans using the postcranial geometric mean as a proxy for size ($r^2 = 0.04$ and $r^2 = 0.02$, respectively). In apes and humans, the length of the spinous process has only a weak association with this

Table 5.19 Posterior/anterior component ratio (posterior geometric mean/centrum geometric mean)

Group	*N*	C2	*N*	C3	*N*	C4	*N*	C5	*N*	C6	*N*	C7	Mean C2–C7
Gorilla	17	0.75 (0.03)	19	0.84* (0.05)	10	0.88* (0.04)	10	0.91* (0.08)	10	0.9* (0.07)	15	0.87 (0.07)	0.86
Pan	18	0.75 (0.04)	19	0.83* (0.05)	9	0.89* (0.04)	9	0.88* (0.05)	10	0.87 (0.05)	13	0.84 (0.05)	0.84
AMH	50	0.77 (0.04)	50	0.76^{h} (0.05)	20	0.79^{h} (0.05)	21	0.8^{h} (0.06)	20	0.81 (0.07)	24	0.83 (0.05)	0.79
KSD	–	–	1	0.96*	–	–	–	–	–	–	–	–	–
A.L. 333	–	–	–	–	–	–	–	–	1	0.91	–	–	–
WT-15K	–	–	–	–	–	–	–	–	–	–	1	0.91	–
Dmanisi	1	0.8	1	0.9	–	–	–	–	–	–	–	–	–
Neand	–	–	1	0.82	–	–	1	0.96	–	–	–	–	–

High values represent greater posterior proportions relative to the centrum. AMH = anatomically modern humans, KSD = KSD-VP-1/1, A.L. 333 = *Au. afarensis*, WT-15 k = KNM-WT-15000 *H. erectus*, Dmanisi = *H. erectus*, Neand = Neandertal. * = different from humans $p < 0.05$, h = different from apes $p < 0.05$. Standard deviations in parentheses

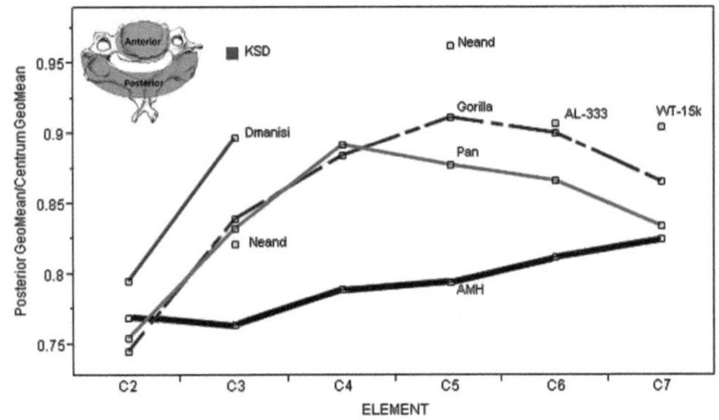

Fig. 5.22 Posterior/anterior component ratio (posterior geometric mean/centrum geometric mean) by element. Note the pronounced mid-cervical peak of the posterior vertebral components in the African great apes. AMH = anatomically modern humans, A.L. 333 = *Au. afarensis,* WT-15k = KNM-WT 15000 *H. erectus,* Dmanisi = *H. erectus*, Neand = Neandertal

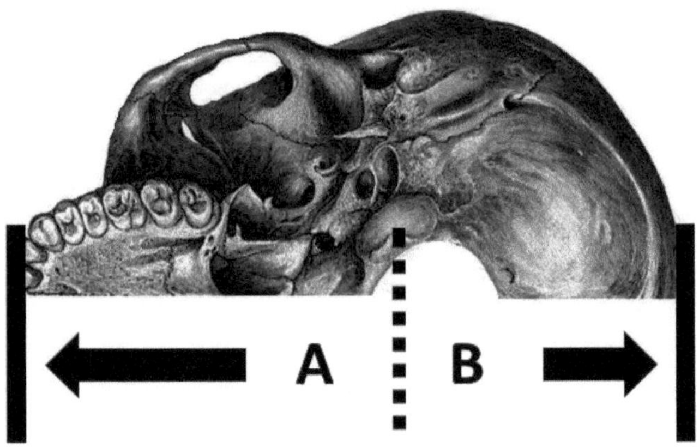

Fig. 5.23 The cranial base balance ratio is A cranial length anterior to occipital condyles/B cranial length posterior to condyles. The cranial base balance ratio expresses the proportion of total cranial base length anterior to the condyles, a measure that correlates well with actual inferior cranial base area ($N = 31$, $r^2 = 0.97$)

ratio ($r^2 = 0.28$ and $r^2 = 0.13$, respectively), similar to the proportion of head mass anterior to the cranial condyles (cranial balance ratio; $r^2 = 0.001$ in humans and $r^2 = 0.07$ in apes). As the posterior–anterior component ratio is a relatively independent isometric variable, it may serve as a salient indicator of relative load vectors in the cervical spine and demonstrate functional differences between modern humans and other hominoid taxa.

Spinous Processes

The cervical spinous processes are the attachment site for the nuchal muscles that insert on the cranium, and their relative sizes reflect the size and importance of the nuchal muscles and their attachments (Moskovich 2001), as well as highlight adaptations in closely related species (O'Higgins et al. 1997). These relationships occur, in part, because the spinous processes follow a slow somatic growth pattern, while other features such as the vertebral canal follow a more rapid neural pattern of growth (Sinclair 1969), allowing environmental influences such as activity to accumulate greater effects during development. Accordingly, spinous processes are extremely informative with respect to cervical functional anatomy and their size in the cervical spine has been effective in partitioning suspensory primates from nonsuspensory taxa (Nalley 2013) whose short spinous processes reflect reduced movements of athletic agility (Straus and Wislocki 1932; Ankel-Simons 1983, 2000; Gebo 1989).

Relatively long spinous processes in the apes are due in part to crania that are ventrally displaced above the cervicothoracic junction (Vidal et al. 1986) and hafted to an obliquely oriented thoracic spine. This translates to different cervical kinematics than humans and the need to counter ventral head pitch through muscular and ligamentous intervention (Niemitz 2002; Dunbar et al. 2008). Data collected for this study corroborate a long line of studies (e.g., Dart 1925; Weidenreich 1943; Le Gros Clark 1954; Nevell and Wood 2008) showing that retrognathism and cranial base remodeling in humans balance crania more vertically above the cervicothoracic junction (Fig. 5.23, Table 5.20), obviating the need for large counterbalancing dorsal musculature (Shapiro and Frankel 1989; Bramble 2000; Bramble and Lieberman 2004; Cailliet 2004). Orthograde posture and bipedal locomotion further mitigate ventral loading relative to pronograde primates, placing the line of axial loading vertically above the hip joint and reduce the need for long spinous processes. Consequently, the spinous processes of humans are shorter than those of African great apes (Ward 1991; Latimer and Ward 1993) and reveal the importance of spinous process dimensions in understanding the nature of head carriage and posture in fossil taxa.

Table 5.20 Cranial base balance ratio (cranial length anterior to occipital condyles/cranial length posterior to condyles)

Group	*N*	Ratio	SD
Gorilla	10	4.52*	1.1
Pan	10	3.74*	0.5
AMH	11	1.36^{h}	0.2
WT-15K	1	1.95	–
Dmanisi	1	1.56	–

This ratio is similar to Weidenreich's (1943) opisthion to cranial base length index indicating the relative position of the posterior margin of the foramen magnum (Schultz 1955). Relative to earlier hominins and the African great apes, the occipital condyles are anteriorly positioned in anatomically modern humans. AMH anatomically modern humans, WT-15 k = KNM-WT-15000 *H. erectus,* Dmanisi = *H. erectus.* * = different from humans $p < 0.05$, h = different from apes $p < 0.05$

Although none of the KSD-VP-1/1 vertebrae preserve a spinous process, C3 and C4 provide adequate morphological information to reconstruct spinous process lengths. Spinous process length was modeled on ten linear dimensions of the centrum, neural arch and lamina in *Pan*, *Gorilla*, and humans, and is reasonably reliable in predicting spinous process length at the C3 ($r^2 = 0.92$, $N = 42$) and C4 vertebral level ($r^2 = 0.95$, $N = 40$) (Table 5.21). The KSD-VP-1/1 C3 and C4 spinous processes are predicted to have been longer than those of humans and close in length to those of *Pan* (Table 5.22). Individually, the predicted values are not significantly different from those of humans, but the predicted 20% spinous process length increase from C3 to C4 in KSD-VP-1/1 is not typically observed in humans where the mean change is 2%. Instead, the pattern seen in KSD-VP-1/1 mirrors the relationship observed in Neandertals (30%) and *Pan* (23%), and, unlike modern humans, the KSD-VP-1/1 vertebral column may have evinced the mid-cervical peak in spinous process length common among *Gorilla*, *Pan*, and Neandertals. By contrast, anatomically modern human mid-cervical spinous processes are of roughly equal size until they increase in length at the two most caudal levels, indicative of lesser force magnitudes acting on the dorsal mid-cervical column to maintain lordosis and head position (Table 5.23).

Because the values for KSD-VP-1/1 are predicted, any functional implications must be considered to be entirely speculative; however, because KSD-VP-1/1 and Neandertals share a common spinous process size, it is likely that long mid-cervical spinous process lengths are partially related to larger dorsal moments in response to proportionately greater craniofacial mass anterior to the cranial condyles rather than signaling locomotor differences with modern humans. This inference is supported by a subset of sample data ($N = 31$) demonstrating the association between relative spinous process lengths and cranial base proportions across taxa, with peak correlation at C5 ($r^2 = 0.79$) and slightly lower correlations cranially and caudally. However, increased epaxial musculature relative to modern humans likely accompanied the long spinous processes of KSD-VP-1/1,

Table 5.21 Spinous process length prediction model parameter estimates ($N = 82$)

Term	Estimate	Std Error	t Ratio	Prob > |t|
Neural canal A–P	2.382087	0.518215	4.60	<0.0001*
Lamina thick	3.7617642	0.98197	3.83	0.0003*
Neural canal WD	−2.609803	0.689612	−3.78	0.0003*
Dorsal centrum HT	3.0455661	1.304902	2.33	0.0222*
Superior centrum A–P	2.7590988	1.268735	2.17	0.0328*
Lamina HT	1.3867064	0.770338	1.80	0.0758
Inferior centrum WD	−1.093233	0.673504	−1.62	0.1087
Superior centrum WD	1.0452635	0.663207	1.58	0.1192
Inferior centrum A–P	−1.298347	1.261156	−1.03	0.3065
Anterior centrum HT	−1.107391	1.113878	−0.99	0.3233

Table 5.22 Spinous process length (mm) by element with predicted values for the KSD-VP-1/1 C3 and C4 vertebrae

	N	C2	*N*	C3	*N*	C4	*N*	C5	*N*	C6	*N*	C7	Ratio C4/C3
Gorilla	19	28.2* (7.5)	20	55.7* (19.5)	10	70.4* (19.9)	10	73.1* (18)	9	74.3* (18.2)	9	69.1* (16.8)	1.26
Pan	18	16.8 (1.8)	19	20.5 (3.7)	10	26.7* (3.1)	10	29.6* (3)	9	31.8 (3.0)	10	34.7 (3.4)	1.30
AMH	48	16.7 (2.3)	47	14.5 (2.2)	24	14.8h (2.7)	23	17.4h (2.4)	23	23.4 (3.2)	23	30.2 (3.4)	1.02
KSD	–	–	1	22	1	27	–	–	–	–	–	–	1.23
A.L. 333	1	13.8	–	–	–	–	–	–	1	28.7	–	–	–
WT-15K	–	–	–	–	–	–	–	–	–	–	1	29.4	–
Dmanisi	1	20.1	1	15.4	–	–	–	–	–	–	–	–	–
ATA	3	17.8 (1.4)	–	–	–	–	–	–	–	–	–	–	–
Neand	1	18.8	1	16.5	1	23.4	1	31.5	2	31.4 (4.4)	2	38 (4.2)	1.42

Predicted values for spinous process length in the KSD-VP-1/1 specimens fall near the mean for *Pan* and exhibit a slope increase similar to both *Pan* and Neandertals that is absent in modern humans. AMH = anatomically modern humans, KSD = KSD-VP-1/1, A.L. 333 = *Au. afarensis*, WT-15 k = KNM-WT-15000 *H. erectus*, Dmanisi = *H. erectus*, ATA = *H. heidelbergensis* (Atapuerca), Neand = Neandertal. * = different from humans $p < 0.05$, h = different from apes $p < 0.05$. Standard deviations in parentheses

Table 5.23 Relative spinous process length (spinous process length/centrum anteroposterior length) by element with predicted values for KSD-VP-1/1 C3 and C4 vertebrae

Group	*N*	C3	*N*	C4	*N*	C5	*N*	C6	*N*	C7	Ratio C4/C3
Gorilla	19	3.23* (0.9)	10	4.24* (0.75)	10	4.65* (0.69)	9	4.26* (0.56)	9	3.64* (0.34)	1.31
Pan	19	1.81* (0.27)	10	2.4* (0.21)	10	2.63* (0.26)	9	2.62* (0.26)	10	2.66* (0.31)	1.33
AMH	47	1.0^{h} (0.16)	24	1.05^{h} (0.2)	23	1.23^{h} (0.2)	23	1.59^{h} (0.22)	23	2.06^{h} (0.25)	1.05
KSD	1	1.51	1	1.71	–	–	–	–	–	–	1.13
A.L. 333	–	–	–	–	–	–	1	2.32	–	–	–
WT-15K	–	–	–	–	–	–	–	–	1	2.14	–
Dmanisi	1	1.21	–	–	–	–	–	–	–	–	–
Neand	1	0.91	–	–	–	–	1	1.85	1	2.56	–

AMH = anatomically modern humans, KSD = KSD-VP-1/1, A.L. 333 = *Au. afarensis,* WT-15 k = KNM-WT-15000 *H. erectus*, Dmanisi = *H. erectus*, Neand = Neandertal. * = different from humans $p < 0.05$, h = different from apes $p < 0.05$. Standard deviations in parentheses. Ratio values were logarithmically transformed for statistical testing

suggesting a point of variance with anatomically modern humans in terms of head carriage and neck flexibility on the sagittal plane.

Importantly, the short spinous processes of modern humans result in a physiological "void" unoccupied by the large trapezius, splenius capitis, and rhomboid muscles of the African great apes. Instead, in humans, bulky muscles are supplanted by a thin dorsal raphe formed from interweaving fibers of these muscles, and a fascial septum consisting of dense connective tissue runs ventrally from the midline raphe to interspinous ligaments connecting the spinous processes (Mercer and Bogduk 2003). This elastic ligamentous band, the nuchal ligament, is a uniquely derived feature in cursorial animals, including genus *Homo*, which connects the cervical spine to the cranium and C7 vertebra and maintains head stability during distance running (Bramble and Lieberman 2004). Intermediate spinous processes in KSD-VP-1/1 imply that the nuchal ligament in *Au. afarensis* was either absent, or at best at an inchoate stage of development. This hypothesis is supported by the absence in australopiths of a strongly everted median nuchal line (the attachment site for the nuchal ligament) and not simply the presence of an external occipital protuberance at the dorsal terminus of the median nuchal line common among hominoid crania but not indicative of the nuchal ligament. All hominin crania except those of genus *Homo* lack the everted median nuchal line, and therefore, may have lacked the nuchal ligament associated with cursorial running (Lieberman 2011).

Neural Canal

Neural Canal Size

In general, modern human neural canals (and the associated spinal cord) are relatively expanded relative to the great apes (Schultz 1930; MacLarnon 1987) and australopiths (MacLarnon 1993; MacLarnon and Hewitt 1999). Despite expectations, mean canal areas in the KSD-VP-1/1 cervical vertebrae are similar to those of modern humans (only 6% smaller than the aggregate mean) and are 29% larger than the mean neural canal area of *Pan* (Table 5.24). At C5, the neural canal of KSD-VP-1/1 actually exceeds the mean for humans. By contrast, neural canal areas in SK 854 and A.L. 333-106 are close to the *Pan* mean and are significantly smaller than humans. *Contra* to Scheiss and Haeusler (2013), the *Pan*-like neural canal area of the KNM-WT 15000 C7 vertebra is consistent with suggestions of pathology (neural stenosis) in the axial spine (Latimer and Ohman 2001; Meyer 2003), especially when compared to the larger canal areas of the Dmanisi *H. erectus* specimens of much smaller body size (Meyer 2005). Unlike KNM-WT 15000 and other australopiths, the subaxial neural canal areas of KSD-VP-1/1 predict a spinal cord size larger than that of *Pan* and similar in cross-sectional area to that of Dmanisi *H. erectus* and modern humans.

Neural canal dimensions in many mammals follow a simple curve paralleling the size of the spinal cord (O'Higgins et al. 1989), which allows for the estimation of spinal cord size in fossil taxa from vertebral remains. The exception to this rule is the C2 vertebra, where the largest neural canal cross-sectional areas in the spine occur, but do not relate to spinal cord sizes, as the spinal cord at this level is 10% smaller than at the mid-cervical levels (Nordquist 1964; Sherman et al. 1990). Although the KSD-VP-1/1 C2 neural canal is below the human mean (but not statistically different), no inferences can be made with respect to spinal cord size. Cross-sectional canal areas at C2 across taxa are probably a function of expanded dorsoventral canal lengths serving to avoid spinal cord impingement, as this vertebra confers the majority of head movement in the sagittal plane.

Neural canal cross-sectional areas were measured from scaled images using the ImageJ NIH software platform and are in agreement with published *in vivo* analyses deriving

Table 5.24 Neural canal area (cm^2) by element

Group	N	C2	N	C3	N	C4	N	C5	N	C6	N	C7	Mean C2-C5, C7
Gorilla	14	3.29* (0.4)	15	2.81* (0.4)	10	2.89* (0.4)	10	3.10* (0.4)	10	2.93* (0.3)	10	2.63* (0.4)	2.90
Pan	16	2.11* (0.4)	17	1.83* (0.3)	11	1.75* (0.2)	11	1.75* (0.2)	11	1.65* (0.2)	11	1.45* (0.2)	1.78
AMH	35	2.78^h (0.4)	36	2.33^h (0.3)	27	2.29^h (0.2)	27	2.38^h (0.3)	29	2.40^h (0.3)	28	2.30^h (0.3)	2.42
KSD	1	2.10	1	2.28	1	2.23	1	2.58	-	-	1	2.28	2.29
A.L. 333	1	2.90	-	-	-	-	-	-	1	1.70*	-	-	-
SK 854	1	1.95*	-	-	-	-	-	-	-	-	-	-	-
WT-15K	-	-	-	-	-	-	-	-	-	-	1	1.44*	-
Dmanisi	1	2.26	1	2.27	-	-	-	-	-	-	-	-	-
ATA	3	2.70 (0.2)	2	2.09 (0.3)	-	-	-	-	-	-	-	-	-
Neand	2	3.23 (0.2)	2	2.62 (0.4)	2	2.68 (0.1)	2	2.93*	2	2.68 (0.2)	2	2.64 (0.4)	2.82

In contrast to expectations, the subaxial neural canal areas of KSD-VP-1/1 predict a spinal cord size similar to that of modern humans. AMH = anatomically modern humans, KSD = KSD-VP-1/1, A.L. 333 = *Au. afarensis*, SK 854 = *P. robustus*, WT-15 k = KNM-WT-15000 *H. erectus*, Dmanisi = *H. erectus*, ATA = *H. heidelbergensis* (Atapuerca), Neand = Neandertal. * = different from humans $p < 0.05$, h = different from apes $p < 0.05$. Standard deviations in parentheses

neural canal dimensions via computed tomography (CT) (e.g., Stanley et al. 1986). Mathematical estimates of neural canal areas are highly erroneous, as the neural canals are not absolute ellipses as formulae require, and, as such, they underestimate chimpanzee values, overestimate human values, and have almost no correlation with gorilla neural canal areas (Meyer 2005).

Neural Canal Shape

Modern human neural canals are relatively wider transversely than those of the African great apes, resulting in a canal shape that is significantly different at each vertebral level (Table 5.25) (MacLarnon and Hewitt 1999). Neural canal shape (transverse canal width/dorsoventral canal diameter × 100) in five of the KSD-VP-1/1 cervical vertebrae was reconstructed based on mirror imaging the preserved canal margins. The results demonstrate that, unlike the apes, the vertebrae of KSD-VP-1/1 exhibit transversely expanded canals. Other *Au. afarensis* cervical vertebrae (A.L. 333-101 and A.L. 333-106) do not share this derived morphology and, like *Paranthropus* (SK-854), have values at the mean for *Pan*.

Dorsoventrally expanded neural arches in the African great apes, especially *Gorilla*, are not representative of a larger spinal cord. Instead, their dorsally elongated mid-cervical canals are associated with long dorsal vertebral structures that offer higher moments and greater surface attachment areas for epaxial musculature acting to counter ventral gravitational loads incumbent to pronogrady. In hominins, this may represent a primitive retention well suited to balance heavy, prognathic crania weighted ventrally to the cervical spine.

Postcranial Neurological Indications

In terms of postcranial neurological evolution, perhaps the most cogent difference between humans and the African great apes is in the dramatic transverse expansion of the human spinal cord for the brachial plexus axons that control the upper limbs. Because nerves to the hands and arms exit the spinal cord at mid-cervical levels associated with the brachial plexus, neural canal size and shape in this cervical region may shed light on neurological potential in the upper extremities, assuming a correlation between spinal cord size distribution and function. Enhanced neural control over the extremities has been hypothesized to explain differences between humans and apes in throwing accuracy and tool-making capabilities (W. Calvin in Walker 1993; Dunsworth et al. 2004; Meyer 2003).

Potentially confounding this discussion is that neural canals in gorillas are larger than those of humans. This may be attributed partly to spinal cord scaling associated with large body sizes, but also due in large part to dorsal expansion of the neural arch as a function of dorsal extension of the laminae for the spinous process lever arm (Schultz 1930). However, the large neural canals in the KSD-VP-1/1 cervical column are not related to dorsoventral expansion as in *Gorilla*; instead, they are derived chiefly from expanded transverse widths as in modern humans.

At every cervical level, the transverse widths (both absolute and size-controlled) of modern humans are

Table 5.25 Neural canal shape index (transverse canal width/dorsoventral canal diameter × 100) by element

Group	*N*	C2	*N*	C3	*N*	C4	*N*	C5	*N*	C6	*N*	C7	Mean C2–C5, C7
Gorilla	20	97* (10)	21	105* (12)	11	107* (15)	11	109* (16)	11	117* (17)	11	128* (27)	109
Pan	23	113* (10)	24	122* (13)	14	128* (15)	14	132* (15)	15	140* (17)	14	154* (17)	130
AMH	60	141^{h} (10)	59	160^{h} (15)	34	172^{h} (16)	32	176^{h} (14)	33	179^{h} (15)	33	173^{h} (17)	164
KSD	1	167^{h}	1	200^{h}	1	170	1	172	–	–	1	179	178
A.L. 333	1	112	–	–	–	–	–	–	1	134	–	–	–
SK 854	1	111	–	–	–	–	–	–	–	–	–	–	–
WT-15K	–	–	–	–	–	–	–	–	–	–	1	206	–
Dmanisi	1	131	1	154	–	–	–	–	–	–	–	–	–
ATA	3	141^{h} (12)	2	153 (19)	–	–	–	–	–	–	–	–	–
Neand	1	139	1	183^{h}	2	180^{h} (28)	2	172^{h} (1)	2	221^{h}* (28)	1	156	166

The high values in the C7 of KNM-WT 15000 are the product of an anomalously small dorsoventral neural canal length. AMH = anatomically modern humans, KSD = KSD-VP-1/1, A.L. 333 = *Au. afarensis*, SK 854 = *P. robustus*, WT-15 k = KNM-WT 15000 *H. erectus*, Dmanisi = *H. erectus*, ATA = *H. heidelbergensis* (Atapuerca), Neand = Neandertal. * = different from humans $p < 0.05$, h = different from apes $p < 0.05$. Standard deviations in parentheses. Ratio values were logarithmically transformed for statistical testing

Table 5.26 Neural canal transverse width (mm) by element

Group	*N*	C2	*N*	C3	*N*	C4	*N*	C5	*N*	C6	*N*	C7	Mean C2–C5, C7
Gorilla	20	19.5* (1.2)	21	19.9* (1.2)	11	20.6* (1.7)	11	21* (1.9)	11	21.7* (1.9)	11	22.1* (3.1)	20.6
Pan	23	16.9* (1.6)	24	17.4* (1.3)	14	18.1* (1.4)	14	18* (1.4)	15	18* (1.3)	14	17.8* (1.3)	17.6
AMH	60	22.4^{h} (1.2)	59	22.4^{h} (1.2)	34	23.7^{h} (1.2)	32	24.5^{h} (1.3)	33	25.1^{h} (1.5)	33	24.1^{h} (1.6)	23.4
KSD	1	20	1	24^{h}	1	23	1	27.5^{h}	–	–	1	26	24.1
A.L. 333	1	20.7	–	–	–	–	–	–	1	17.7*	–	–	–
SK 854	1	16.9*	–	–	–	–	–	–	–	–	–	–	–
WT-15K	–	–	–	–	–	–	–	–	–	–	1	21.3	–
Dmanisi	1	18.9	1	19.7	–	–	–	–	–	–	–	–	–
ATA	3	23.4 (0.5)	2	20.9 (0.5)	–	–	–	–	–	–	–	–	–
Neand	2	23.8^{h} (0.4)	2	24.5^{h} (3.3)	2	25.4^{h} (2.8)	2	25.9^{h} (0.6)	2	28.5^{h} (0.8)	2	23.3 (0.2)	24.6

Throughout the cervical spine, the transverse widths of modern human neural canals are significantly wider than those of the African great apes. AMH = anatomically modern humans, KSD = KSD-VP-1/1, A.L. 333 = *Au. afarensis*, SK 854 = *P. robustus*, WT-15 k = KNM-WT 15000 *H. erectus*, Dmanisi = *H. erectus*, ATA = *H. heidelbergensis* (Atapuerca), Neand = Neandertal. * = different from humans $p < 0.05$, h = different from apes $p < 0.05$. Standard deviations in parentheses

significantly wider than those of the African great apes (Tables 5.26 and 5.27). In modern humans, transverse enlargement of the neural canals for the cervical spinal cord bulge is most pronounced from levels C4 to C6 with no increase in the dorsoventral dimension (Elliott 1945; Sherman et al. 1990; Fountas et al. 1998). The transverse widths of the KSD-VP-1/1 neural canals are at or above the canal widths of anatomically modern humans in all but the C2 vertebral level, and, like modern humans, the C4, C5, and C7 neural canals of KSD-VP-1/1 exhibit transverse enlargement for the cervical cord bulge associated with the brachial plexus. By contrast, the isolated vertebrae for SK 854 and A.L. 333-106 are smaller, falling at the mean for *Pan*. Despite the fact that A.L. 333-106 is an isolated C6 vertebra, single neural canal segments below the C2 level are representative of the cross-sectional area of the cervical spinal cord (Kameyama et al. 1994, 1996; Fourie and Kirberger 1999), signifying that A.L. 333-106, a putatively female *Au. afarensis*, did not share the derived cervical bulge for the brachial plexus with KSD-VP-1/1 (a large male) and modern humans.

The observation that KSD-VP-1/1 possessed the transverse cervical bulge does not necessarily mean, however, that KSD-VP-1/1 was neurologically equivalent to modern

Table 5.27 Relative neural canal transverse width (neural canal transverse width/centrum geometric mean)

Group	*N*	C2	*N*	C3	*N*	C4	*N*	C5	*N*	C6	*N*	C7	Mean C3–C5
Gorilla	17	0.95* (0.1)	19	1.29* (0.1)	10	1.32* (0.1)	10	1.33* (0.1)	10	1.26* (0.1)	10	1.22* (0.1)	1.31
Pan	18	1.07* (0.1)	19	1.48 (0.1)	10	1.55* (0.1)	10	1.50* (0.1)	10	1.37* (0.1)	10	1.28* (0.1)	1.51
AMH	52	1.18^{h} (0.1)	50	1.56 (0.1)	22	1.70^{h} (0.1)	20	1.76^{h} (0.1)	21	1.72^{h} (0.2)	21	1.53^{h} (0.1)	1.67
KSD	–	–	1	1.77	1	1.46	1	1.65	–	–	–	–	1.63
A.L. 333	–	–	–	–	–	–	–	–	1	1.35	–	–	–
WT-15K	–	–	–	–	–	–	–	–	–	–	1	1.62^{h}	–
Dmanisi	1	1.11	1	1.84^{h}	–	–	–	–	–	–	–	–	–
ATA	–	–	2	1.71 (0.1)	–	–	–	–	–	–	–	–	–
Neand	–	–	2	1.65^{h} (0.1)	–	–	1	1.99	–	–	1	1.63^{h}	–

AMH = anatomically modern humans, KSD = KSD-VP-1/1, A.L. 333 = *Au. afarensis,* WT-15 k = KNM-WT 15000 *H. erectus,* Dmanisi = *H. erectus,* ATA = *H. heidelbergensis* (Atapuerca), Neand = Neandertal. * = different from humans $p < 0.05$, h = different from apes $p < 0.05$. Standard deviations in parentheses. Ratio values were logarithmically transformed for statistical testing

humans in terms of manual dexterity. The transverse expansion of the KSD-VP-1/1 mid-cervical neural canals results in a 'shouldered' appearance at the ventrolateral neural canal margins (Fig. 5.1), which corresponds with the increase of motor pools in the ventral horn of the spinal cord (Romanes 1964) as well as the ventrolateral component of the vestibulospinal tract. This tract is the conduit for the vestibulospinal reflex, a feedback loop that stimulates extensor motor neurons in lower limbs and the trunk that coordinates postural adjustments to keep the body balanced along the midline in bipedality. It is possible to speculate that instead of enhanced manual coordination, the increased area afforded by the "shouldered" morphology might most parsimoniously relate to an increase in the ability to maintain posture and head stabilization in bipedalism.

An additional hypothesis is that the transversely wide neural canals in KSD-VP-1/1 are independent of spinal cord size and simply reflect the progressively widening interfacet distances in the cervical column, as is the case in the hominin lumbar region for postural stability in bipedalism (Sanders 1990; Latimer and Ward 1993). However, this last hypothesis is unlikely as subaxial cervical spinal cord size and cervical neural canal dimensions across primates are closely linked (MacLarnon 1987).

Pathological Analysis

The cost of transitioning the bipedal spine from a horizontal suspension strut to a vertical column has been endemic pathology (Krogman 1951; Jurmain 1989; Gracovetsky 1996; Tatarek 2005). Australopiths exhibit a staggering incidence of vertebral pathology, with nearly every individual affected to some degree (Sanders 1998). Even the iconic *Au. afarensis* specimen, A.L. 288-1 ("Lucy"), exhibits pronounced thoracic kyphosis and new bone formation on its ventral surfaces and Scheuermann's disease at the T8 level (Cook et al. 1983). This deformity results from avascular necrosis of the vertebral ring apophysis and is characterized by compressed and wedged vertebral centra with end-plate irregularities (Scheuermann 1920, 1921). Its etiology is still a subject of discussion, having been suggested to be a genetic predisposition, the product of a hormonal disturbance, but most often described as the result of mechanical trauma (Nathan 1962; Boachie-Adjei and Sarwahi 2003; Damborg et al. 2011). Cook and colleagues (1983) argued for the latter explanation in A.L. 288-1 that vertebral pathology in this small-bodied female was related to lifting, climbing, or acrobatic activities.

Because osteophytic lesions are so commonplace among other australopith and paranthropine vertebral remains, the shift to orthograde posture may have predisposed australopiths to compressive load injuries before their centra were capable withstanding the high magnitudes of vertical loading incumbent to bipedalism (Meyer 2012), although the *Paranthropus robustus* specimen, Stw 431, may be an exception to this pattern (Barrickman 2003). High incidence of pathology in the australopiths may partly be a function of a center of gravity located more ventrally than in modern humans (Trinkaus 1987; Hausler and Berger 2001), but in a general sense, their small centra and overall spinal column appear to be better structured for suspension than compressive loading, sharing much in common with the vertebral column of pronograde apes, best suited to act as a bridge to support suspended viscera (Cartmill et al. 1987). The prevalence of vertebral pathology in australopiths is analogous to the inexorable structural failure that would result if the Brooklyn Bridge was reoriented to stand on one end, and provides a classic example of "behavior preceding biology," as the ape spine was exapted for vertical posture despite being ill prepared for the task.

The KSD-VP-1/1 cervical vertebrae are no exception to this observation. As in many other australopith vertebrae, the C4, C5, and C6 centra in the KSD-VP-1/1 cervical spine exhibit spondylosis deformans, with osteophytes originating from the centrum margins extending across the intervertebral space toward the adjacent centrum. Although the caudal centra are affected, there is a complete lack of pathological involvement of the corresponding zygapophyseal joints. The KSD-VP-1/1 C4 vertebra has a small degree of osteophytosis on the periphery of the right ventral margin that would be classified as first degree lipping according to the scoring system of Nathan (1962). More pronounced is the thin outgrowth of bone on the right inferior centrum margin of the C5, forming a small demi-ring of bone extending about 1.5 mm anteroinferiorly at its maximum, scored as second degree osteophytic lipping.

The KSD-VP-1/1 C6 vertebra exhibits pathological indications affecting both the superior and inferior centrum surfaces. The superior centrum's right ventral margin presents a second degree osteophytic lipping commensurate to that of the superjacent C5 inferior surface. The inferior surface of the C6 is marked by two significant pathologies – moderately severe spondylosis deformans and a probable Schmorl's node. Spondylosis deformans of the right inferior centrum margin is moderately severe, with third degree osteophytic lipping projecting approximately 5 mm ventrolaterally at its maximum (Fig. 5.24). The left side of the inferior centrum is damaged and missing aspects of the posterolateral surface, making it uncertain whether this aspect of the vertebra was similarly affected. Spondylosis deformans increases with advancing age, yet only approximately 10% of individuals experience negative clinical symptoms (Smith 1960; Rydevik et al. 2001), rendering it unlikely that osteophytosis in the KSD-VP-1/1 cervical spine incurred debilitating pain or an appreciable loss of spinal function. Although there is obfuscating damage to the inferior centrum surface, a relatively shallow, irregularly shaped Schmorl's node appears to be localized slightly dorsal to the midpoint of the inferior centrum surface. Schmorl's nodes typically accompany the degree of osteophytic lipping observed in the KSD-VP-1/1 C6 vertebra (Nathan 1962).

Pathogenesis of spondylosis deformans follows degeneration of the anulus fibrosis and affects all three precaudal vertebral regions in many taxa (Morgan et al. 1967). The cervical spine is susceptible to this degenerative change because of its complexity and wide range of motion, and its incidence increases with the degenerative effects of aging as the structure of the annulus fibrosis loses its capacity to rebound subsequent to compression. Spondylosis deformans in humans is most pronounced on the ventral centrum margins (Shimoda et al. 2011) between the C5 and C7 levels, which serve as the pivot through which movements of the head and neck are mediated (Boyd-Clark et al. 2002; Van der Merwe et al. 2006). Spinal degenerative changes result in reduced disc height, loss of cervical lordosis, and increase the risk for disc herniation (Schmorl's nodes) (Rydevik et al. 2001). However, rather than viewed simply as a negative pathological consequence, the formation of osteophytes on the centrum effectively renovates centra into more compression-resistant pillars, similar to how capitals and bases are employed in architecture to increase the resistance of pillars to vertical compression (Nathan 1962).

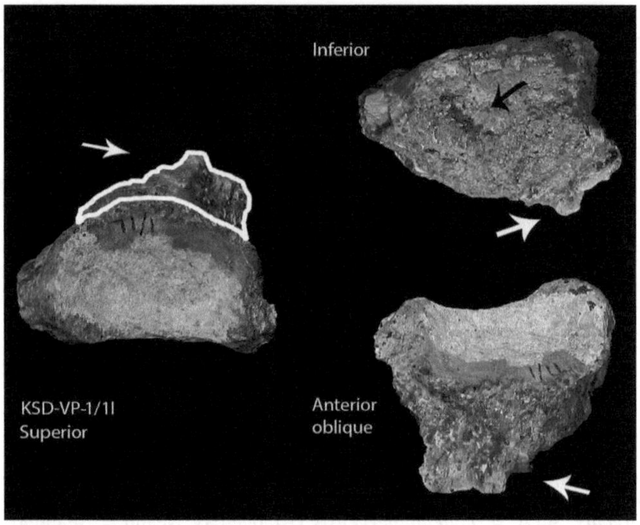

Fig. 5.24 Superior, inferior, and anterior oblique views of KSD-VP 1/11 (C6 centrum) highlighting the extent of pathology. White arrows point to area of spondylosis deformans on the inferior, and the black arrow points to the Schmorl's node on the inferior centrum surface. In the superior view, the white solid line outlines the area of osteophytic lipping ventral to the anterior centrum margin

Compared with modern humans, other hominoids exhibit significantly less degenerative spinal changes (Jurmain 2000). For example, a sample of 344 wild chimpanzees displayed a complete lack of osteophytosis and no evidence of degenerative spinal disease (Jurmain 1989). Although these conditions are pervasive in adult humans, even the oldest chimpanzees in the sample were completely free from spinal degeneration. Osteophytosis in wild gorilla cervical vertebrae is also rare (Lovell 1990; Jurmain 2000), although a higher incidence of pathology was observed in captive male gorillas (Nichols and Zilhlman 2002). Polyarticular spinal degeneration has been documented in rhesus macaques, including spinal osteophytosis, possibly because they habitually sit upright, which vertically loads their spines (Bailey et al. 2013). Yet, the systematic prevalence of vertebral pathology,

Table 5.28 Frequencies of spondylosis deformans (osteophytic lipping of the centrum) by element

Group	*N*	C2	C3	C4	C5	C6	C7
AMH	52	0.02	0.13	0.25	0.27	0.29	0.21
Neandertal	2	0.0	0.0	0.0	0.50	100.0	0.50
Gorilla	20	0.0	0.18	0.07	0.0	0.0	0.0
Pan	20	0.0	0.0	0.0	0.0	0.0	0.0

Note the caudal pattern of pathological expression in anatomically modern humans (AMH) and Neandertals in comparison to the more cranial pattern in *Gorilla*. The human sample consists of individuals over age 40 (mean age 52; median age 50; range 40–70 years)

including cervical pathology, in hominins is unparalleled, extending from the australopiths through large-bodied *Homo heidelbergensis* (Perez et al. 1997) and Neandertals (Boule 1911; Vallois 1937; Straus and Cave 1957). Thus, the presence of human-like pathology may serve as a biomechanical signal of human-like posture and locomotion.

Spondylosis deformans might be the expected reaction in the cervical column of a highly prognathic and large-bodied adult biped, but especially if KSD-VP-1/1 was afflicted with kyphosis as is the case in the smaller bodied A.L. 288-1. This is because the "hunchback" posture associated with Scheuermann's kyphosis initiates a compensatory response in the lumbar and cervical spine which causes the head to protrude forward, or gooseneck, resulting in a negative sagittal balance with the C7 oriented dorsal to the sacral promontory (Boachie-Adjei and Sarwahi 2003). Because there are no thoracic vertebrae associated with the KSD-VP-1/1 cervical fossils, it is speculation that the cervical pathology described here is the result of compensation for Scheuermann's kyphosis, but such a scenario is plausible.

In humans high-load magnitudes (Bridges 1994), bipedalism (Van der Merwe et al. 2006) and advanced age (Rydevik et al. 2001; Shimoda et al. 2011) increase the frequency of cervical osteophytosis, supported by data collected for this study showing almost 1/3 of human adults over the age of 40 with spondylosis deformans in the mid-cervical region (Table 5.28). Because lateral bending in human orthogrady is maximal at the C6–C7 junction (Lang 1972; Wilke et al. 1997), this differentially predisposes the inferoventral aspect of the human C6 centrum to potential trauma and intervertebral disc degeneration resulting in spondylosis deformans (Farfan et al. 1970). Moreover, in humans, the incidence of osteophytic lipping is greatest near the peaks of spinal curvature away from the line of gravity, whereas little or no pathology occurs where the line of gravity crosses the spine (Nathan 1962). Because the KSD-VP-1/1 pattern of caudal pathology with a peak expression at the C6 level matches the pattern exclusive to anatomically modern humans (and Neandertals), this implies that KSD-VP-1/1 shared a common loading regime and cervical posture with modern humans.

Discriminant Function Analysis

Stepwise discriminant analysis was performed to select the most correlated predictors from preserved variables in the KSD-VP-1/1 vertebrae to classify sample group affinity for each vertebral specimen. Each stepwise discriminant analysis was performed to select the most correlated predictors from preserved dimensions in the corresponding KSD-VP-1/1 vertebra. Following Cokluk and Buyukozturk (2008), the number of predictor variables was fewer than the sample size of the smallest group. Most stepwise models predicted group classification with 100% accuracy, with the lowest model accuracy of 96% at the C6 level.

For the C2 level, 11 dimensions were selected from the neural canal, inferior articular facet, pedicle, lamina, pars interarticularis, and inferior centrum. For the C3 level, 18 dimensions were selected from the centrum, neural canal, superior articular facets, inferior articular facets, pedicle, lamina, and pars interarticularis. Nine dimensions were selected at the C4 level from the centrum, neural canal, articular facets, pedicle, lamina, and pars interarticularis. For the C5 level, nine dimensions were selected from the centrum, neural canal, and superior articular facets. Six dimensions at the C6 level were selected from the centrum and six dimensions for the C7 level were selected from the superior articular facets, lamina, neural canal, and root of the transverse process.

All of the subaxial KSD-VP-1/1 cervical vertebrae were classified as anatomically modern human (AMH) with high prediction values (Tables 5.29 and 5.30). However, the KSD-VP-1/1 C2 vertebra classified as *Gorilla*, as did A.L. 333-101. SK 854 and A.L. 333-106 both classified as *Pan*. The more cranial vertebral levels in the Dmanisi *H. erectus* specimens and Neandertals also tended to classify with less affinity to anatomically modern humans. The pathological C7 vertebra of KNM-WT 15000 (*H. erectus*) only weakly classified with humans, although when centrum measures were incorporated into the analysis, classification probability increased to 1.0.

Table 5.29 Summary of discriminant function classifications in hominin cervical vertebrae (prediction value in parentheses)

	KSD	A.L. 333	SK 854	Dmanisi	WT-15k	Neandertal	Model accuracy (%)	2LogLikelihood
C2	*Gorilla* (0.98)	*Gorilla* (0.99)	*Pan* (0.99)	AMH (0.52)	–	–	99	3.163
C3	AMH (0.99)	–	–	AMH (1.0)	–	*Gorilla* (0.98)	100	0
C4	AMH (1.0)	–	–	–	–	*Gorilla* (0.97)	100	0.03
C5	AMH (1.0)	–	–	–	–	AMH (1.0)	100	0.0002
C6	AMH (0.97)	*Pan* (0.90)	–	–	–	–	96	4.137
C7	AMH (1.0)	–	–	–	AMH (0.54)	–	100	0.917

All except the C2 vertebra in KSD-VP-1/1 classifies with anatomically modern humans (AMH). Because A.L. 333-101 does not preserve the same element parts as KSD-VP-1/1, a separate stepwise discriminant function analysis was performed using ten preserved dimensions. When dorsoventral neural canal dimensions are removed from analyses (reconstructed in the KSD-VP-1/1 specimens), the results are nearly identical. AMH = anatomically modern humans, KSD = KSD-VP-1/1, A.L. 333 = *Au. afarensis,* SK 854 = *P. robustus,* WT-15k = KNM-WT 15000 *H. erectus,* Dmanisi = *H. erectus*

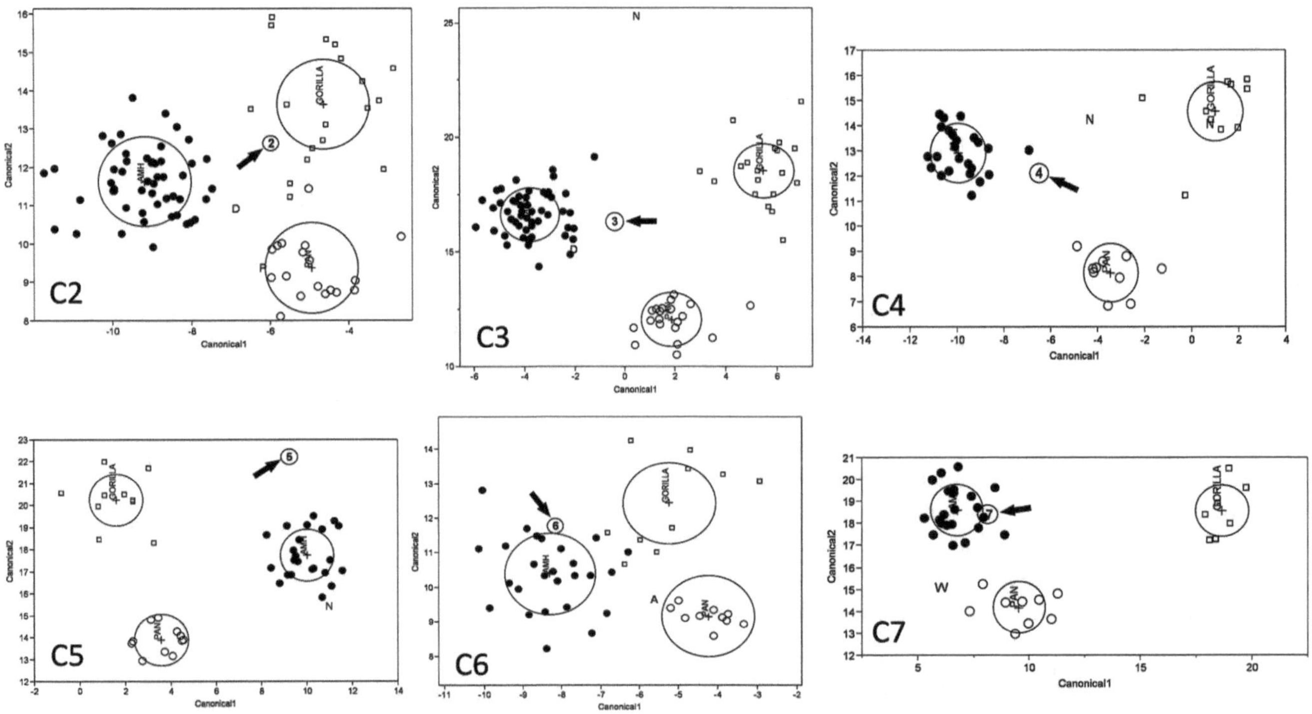

Fig. 5.25 Stepwise discriminant function analysis canonical scatterplots by element. Normal 50% contours shown. KSD = numerals indicating vertebral level, modern humans = solid points, *Pan* = open circles, *Gorilla* = open squares, A = A.L. 333-106 *Au. afarensis,* P = SK 854 *P. robustus,* D = Dmanisi *H. erectus,* W = KNM-WT 15000 *H. erectus,* N = Neandertals

Table 5.30 Summary of principal component analysis groupings in hominin cervical vertebrae

	KSD	A.L. 333	SK 854	Dmanisi	WT-15k	Neandertal	PC1 and PC2 % of variab. (%)	PC2 and PC3 % of variab. (%)
C2	AMH/*Gorilla*	–	*Pan*/*Gorilla*	Intermediate	–	–	68	27
C3	Intermediate	–	–	AMH	–	Outlier	70	26
C4	AMH	–	–	–	–	–	72	29
C5	AMH	–	–	–	–	AMH	79	29
C6	AMH	*Pan*/*Gorilla*	–	–	–	–	94	26
C7	AMH	–	–	–	AMH	–	73	27

Below the C2 and C3 levels, KSD-VP-1/1 cervical vertebrae fall exclusively within the distribution of anatomically modern humans (AMH). The most superior cervical vertebrae across hominins are more primitive than those of modern humans, likely reflecting a craniocervical developmental link. Note that unlike the large male KSD-VP-1/1, the smaller female A.L. 333 is more primitive in the lower cervical column. AMH = anatomically modern humans, KSD = KSD-VP-1/1, A.L. 333 = *Au. afarensis,* SK 854 = *P. robustus,* WT-15 k = KNM-WT 15000 *H. erectus,* Dmanisi = *H. erectus*

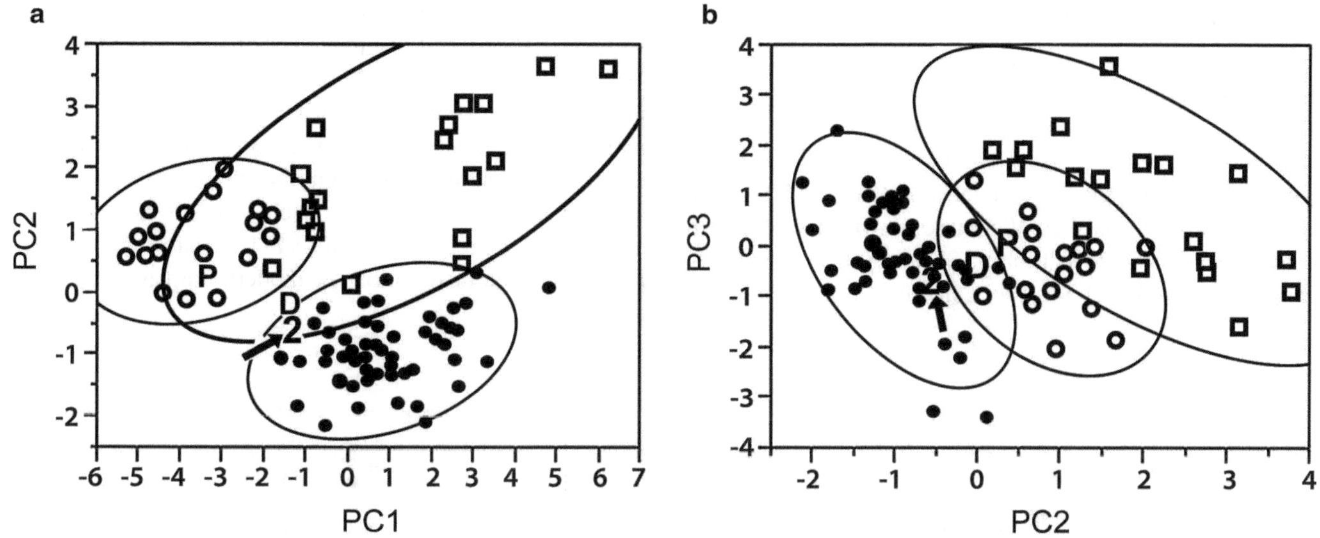

Fig. 5.26 **a** Scatterplot of C2 vertebrae first principal component by second principal component. The KSD-VP-1/1 C2 vertebra (indicated by the number 2) falls within the 95% confidence ellipse for both anatomically modern humans and *Gorilla*, whereas the Dmanisi *H. erectus* C2 (letter D) falls outside the human distribution, intermediate to humans and *Gorilla*. The *P. robustus* C2 vertebra SK 854 (letter P) falls in the distribution of *Pan*. **b** Scatterplot of C2 second principal component by third principal component. The KSD-VP-1/1 C2 vertebra (number 2) falls within the 95% confidence ellipse for modern humans. The *P. robustus* C2 vertebra SK 854 (letter P) falls outside the human distribution and within the distribution of *Pan*. Modern humans = solid circles (N = 64), *Pan* = open circles (N = 18), *Gorilla* = open squares (N = 17)

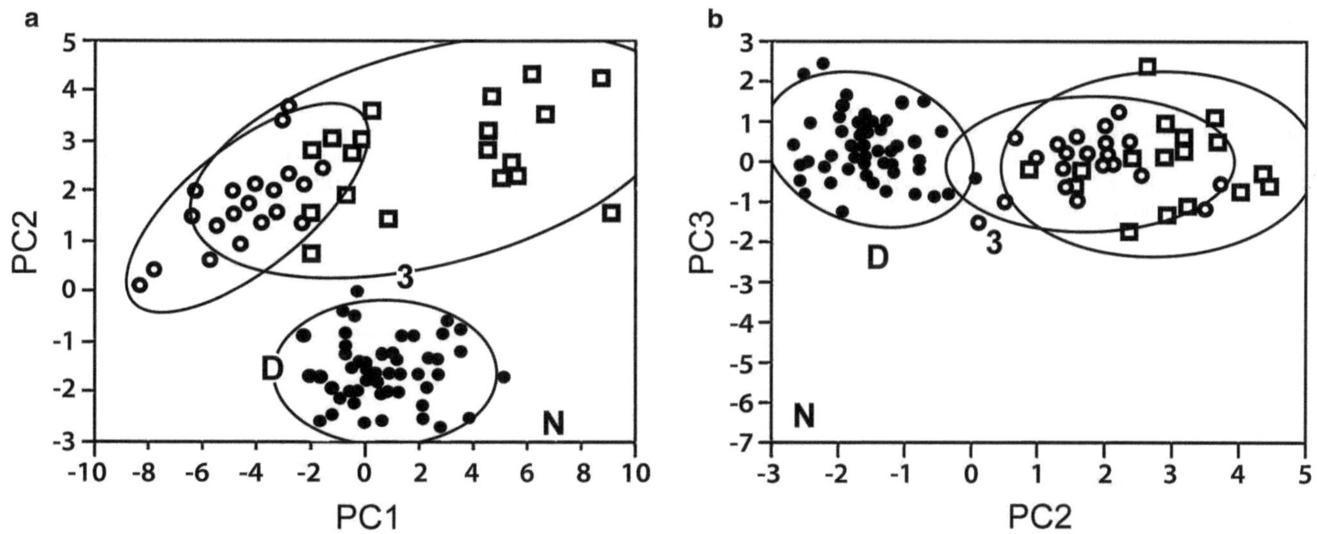

Fig. 5.27 **a** Scatterplot of C3 vertebrae first principal component by second principal component. The KSD-VP-1/1 C3 vertebra (indicated by the number 3) falls between the 95% confidence ellipses for modern humans and *Gorilla*, whereas the Dmanisi *H. erectus* C3 (letter D) falls on the margin of the human distribution. The La Ferrassie Neandertal C3 vertebra (letter N) falls outside of the human distribution. **b** Scatterplot of C3 second principal component by third principal component. The KSD-VP-1/1 C3 vertebra (number 3) falls outside the 95% confidence ellipse for humans, *Pan*, and *Gorilla*. The Dmanisi *H. erectus* C3 (letter D) and La Ferrassie Neandertal (letter N) also fall outside of the extant sample distributions. Modern humans = solid circles (N = 65), *Pan* = open circles (N = 20), *Gorilla* = open squares (N = 19)

Principal Components

In each of the principal component analyses, only dimensions from preserved elements in the KSD-VP-1/1 vertebrae were incorporated. As discussed above, since the first principal component typically correlates positively with size, to eliminate size effects with allometric growth, the first component was discarded (sheared). Shearing did not radically change the results of analyses, probably because the first component contains a degree of allometrically related shape variation (Healy and Tanner 1981; James and McCulloch 1990), but both standard PCA and sheared PCA results are presented here for rigor. As was the case with the discriminant function analyses, principal component analyses (PCA) demonstrate most hominins to be somewhat more ape-like at the most cranial levels. Unlike A.L. 333-106, each of the vertebrae of KSD-VP-1/1 below the C4 level falls within the distribution for modern humans (Fig. 5.25).

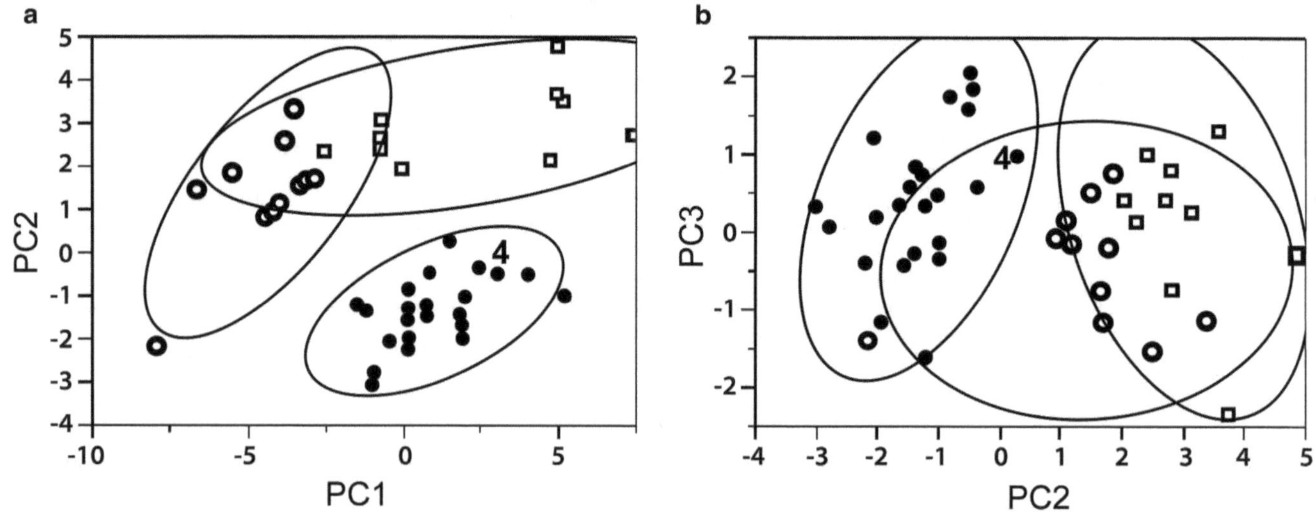

Fig. 5.28 **a** Scatterplot of C4 vertebrae first principal component by second principal component. KSD-VP-1/1 (number 4) falls within the 95% confidence ellipse for anatomically modern humans. **b** Scatterplot of C4 second principal component by third principal component. KSD-VP-1/1 (number 4) falls within the 95% confidence ellipse for anatomically modern humans. Modern humans = solid circles (N = 23), *Pan* = open circles (N = 10), *Gorilla* = open squares (N = 10)

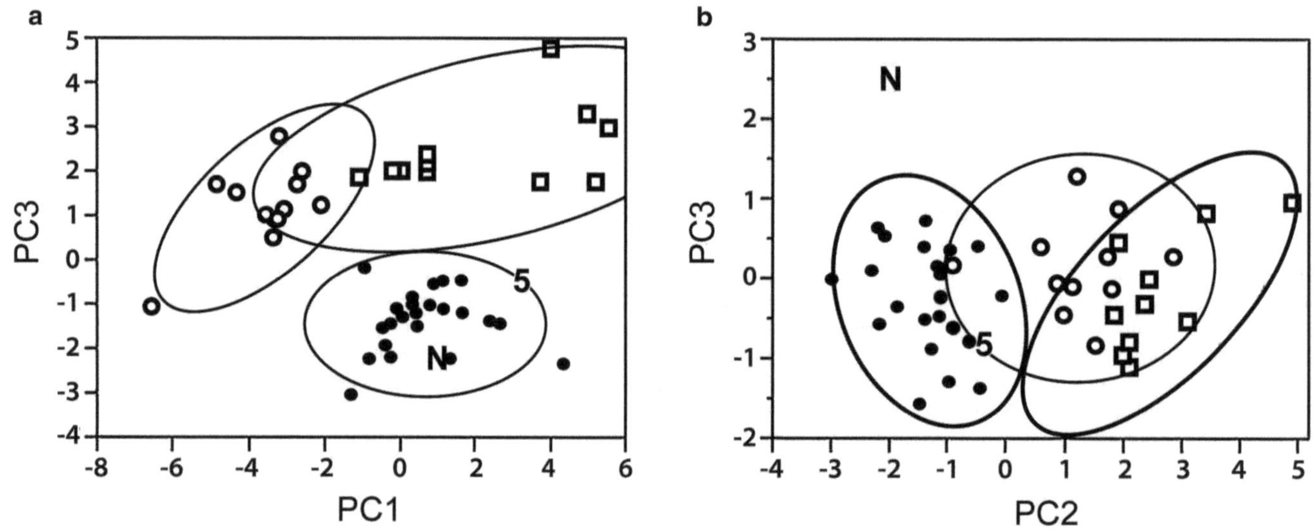

Fig. 5.29 **a** Scatterplot of C5 vertebrae first principal component by second principal component. The KSD-VP-1/1 C5 vertebra (number 5) falls within the 95% confidence ellipse for anatomically modern humans as does the Kebara Neandertal (letter N) specimen. **b** Scatterplot of C5 second principal component by third principal component. KSD-VP-1/1 (number 5) falls within the 95% confidence ellipse for anatomically modern humans and *Pan*. Modern humans = solid circles (N = 23), *Pan* = open circles (N = 11), *Gorilla* = open squares (N = 10)

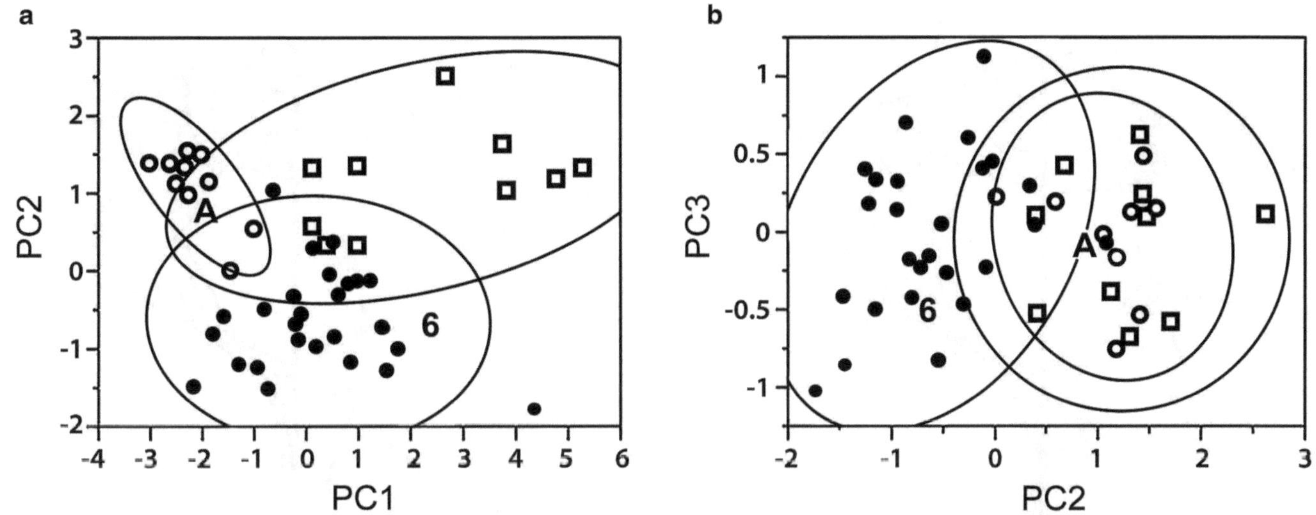

Fig. 5.30 **a** Scatterplot of C6 vertebrae first principal component by second principal component. The KSD-VP-1/1 C6 vertebra (number 6) falls within the 95% confidence ellipse for anatomically modern humans, whereas A.L. 333-106 (letter A) falls outside the human distribution and within the distribution of *Pan*. **b** Scatterplot of C6 second principal component by third principal component. KSD-VP-1/1 (number 6) falls within the 95% confidence ellipse for anatomically modern humans. A.L. 333-106 (letter A) falls outside the human distribution and within the distribution of *Pan* and *Gorilla*. Modern humans = solid circles (N = 26), *Pan* = open circles (N = 10), *Gorilla* = open squares (N = 10)

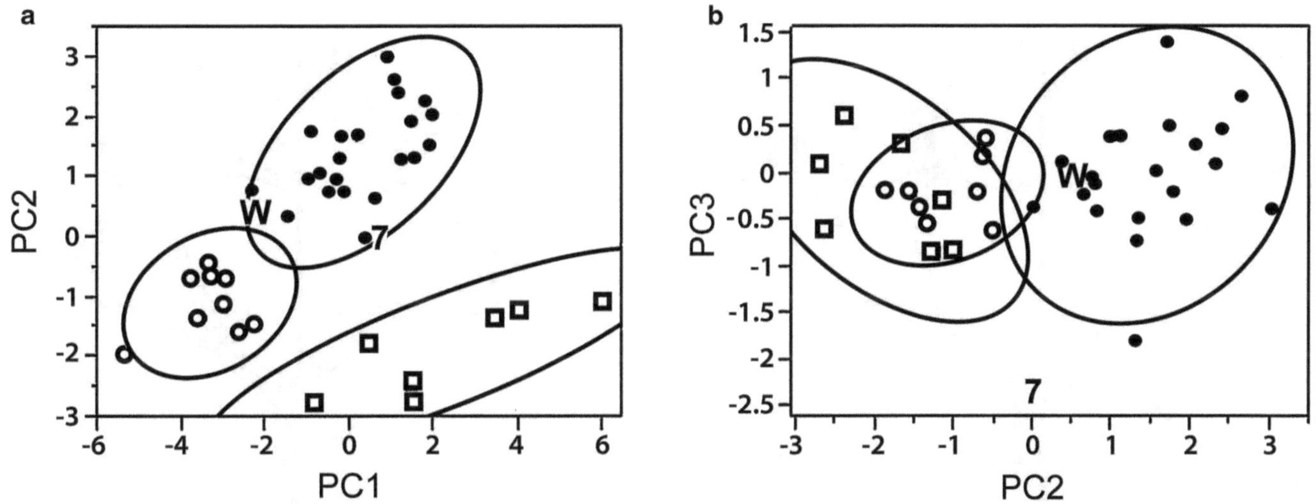

Fig. 5.31 **a** Scatterplot of C7 vertebrae first principal component by second principal component. The KSD-VP-1/1 C7 vertebra (number 7) falls just within the 95% confidence ellipse for anatomically modern humans, as does the KNM-WT 15000 (letter W) C7 vertebra. **b** Scatterplot of C7 second principal component by third principal component. The KSD-VP-1/1 C7 vertebra (number 7) falls outside the 95% confidence ellipse for anatomically modern humans. Modern humans = closed circles (N = 22), *Pan* = open circles (N = 9), *Gorilla* = open squares (N = 7)

For the C2 vertebrae PCA (Fig. 5.26), 11 dimensions were incorporated from the neural canal, inferior articular facet, pedicle, lamina, pars interarticularis, and inferior centrum. For the C3 vertebrae PCA (Fig. 5.27), 22 dimensions from the centrum, neural canal, superior and inferior articular facets, pedicle, lamina, and pars interarticularis were incorporated. For the C4 vertebrae PCA (Fig. 5.28), 21 preserved dimensions were employed from centrum, neural canal, superior and inferior articular facets, pedicle, lamina, and pars interarticularis. Fourteen dimensions incorporated for the C5 vertebrae PCA (Fig. 5.29) were from the centrum, neural canal, and superior articular facets. For the C6 vertebrae PCA (Fig. 5.30), six dimensions were employed from the centrum. For the C7 vertebrae PCA (Fig. 5.31), 14 dimensions were incorporated from the superior and inferior articular facets, neural canal, lamina, pars interarticularis, and root of the transverse process.

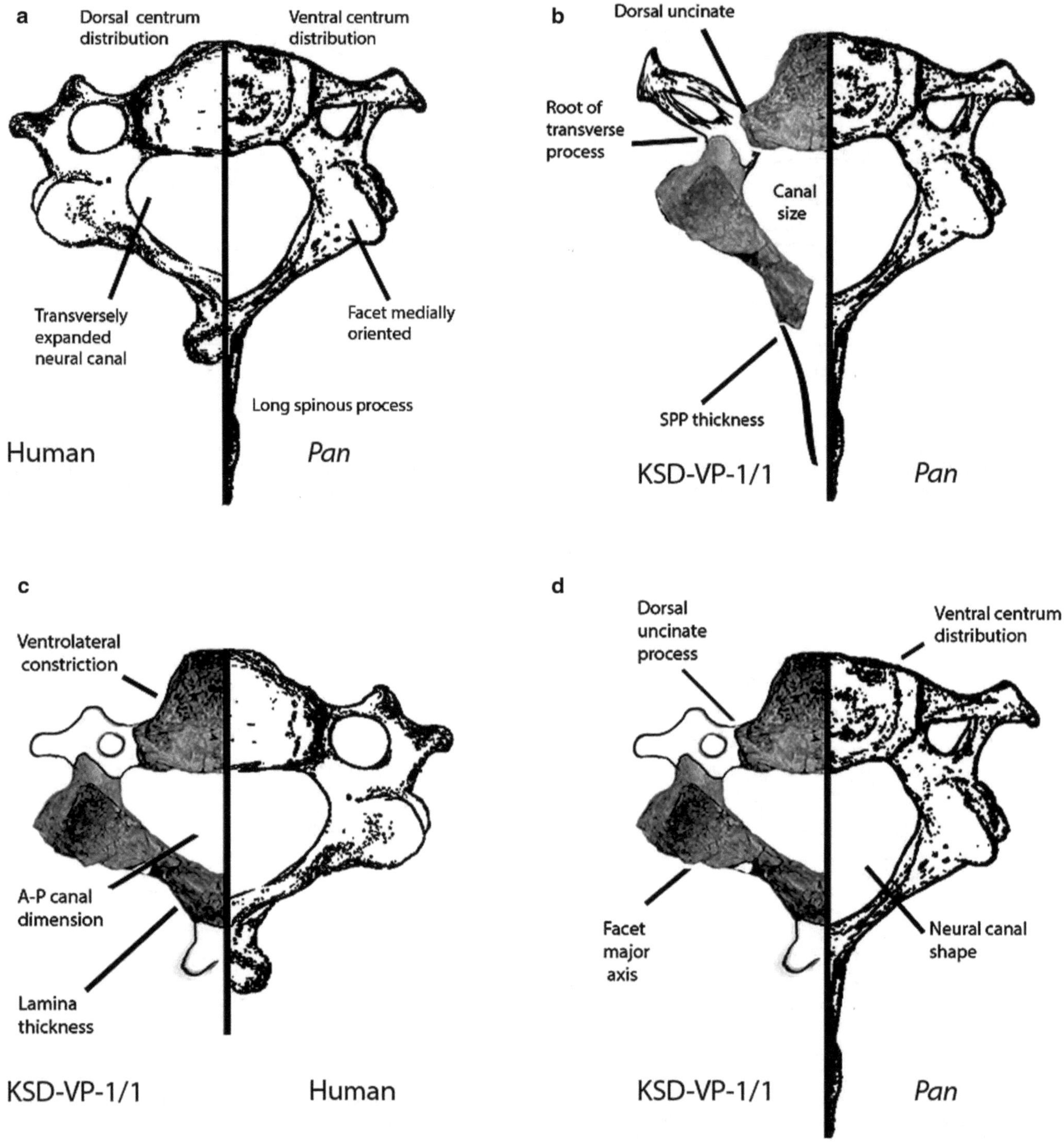

Fig. 5.32 **a** Modern human (left) versus *Pan* C3 vertebra (right) scaled to match midsagittal dorsoventral centrum length. The human vertebra features a dorsally eccentric centrum and uncinate process, obliquely oriented transverse processes, and overall transverse expansion compared to dorsoventral expansion in *Pan*. **b** KSD-VP-1/1 C3 vertebra (left) reconstructed following *Pan* morphology (right). Reconstruction of the KSD vertebra using *Pan* as a model is anatomically untenable, leaving no space for the transverse process, an abnormally compressed neural canal for a primate, and spinous process morphology unknown among hominoids. **c** KSD-VP-1/1 C3 vertebra (left) reconstructed following modern human morphology (right). Despite the highlighted differences, the vertebrae are similar in overall morphology. **d** KSD-VP-1/1 C3 vertebra (left) reconstructed following modern human morphology versus *Pan* (right). KSD-VP-1/1 differs from *Pan* in its dorsally eccentric centrum and uncinate process, transverse process orientation, and transversely expanded neural canal and articular facets

Principal component analyses result in a distribution of taxa very similar to that seen in the discriminant function analyses of canonical variates, demonstrating that assigning group membership *a priori* has little effect on the pattern of shape similarities. Both sets of analyses load heavily on features on the transverse plane, where humans are expanded (especially at levels of the cervical bulge) and apes are transversely compressed. Like humans, KSD-VP-1/1 centra feature dorsally eccentric centra and uncinate processes, obliquely oriented transverse processes, and overall transverse expansion, and lack the archetypal dorsoventral expansion seen in *Pan* (Fig. 5.32). However, because the KSD-VP-1/1 vertebrae are fragmentary, it is necessary to reconstruct the orientation of the neural arches. There is a highly conserved relationship in humans and the apes in the placement of the medial articular facet margins relative to the dorsal centrum margin which results in high confidence for the reconstruction of ventrolateral margins of the neural canal. Variability between apes and humans in orientation of the lamina potentially confounds neural canal dorsoventral length reconstruction, as the lamina constitutes the posterior border of the neural arch. But reconstruction of the KSD-VP-1/1 neural canal using the laminar orientation of *Pan* results in untenable vertebral anatomy, as the result eliminates sufficient area for the transverse process to emerge, an abnormally compressed neural canal, and a spinous process thickness and morphology unknown among hominoids (Fig. 5.32b). Only when the KSD-VP-1/1 C3 vertebrae are reconstructed following a laminar orientation closer to that of humans, are the cervical vertebrae physiologically viable. Using the human model is further warranted by systematic morphological similarities between KSD-VP-1/1 and humans across the centra and articular facets. The statistical analyses employed here underscore morphological observations pointing to the similarity of KSD-VP-1/1 cervical vertebrae to those of modern humans and illustrate systematic differences with those of the African great apes.

Discussion

It is important to recognize that many aspects of morphology do not directly relate to function (Begun 2004), a principle emphasized by the fact that some species of goats climb trees (Coblentz 1978; Lu 1988) despite the absence of a functional skeletal signal suggesting a capacity for this behavior (the goat principle). Yet, axial skeletal remains can reveal adaptations to load frequency and magnitude, constraints on possible ranges of behavior (Ward 1991; Nalley 2013), and epigenetic modifications that may reveal patterns of habitual activity (Ward 2002). Despite the goat principle, vertebral size, shape, geometry, and especially epigenetically sensitive traits offer a conservative functional link between morphology and behavior in KSD-VP-1/1.

Vertebral Number

Although there is considerable disagreement on the number of lumbar vertebrae in early hominins (i.e., Robinson 1972; Latimer and Ward 1993; Haeusler et al. 2002; Ward 2002) the cervical columns of early hominins undoubtedly comprised seven vertebrae, as is the case for all primates and nearly all mammals (Williams 2011). Thus, there is no question that the KSD-VP-1/1 vertebrae represent six levels of a cervical column that comprised a total seven vertebrae, and unlike questions concerning the evolution of the hominin lumbar column, variance in cervical vertebral number is not at issue (Table 5.30).

The Neural Canals

In general, human vertebral canals (and the associated spinal cord) are relatively expanded relative to the great apes (Schultz 1930; MacLarnon 1987), and because cervical neural canal dimensions mirror the size of the spinal cord below the C2 level, they can be used to infer spinal cord size. The analyses here show that despite their antiquity, the size and shape of the subaxial KSD-VP-1/1 neural canals match those of modern humans, implying that KSD-VP-1/1 possessed a fully human-like spinal cord. Meyer (2005) showed that human-like neural canal dimensions evolved in the Dmanisi *H. erectus* specimens at 1.8 Ma, indicating that the evolution of the human spinal cord was completed prior to the evolution of human brain size. Now, the capacious cross-sectional areas in KSD-VP-1/1 appear to push back this development to 3.6 Ma. Moreover, the neural canals of KSD-VP-1/1 exhibit the human-like pattern of increased transverse enlargement in the caudal direction for the cervical cord bulge associated with the brachial plexus, a region of the spinal cord that supplies nerves to the hands and arms. Although African great apes and all other australopiths exhibit only a modest cervical bulge in an otherwise smaller spinal cord, assuming a correlation between spinal cord size and function, the larger spinal cord and exaggerated transverse bulge in KSD-VP-1/1 might indicate that despite an ape-sized brain, it had derived enhanced manual control relative to the African great apes.

Alternatively, because the size increase corresponds with a region of motor pool increase in the ventral horn of the spinal cord and ventrolateral component of vestibulospinal tract (the conduit for the vestibulospinal feedback reflex for balance), the putatively increased spinal cord dimensions of KSD-VP-1/1 could simply relate to enhanced coordination to keep the body balanced for bipedality. Transverse spinal cord expansion with concomitant neural canal expansion might be expected in obligate bipeds to manage the difficulties of negotiating increased sway amplitudes around a vertical axis relative to quadrupedalism (see Hirasaki et al. 1999; Xiang et al. 2008), as robotics engineers are well aware (Sellers et al. 2004; Rummel et al. 2010; Klein and Lewis 2012). Compensatory reflex mechanisms for head and body stabilization in space are of considerable importance in maintaining gaze and heading (Patla et al. 1999), but are not associated with brain size, as even decerebrate animals maintain proprioceptive control of the limbs (Wilson et al. 1986), highlighting the independence of spinal cord and brain sizes and how spinal cord enlargement could have preceded brain expansion in hominins.

As discussed above, because subaxial cervical spinal cord size and cervical neural canal dimensions across primates are closely matched (MacLarnon 1987), it is unlikely that the canals expanded independently of the spinal cord for structural stability. While this is certainly the case for the hominin lumbar column where progressive widening of the neural canals follow increasing interfacet distances for mitigating incrementally increasing loads (Sanders 1990; Latimer and Ward 1993), unmatched cord and canal dimensions are unknown for the cervical column in primates. These hypotheses are speculative, untested, and not exhaustive; notwithstanding, the large and transversely expanded neural canals in KSD-VP-1/1 appear reflective of spinal cord expansion previously thought to be a synapomorphism in Pleistocene hominins and modern humans.

Lordosis, Pathology, and Cervical Posture

Lordosis in the lumbar spine is exclusive to hominins, but in the cervical spine lordosis is present across the entire extant primate sample. Despite the fact that the mid-cervical centra tend to be ventrally wedged, unlike other vertebral regions, the cervical intervertebral discs (which do not preserve) produce a substantial degree of lordosis. In terms of skeletal anatomy, the aggregate human cervical column is only slightly more lordotic than those of apes by virtue of a diagnostic C7 vertebra that has significantly more dorsal wedging. Unfortunately, the diagnostic C7 centrum for KSD-VP-1/1 was not recovered, but its axial remains, and those of other fossil hominins, nonetheless exhibit kinematic signatures consistent with cervical lordosis. Dorsally inflected inferior articular facet angles that facilitate lordosis in the cervical and lumbar columns in australopiths fall within the human distribution, indicating the presence of cervical lordosis. Superior articular processes of KSD-VP-1/1 are also essentially human-like in morphology and orientation, indicating similar relationships between motion segments in the cervical spine. As in the lordotic lumbar spine, pyramidal increases in transverse centrum and articular process widths are a functional signal of lordosis and are present across the sample, including KSD-VP-1/1. While the KSD-VP-1/1 C2 vertebra suffers from considerable damage, its inferior centrum surface indicates the typical dorsal wedging angle that is a cogent skeletal factor in producing cervical lordosis.

Although the subaxial KSD-VP-1/1 centra are ventrally wedged (more so than in humans), further evidence for lordosis in the KSD-VP-1/1 cervical spine comes from the dorsally eccentric centra. This signals that the nucleus pulposus at each level in KSD-VP-1/1 was skewed toward the dorsal centrum as in humans, indicating that load mitigation in the KSD-VP-1/1 cervical spine was comparable to modern humans. Since dorsal eccentricity of the centrum is systematically dissimilar from more ventral centrum distribution of the apes, a human degree of cervical lordosis appears likely. Yet, the pathology of KSD-VP-1/1 may serve as the most informative biomechanical signal pertaining to lordosis and head carriage.

Although humans and African great ape cervical columns are both lordotic, the human cranium is not cantilevered on the cervical spine as in the African great apes. As a result, in static posture, the center of gravity in human and ape crania are situated differently causing the line of gravity to cross the peak of spinal curvature at different cervical levels. Load-induced osteophytic lipping typically occurs at the peak of spinal curvature away from the line of gravity, whereas little or no pathology occurs where the line of gravity crosses the spine (Nathan 1968; Farfan et al. 1970). In humans, cervical lordosis produces a maximal distance from the line of gravity at the C6–C7 junction (Lang 1972; Wilke et al. 1997), resulting in pathology at the C6 level unobserved in apes. Because the pathology of KSD-VP-1/1 exhibits peak expression at the C6 level as in humans, this implies that the line of gravity of the KSD-VP-1/1 cranium crossed the peak of spinal curvature at the same C6 level and that the overall degree of curvature of the cervical spine was similar, if not identical, in humans and KSD-VP-1/1.

Head Carriage

The cantilevered cranium of the African great apes situates a center of gravity relatively more ventral to the cranial condyles than in humans despite the shared lordotic cervical columns, mandating greater mid-cervical dorsal moments and constant support from large epaxial (i.e., nuchal) muscles to counter the ventral forces of gravity (Lieberman 2011). This results in a high, steeply angled, robustly flaring nuchal crest in the apes, a different mode of head carriage, and a cervical column that endures constant ventral loading. This suite of demands results in ape cervical vertebrae with long spinous processes hafted to robust mid-cervical lamina and pedicles, dorsoventrally expanded neural canals to counter ventral axial loads along with tall, ventrally eccentric centra, and tall, acutely angled articular facet on the sagittal plane for stability and ventral shear resistance. The KSD-VP-1/1 vertebrae exhibit many of these primitive traits, especially in the posterior vertebral components. For example, the mid-cervical spinous process length reconstructions for KSD-VP-1/1 are intermediate to humans and *Pan*, the lamina are taller and thicker than the human mean, the articular processes are tall, and the pedicles are most similar to *Gorilla* in their size and proportion. For the reasons discussed above, the posterior components in the apes are larger, reflecting the transfer of proportionately greater load magnitudes through these elements compared to modern humans who transfer proportionally greater axial loads through the centra (Sanders 1990, 1998). Considered individually, nearly all mid-cervical posterior components of the KSD-VP-1/1 spine are more similar to the apes than to humans (as are other early hominins), and suggest an ape-like pattern of load transfer, and by extension, significant differences with human head carriage.

Additional hints that head carriage in KSD-VP-1/1 differed from modern humans come from strongly angled atlanto-occipital articular surfaces on *Au. afarensis* crania (A.L. 333-45 and A.L. 822-1; Kimbel and Rak 2010) and steep angulation in the articular facets of the fragmented *Au. afarensis* C1 vertebra (A.L. 333-83; Lovejoy et al. 1982; Kimbel et al. 1994). Similarly, the primitive rounded slope morphology of the superior apophyseal surface of the KSD-VP-1/1 C3 centrum discussed above suggests a loading pattern at variance with humans.

However, when the KSD-VP-1/1 cervical spine is examined as a whole, rather than as separate isolated elements, a more human-like pattern emerges. Despite being taller and thicker than the human mean, lamina heights in the KSD-VP-1/1 mid-cervical column parallel the incremental human pattern and are unlike the apes where mid-cervical lamina heights increase dramatically. This implies that dorsal loads channeled through each lamina level in humans and KSD-VP-1/1 in similar proportions and along similar vectors, and loads were not progressively cumulative in the caudal direction as in the apes. Similarly, the large cantilevered head of the ape produces progressively greater shear stresses in the caudal direction resulting in the largest subaxial articular processes at the C6 and C7 levels. Although the articular facets of KSD-VP-1/1 are large, their sizes do not progressively increase, and they maintain consistent facet areas throughout the cervical column as in humans.

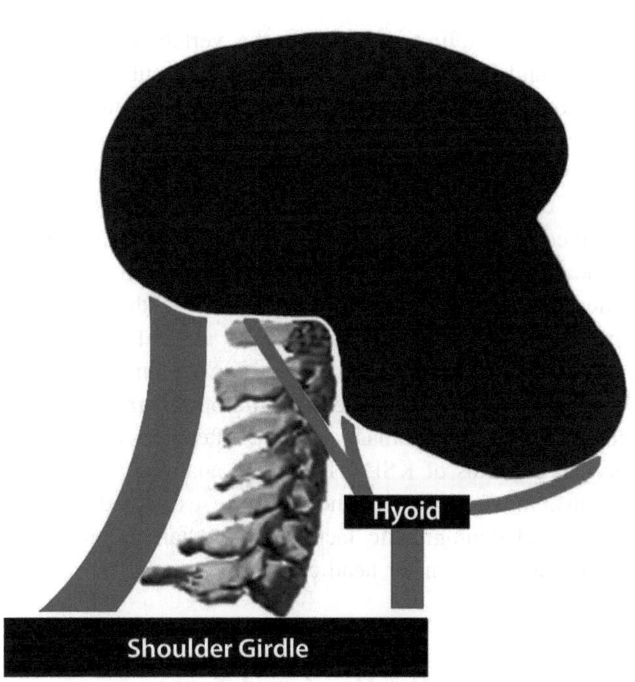

Fig. 5.33 Schematic of superficial epaxial (dorsal) and hypaxial (ventral) muscles acting on the cervical spine. Even with basicranial remodeling and an anteriorly repositioned foramen magnum, large masticatory structures (craniodental and hypaxial musculature) in *Au. afarensis* act to shift the center of cranial mass more ventrally to the cranial condyles than in modern humans, mandating relatively greater antagonistic epaxial muscle recruitment to maintain lordosis and head position

The forward shift of the foramen magnum and cranial condyles in humans combined with a retrognathic cranium better balance the head above a vertical spine and obviates the need for large epaxial musculature (Aiello and Dean 1990; Cartmill and Smith 2009). As a result, the attendant levers for epaxial muscles, the spinous processes, are significantly shorter in humans, and the homolog of the large ape nuchal crest is the relatively slight superior nuchal line. The australopiths present a morphological compromise between these features because despite being bipedal and having a more anteriorly positioned foramen magnum and cranial condyles than apes, enlarged craniodental structures and hypaxial musculature would produce higher load magnitudes on the ventral cervical spine than in humans (Fig. 5.33). Thus, intermediate spinous process lengths are entirely expected in prognathic bipeds that engage in heavy masticatory or paramasticatory activities – a contention supported by common spinous process lengths in KSD-VP-1/1 and Neandertals.

Although many features in the KSD-VP-1/1 cervical column are larger than those of humans, principal component analyses and discriminant function analyses both indicate close affinities between the vertebrae of modern humans and KSD-VP-1/1, especially at the subaxial levels. The subaxial centra are relatively shorter craniocaudally than the apes, lack a ventral keel, have low sloping uncinate processes, and exhibit a shift from the ventral axial loading regime characteristic of primates to the more vertical and dorsal cervical loading regime of orthograde bipedalism. The cervical vertebrae of KSD-VP-1/1, together with postcranial indications of a clavicular, scapular, and thoracic form similar to humans (Haile-Selassie et al. 2010), reinforce cranial evidence for *Au. afarensis* (Kimbel and Rak 2010) that imply the head carriage of australopiths greatly resembled that of modern humans. Despite differences between the cervical columns of KSD-VP-1/1 and anatomically modern humans, most of the variance may be explained by basicranial and dentognathic factors rather than differences in posture, locomotion, or head carriage.

Decoupling of the Head and Torso

The nervous system requires at least one stabilized axial segment to determine and maintain whole-body spatial orientation (Dunbar et al. 2004). In primates, either the head or the trunk can provide a stable reference for this proprioceptive signal depending on the pattern of gait. For example, during quadrupedal walking the primate trunk remains stable while the head pitches, whereas in quadrupedal running the trunk pitches on a stabilized head (Dunbar 2004; Dunbar et al. 2004, 2008). By contrast, in both bipedal running and walking the human trunk remains stable on the sagittal plane, providing the platform for movement of the head and neck (Cromwell et al. 2001). As a result, humans have uniquely derived postcranial features for trunk stability (Lieberman et al. 2006) and head and neck that are effectively "decoupled" from the trunk.

Larger and more cranial articulations of the great ape nuchal musculature (i.e., trapezius, rhomboideus) limit the independence of the head and neck segments from the trunk, and unlike humans, head position in the apes is more dependent upon cervical spine orientation (Aiello and Dean 1990; Dunbar et al. 2008; Lieberman 2011). Tall uncinate processes, acutely oriented articular facets, and long spinous processes in apes additionally limit cervical segmental motion and further "couple" head and neck movements to the torso (Clausen et al. 1997; Moskovich 2001) as indicated by the typical stiff-necked appearance of apes. Meyer (2005, 2008) posited that shallower uncovertebral joints in the cervical spine greatly increase the range of cervical rotation and lateral bending and contribute to fluid contralateral head and neck movement (decoupling) in early *Homo*. Moreover, as discussed above, obliquely oriented articular facets on the sagittal plane derived in humans allow for greater range of side-to-side movement (roll) and rotation (yaw) across cervical motion segments. Conversely, acute articular facet angles in apes act like lumbar facet angles, conferring increased stability and ventral shear resistance but severely limits roll and yaw movements.

Oblique articular facet angles in the KSD-VP-1/1 cervical column combined with shallow uncinate processes indicate that these two important components of functional morphology for independent torso and head segmental movement was in place in this large male individual of *Au. afarensis*. The range of coupled movements such as axial rotation and lateral bending in the KSD-VP-1/1 cervical spine would have been increased in this individual relative to apes. This hypothesis is supported by crania of *Au. afarensis* (particularly males) that indicate reduced nuchal musculature relative to *Pan* and *Gorilla* (Aiello and Dean 1990; Kimbel and Rak 2010), implying a degree of cranial independence from the torso in KSD-VP-1/1 *at least* intermediate to humans and the apes.

Locomotor Indications

Although shallow uncovertebral joints and obliquely oriented articular facets are part of the functional anatomy derived in genus *Homo*, it is unclear whether *Au. afarensis* had evolved the requisite nuchal musculature reduction that permits inertia of the head implemented in human running. The relatively anterior position of the foramen magnum and occipital condyles in australopiths combined with a relatively small horizontal nuchal plane with a low compound temporal/nuchal crest (Kimbel and Rak 2010) provide anatomical support for Le Gros Clark's (1947, 1950) contention that nuchal musculature in *Australopithecus* was greatly reduced, and its head carriage greatly resembled that of modern humans. The transversely large centrum surface areas, dorsal eccentricity, and short superoinferior heights of the KSD-VP-1/1 cervical centra appear to corroborate this suggestion, as the KSD-VP-1/1 cervical spine indicates a functional anatomy very similar to that of the genus *Homo*, known to be specialized for vigorous distance locomotion. However, relatively long spinous processes lengths in A.L. 333-106 and those reconstructed for the cervical vertebrae of KSD-VP-1/1 denote cross-sectional areas for the nuchal musculature intermediate between humans and African great apes. In addition, a projected 20% caudal increase in spinous process length from the C3 to C4 level would have created a mid-cervical dorsal peak characteristic of the apes not observed in humans.

Bipedalism, retrognathism, and cranial base remodeling explain reduced nuchal musculature and the commensurate reduction of the mid-cervical spinous processes in humans, as gravitational hypaxial loads and antagonistic muscular epaxial loads approach equivalency. Hominins more prognathic than anatomically modern humans would incur higher antagonistic dorsal loads on the mid-cervical vertebrae and be expected to have longer spinous processes. But the predicted lengths of the KSD-VP-1/1 spinous processes and pattern of mid-cervical dorsal increase would have effectively occupied the anatomic pathway and line of action of the nuchal ligament, resulting in either an inchoate form of nuchal ligament, or none as in the African great apes. Despite recent research on limb lengths showing that locomotor performance in *Australopithecus* may have been equivalent to that of early *Homo* (Pontzer 2012), cervical spine of KSD-VP-1/1 seems to indicate that the nuchal ligament, a significant component of the derived human axial anatomy for cursorial distance locomotion, had probably not yet evolved in early *Au. afarensis*.

Conclusion

Although the suite of characteristics in the KSD-VP-1/1 centra is consistent with human-like orthograde posture and head carriage, the mosaic of derived anatomy with transitional nuchal musculature and inchoate stage of nuchal ligament development would be consistent with the locomotor pattern seen in primates during bouts of bipedal running where the whole-body aerial phase of humans seems absent (see Demes and O'Neill 2012). The relatively small and often pathological thoracic and lumbar centra typical of *Au. afarensis* appear to corroborate the hypothesis that they were ill suited for high-peak vertical loads incurred with human-like, dynamic long-distance bipedal locomotion (Jungers 1988; Berge 1994; Bramble and Lieberman 2004).

Overall, the KSD-VP-1/1 vertebrae exhibit an axial functional structure ideally suited for a transitional locomotor phase lacking the springy gait and whole-body aerial phase of modern humans. But the "goat principle" (where behavior precedes biology) should be evoked here, as the near-human kinematic signature of the KSD-VP-1/1 centra and articular facets combined with the prevalence of load-induced pathology in australopiths might indicate that australopiths disregarded their inadequacies, and, with a somewhat compromised axial structure, intermittently engaged in aerial phase locomotion, setting the stage for subsequent selection for cursorial behavior.

Acknowledgements The author would like to extend thanks to Y. Haile-Selassie for the opportunity to study the cervical vertebrae of KSD-VP-1/1, and to Y. Haile-Selassie, B. Latimer, L. Jellema, D. Lordkipanidze, F. Thackeray, C. Menter, M. Morgan, E. Mbua, M. Leakey, and I. Tattersall for access to skeletal materials in their care. Gratitude is also extended to two anonymous reviewers for their comments, and to M. Fong and D. Lieberman for very useful discussions, and to B. McIndoe, D. Dominguez, and K. Tuosto for their dedicated assistance.

References

Abbie, A. A. (1957). Metrical characters of a Central Australian tribe. *Oceania, 27*, 220–243.

Aiello, L., & Dean, C. (1990). *An introduction to human evolutionary anatomy*. London: Academic Press.

Alemseged, Z., Spoor, F., Kimbel, W. H., Bobe, R., Geraads, D., Reed, D., & Wynn, J. G. (2006). A juvenile early hominin skeleton from Dikika, Ethiopia. *Nature, 443*, 296–301.

Ankel-Simons, F. (1983). *A survey of living primates and their anatomy*. New York: Macmillan.

Ankel-Simons, F. (2000). *Primate anatomy*. San Diego: Academic Press.

Bailey, J. F., Liebenberg, E., Fields, A. J., Mattison, J. A., Lotz, J. C., & Kramer, P. A. (2013). Relating posture to spinal osteoarthritis: Histological evidence. *American Journal of Physical Anthrolopogy, 150*, 71–72.

Barrickman, N. L. (2003). *Morphological variation of the lumbar vertebrae of A. africanus: Implications for locomotor differences between small and large individuals*. Paper presented at the AAPA 73rd Annual Meeting, Tempe, AZ.

Begun, D. R. (2004). The three "Cs" of behavioral reconstruction in fossil primates. *Journal of Human Evolution, 46*, 497–505.

Berge, C. (1994). How did the australopithecines walk? *Journal of Human Evolution, 26*, 259–273.

Berger, L. R., de Ruiter, D. J., Churchill, S. E., Schmid, P., Carlson, K. J., Dirks, P. H., & Kibii, J. M. (2010). *Australopithecus sediba*: A new species of *Homo*-like australopith from South Africa. *Science, 328*, 195–204.

Blackith, R. E., & Reyment, R. A. (1971). *Multivariate morphometrics*. London, New York: Academic Press.

Boachie-Adjei, O., & Sarwahi, V. (2003). Scheuermann's Kyphosis. In V. Arlet, A. L. Carl, & R. L. DeWald (Eds.), *Spinal deformities: The comprehensive text* (pp. 777–805). New York: Thieme.

Boszczyk, B. M., Boszczyk, A. A., & Putz, R. (2001). Comparative and functional anatomy of the mammalian lumbar spine. *Anatomical Record, 264*, 157–168.

Boule, M. (1911). L'homme fossile de La Chapelle-aux-Saints. *Annales de Paléontologie, 6*, 1–64.

Boyd-Clark, L. C., Briggs, C. A., & Galea, M. P. (2002). Muscle spindle distribution, morphology, and density in longus colli and multifidus muscles of the cervical spine. *Spine, 27*, 694–701.

Bramble, D. M. (2000). Head stabilization in human running: Implications for hominid evolution. *American Journal of Physical Anthropology, 30*, 111.

Bramble, D. M., & Lieberman, D. E. (2004). Endurance running and the evolution of *Homo*. *Nature, 432*, 345–352.

Breit, S. (2002). Functional adaptations of facet geometry in the canine thoracolumbar and lumbar spine (Th10-L6). *Annals of Anatomy, 184*, 379–385.

Breit, S., & Kunzel, W. (2002). Shape and orientation of articular facets of cervical vertebrae (C3-C7) in dogs denoting axial rotational ability: An osteological study. *European Journal of Morphology, 40*, 43–51.

Bridges, P. S. (1994). Vertebral arthritis and physical activities in the prehistoric southeastern United States. *American Journal of Physical Anthropology, 93*, 83–93.

Broom, R., & Robinson, J. T. (1949). A new type of fossil man. *Nature, 164*, 322–323.

Brown, F., Harris, J., Leakey, R., & Walker, A. (1985). Early *Homo erectus* skeleton from west Lake Turkana, Kenya. *Nature, 316*, 788–792.

Byrd, S. E., & Comiskey, E. M. (2007). Postnatal maturation and radiology of the growing spine. *Neurosurgery Clinics of North America, 18*, 431–461.

Cailliet, R. (2004). *The illustrated guide to functional anatomy of the musculoskeletal system*. Chicago: AMA Press.

Cartmill, M., & Brown, K. (2013). The relative effects of locomotion and posture on vertebral scaling. *American Journal of Physical Anthrolopogy, 150*, 95.

Cartmill, M., & Smith, F. H. (2009). *The human lineage*. Hoboken, N.J: Wiley-Blackwell.

Cartmill, M., Hylander, W. L., & Shafland, J. (1987). *Human structure*. Cambridge: Harvard University Press.

Chapman, A. E. (2008). *Biomechanical analysis of fundamental human movements*. Champaign, IL: Human Kinetics.

Clausen, J. D., Goel, V. K., Traynelis, V. C., & Scifert, J. (1997). Uncinate processes and Luschka joints influence the biomechanics of the cervical spine: Quantification using a finite element model of the C5-C6 segment. *Journal of Orthopaedic Research, 15*, 342–347.

Coblentz, B. E. (1978). The effects of feral goats (*Capra hircus*) on island ecosystems. *Biological Conservation, 13*, 279–286.

Cokluk, B., & Buyukozturk, S. (2008). Discriminant function analysis: Concept and application. *Eğitim araştırmaları dergisi, 33*, 73–92.

Cook, D. C., Buikstra, J. E., DeRousseau, C. J., & Johanson, D. C. (1983). Vertebral pathology in the Afar australopithecines. *American Journal of Physical Anthropology, 60*, 83–101.

Cromwell, R. L., Aadland-Monahan, T. K., Nelson, A. T., Stern-Sylvestre, S. M., & Seder, B. (2001). Sagittal plane analysis of head, neck, and trunk kinematics and electromyographic activity during locomotion. *Journal of Orthopaedics and Sports Physical Therapy, 31*, 255–262.

Damborg, F., Engell, V., Nielsen, J., Kyvik, K. O., Andersen, M. O., & Thomsen, K. (2011). Genetic epidemiology of Scheuermann's disease. *Acta Orthopaedica, 82*(5), 602–605.

Darroch, J. N., & Mosimann, J. E. (1985). Canonical and principle components of shape. *Biometrika, 72*, 241–252.

Dart, R. (1925). *Australopithecus africanus*. The man-ape of South Africa. *Nature, 115*, 195–199.

Day, M. H., & Leakey, R. E. (1974). New evidence of the genus *Homo* from East Rudolf, Kenya (III). *American Journal of Physical Anthrolopogy, 41*, 367–380.

Dean, M. C., & Wood, B. A. (1981). Metrical analysis of the basicranium of extand hominoids and Australopithecus. *American Journal of Physical Anthrolopogy, 54*, 63–71.

Dean, M. C., & Wood, B. A. (1982). Basicranial anatomy of Plio-Pleistocene hominids from East and South Africa. *American Journal of Physical Anthrolopogy, 59*, 157–174.

Deane, A., Organ, J. M., & Muchlinski, M. (2012). Can caudal vertebral body articular surface shape discriminate among prehensile and non-prehensile tailed anthropoids? *American Journal of Physical Anthrolopogy, 147*, 127.

Delattre, A. (1924). Essai sur l'anatomie comparee et la mechanique fonctionnelle de l'axis de Mammiferes. *Armentieres, Waterlot edition*, 129p.

Demes, B., & O'Neill, M. C. (2012). Ground reaction forces and center of mass mechanics of bipedal capuchin monkeys: Implications for the evolution of human bipedalism. *American Journal of Physical Anthropology, 150*, 76–86.

Digiovanni, B. F., Scoles, P. V., & Latimer, B. M. (1989). Anterior expansion of the thoracic vertebral bodies in Scheuermann's kyphosis: An anatomic study. *Spine, 14*, 712–716.

Dunbar, D. C. (2004). Stabilization and mobility of the head and trunk in vervet monkeys (*Cercopithecus aethiops*) during treadmill walks and gallops. *Journal of Experimental Biology, 207*, 4427–4438.

Dunbar, D. C., Badam, G. L., Hallgrimsson, B., & Vieilledent, S. (2004). Stabilization and mobility of the head and trunk in wild monkeys during terrestrial and flat-surface walks and gallops. *Journal of Experimental Biology, 207*, 1027–1042.

Dunbar, D. C., Macpherson, J. M., Simmons, R. W., & Zarcades, A. (2008). Stabilization and mobility of the head, neck and trunk in horses during overground locomotion: Comparisons with humans and other primates. *Journal of Experimental Biology, 211*, 3889–3907.

Dunsworth, H., Challis, J. H., & Walker, A. (2004). Throwing and bipedalism: A new look at an old idea. In J. L. Franzen, M. Dohler, & S. Moya-Sola (Eds.), *Upright walking*. Senckenberg Institute: Frankfurt.

Elliott, H. C. (1945). Cross-sectional diameters and areas of the human spinal cord. *The Anatomical Record, 93*, 287–293.

Farfan, H. F., Cossette, J. W., Robertson, G. H., Wells, R. V., & Kraus, H. (1970). The effects of torsion on the lumbar intervertebral joints: The role of torsion in the production of disc degeneration. *The Journal of Bone and Joint Surgery, 52*, 468–497.

Fountas, K. N., Kapsalaki, E. Z., Jackson, J., Vogel, R. L., & Robinson, J. S. J. (1998). Cervical spinal cord—smaller than considered? *Spine, 23*, 1513–1516.

Fourie, S. L., & Kirberger, R. M. (1999). Relationship of cervical spinal cord diameter to vertebral dimensions: A radiographic study of normal dogs. *Veterinary Radiology and Ultrasound, 40*, 137–143.

Gebo, D. (1989). Postcranial adaptation and evolution in Lorisidae. *Primates, 30*, 347–367.

Gogen, A., Huang, Y., Sarmiento, E. E., Gidansky, I. H., & Marquez, S. (2007). The relative contribution of intervertebral disc to the curvatures of the vertebral column. *FASEB Journal, 21*, A969-d-970.

Gomez-Olivencia, A., Carretero, J. M., Arsuaga, J. L., Rodriguez-Garcia, L., Garcia-Gonzalez, R., & Martinez, I. (2007). Metric and morphological study of the upper cervical spine from the Sima de los Huesos site (Sierra de Atapuerca, Burgos, Spain). *Journal of Human Evolution, 53*, 6–25.

Gommery, D. (2000). Superior cervical vertebrae of a Miocene Hominoid and a Plio-Pleistocene hominid from southern Africa. *Palaeontologia Africana, 36*, 139–145.

Gordon, A. D., Green, D. J., & Richmond, B. G. (2008). Strong postcranial size dimorphism in *Australopithecus afarensis*: Results from two new resampling methods for multivariate data sets with missing data. *American Journal of Physical Anthropology, 135*, 311–328.

Gracovetsky, S. A. (1996). Function of the spine from an evolutionary perspective *The Lumbar Spine* (2nd ed., pp. 259–269). Philadelphia: W.B. Saunders Company.

Graf, W., de Waele, C., & Vidal, P. P. (1995). Functional anatomy of the head-neck movement system of quadrupedal and bipedal mammals. *Journal of Anatomy, 186*, 55–74.

Grave, B., Brown, T., & Townsend, G. (1999). Comparison of cervicovertebral dimensions in Australian Aborigines and Caucasians. *European Journal of Orthodontics, 21*, 127–135.

Haeusler, M., Martelli, S. A., & Boeni, T. (2002). Vertebrae numbers of the early hominid lumbar spine. *Journal of Human Evolution, 43*, 621–643.

Haile-Selassie, Y., Latimer, B. M., Alene, M., Deino, A. L., Gibert, L., & Melillo, S. M., et al. (2010). An early *Australopithecus afarensis* postcranium from Woranso-Mille, Ethiopia. *Proceedings of the National Academy of Sciences U.S.A, 107*, 12121–12126.

Hardy, A. (1960). Was man more aquatic in the past? *New Scientist, 7*, 642–645.

Harrison, D. D., Janik, T. J., Troyanovich, S. J., & Holland, B. (1996). Comparisons of lordotic cervical spine curvatures to a theoretical ideal model of the static sagittal cervical spine. *Spine, 21*, 667–675.

Harrison, D. D., Harrison, D. E., Janik, T. J., Cailliet, R., Ferrantelli, J. R., Haas, J. W., & Holland, B. (2004). Modeling of the sagittal cervical spine as a method to discriminate hypolordosis: Results of elliptical and circular modeling in 72 asymptomatic subjects, 52 acute neck pain subjects, and 70 chronic neck pain subjects. *Spine, 29*, 2485–2492.

Hausler, M., & Berger, L. (2001). Stw 441/465: A new fragmentary ilium of a small-bodied *Australopithecus africanus* from Sterkfontein, South Africa. *Journal of Human Evolution, 40*, 411–417.

Healy, M. J. R., & Tanner, J. M. (1981). Size and shape in relation to growth and form. *Symposia of the Zoological Society of London, 46*, 19–35.

Hession, P. R., & Butt, W. P. (1996). Imaging of spondylolysis and spondylolisthesis. *European Radiology, 6*, 284–290.

Hinck, V. C., Hopkins, C. E., & Clark, W. M. (1966). Sagittal diameter of the lumbar spinal canal in children and adults. *Radiology*, 929–937.

Hirasaki, E., Moore, S. T., Raphan, T., & Cohen, B. (1999). Effects of walking velocity on vertical head and body movements during locomotion. *Experimental Brain Research, 127*, 117–130.

Hoffmann, F. (2006). *The influence of sitting position on (ventro-) dorsal head translation.* (Master Thesis), University Hospital of Neurology Innsbruck, Innsbruck.

Huggare, J. (1992). Population differences in the morphology of the first cervical vertebra. *American Journal of Physical Anthropology, 88*, 197–201.

Huggare, J. A., & Raustia, A. M. (1992). Head posture and cervicovertebral and craniofacial morphology in patients with craniomandibular dysfunction. *Cranio, 10*, 173–177; discussion 178–179.

Humzah, M. D., & Soames, R. W. (1988). Human intervertebral disc: Structure and function. *Anatomical Record, 220*, 337–356.

Iai, H., Moriya, H., Goto, S., Takahashi, K., Yamagata, M., & Tamaki, T. (1993). Three-dimensional motion analysis of the upper cervical spine during axial rotation. *Spine, 18*, 2388–2392.

James, F. C., & McCulloch, C. E. (1990). Multivariate analysis in ecology and systematics: Panacea or Pandora's Box? *Annual Review of Ecology and Systematics, 21*, 129–166.

Jeyapalan, K., & Chavda, S. V. (1994). Case report 868. *Skeletal Radiology, 23*, 580–582.

Johnson, G. M. (1998). The correlation between surface measurement of head and neck posture and the anatomic position of the upper cervical vertebrae. *Spine, 23*, 921–927.

Jolicoeur, P., & Mosimann, J. E. (1960). Size and shape variation in the painted turtle: A principal component analysis. *Growth, 7*, 179–188.

Jungers, W. L. (1988). Relative joint size and hominid locomotor adaptations with implications for the evolution of hominid bipedalism. *Journal of Human Evolution, 17*, 247–265.

Jungers, W., Falsetti, A., & Wall, C. E. (1995). Shape, relative size, and size-adjustments in morphometrics. *Yearbook of Physical Anthropology, 38*, 137–161.

Jurmain, R. (1989). Trauma, degenerative disease, and other pathologies among the Gombe chimpanzees. *American Journal of Physical Anthropology, 80*, 229–237.

Jurmain, R. (2000). Degenerative joint disease in African great apes: An evolutionary perspective. *Journal of Human Evolution, 39*, 185–203.

Kameyama, T., Hashizume, Y., Ando, T., & Takahashi, A. (1994). Morphometry of the normal cadaveric cervical spinal cord. *Spine, 19*, 2077–2081.

Kameyama, T., Hashizume, Y., & Sobue, G. (1996). Morphologic features of the normal human cadaveric spinal cord. *Spine, 21*, 1285–1290.

Kikuchi, Y., Nakano, Y., Nakatsukasa, M., Kunimatsu, Y., Shimizu, D., Ogihara, N., & Ishida, H. (2012). Functional morphology and anatomy of cervical vertebrae in Nacholapithecus kerioi, a middle Miocene hominoid from Kenya. *Journal of Human Evolution, 62*, 677–695.

Kimbel, W. H., & Delezene, L. K. (2009). "Lucy" redux: A review of research on *Australopithecus afarensis*. *American Journal of Physical Anthropology, 140*, 2–48.

Kimbel, W. H., & Rak, Y. (2010). The cranial base of *Australopithecus afarensis*: New insights from the female skull. *Philosophical Transactions of the Royal Society of London B, Biological Sciences, 365*, 3365–3376.

Kimbel, W. H., White, T. D., & Johanson, D. C. (1984). Cranial morphology of *Australopithecus afarensis*: A comparative study based on a composite reconstruction of the adult skull. *American Journal of Physical Anthropology, 64*, 337–388.

Kimbel, W. H., Johanson, D. C., & Rak, Y. (1994). The first skull and other new discoveries of *Australopithecus afarensis* at Hadar, Ethiopia. *Nature, 368*, 449–451.

Kirpalani, D., & Mitra, R. (2008). Cervical facet joint dysfunction: A review. *Archives of Physical Medicine and Rehabilitation, 89*, 770–774.

Klein, T. J., & Lewis, M. A. (2012). A physical model of sensorimotor interactions during locomotion. *Journal of Neural Engineering, 9*, 046011.

Krogman, W. M. (1951). The scars of human evolution. *Scientific American, 185*, 54–57.

Lang, B. (1972). *Bewegungsmessungen an der Wirbelsaeule von Hund und Katze*. Giessen: Justus-Liebig University.

Latimer, B., & Ohman, J. C. (2001). Axial displasia in *Homo erectus*. *Journal of Human Evolution, 40*, A12.

Latimer, B., & Ward, C. (1993). The thoracic and lumbar vertebrae. In A. Walker & R. Leakey (Eds.), *The Nariokotome Homo erectus skeleton* (pp. 266–293). Cambridge: Harvard University Press.

Le Gros Clark, W. E. (1947). Observations on the anatomy of the fossil Australopithecinae. *Journal of Anatomy, 81*, 300–333.

Le Gros Clark, W. E. (1950). New palaeontological evidence bearing on the evolution of the Hominoidea. *Quarterly Journal of the Zoological Society, 105*, 225–264.

Le Gros Clark, W. E. (1954). Reason and fallacy in the study of fossil man. *Advancement of Science London, 43*, 1–13.

Lieberman, D. (2011). *The evolution of the human head.* Cambridge, MA: Belknap Press of Harvard University Press.

Lieberman, D. E., Raichlen, D. A., Pontzer, H., Bramble, D. M., & Cutright-Smith, E. (2006). The human gluteus maximus and its role in running. *Journal of Experimental Biology, 209*, 2143–2155.

Lleonart, J., Salat, J., & Torres, G. J. (2000). Removing allometric effects of body size in morphological analysis. *Journal of Theoretical Biology, 205*, 85–93.

Lockwood, C. A., Richmond, B. G., Jungers, W. L., & Kimbel, W. H. (1996). Randomization procedures and sexual dimorphism in *Australopithecus afarensis*. *Journal of Human Evolution, 31*, 537–548.

Lordkipanidze, D., Jashashvili, T., Vekua, A., Ponce de Leon, M. S., Zollikofer, C. P., Rightmire, G. P., & Rook, L. (2007). Postcranial evidence from early Homo from Dmanisi, Georgia. *Nature, 449*, 305–310.

Lovejoy, C. O., Johanson, D. C., & Coppens, Y. (1982). Elements of the axial skeleton recovered from the Hadar formation: 1974–1977 collections. *American Journal of Physical Anthropology, 57*, 631–635.

Lovell, N. C. (1990). Skeletal and dental pathology of free-ranging mountain gorillas. *American Journal of Physical Anthropology, 81*, 399–412.

Lu, C. D. (1988). Grazing behavior and diet selection of goats. *Small Ruminant Research, 1*, 205–216.

MacLarnon, A. (1987). *Size Relationships and the spinal cord and associated skeleton in primates.* University of London.

MacLarnon, A. (1993). The vertebral canal. In A. Walker & R. Leakey (Eds.), *The Nariokotome Homo erectus skeleton* (pp. 359–390). Berlin: Springer.

MacLarnon, A. M., & Hewitt, G. P. (1999). The evolution of human speech: The role of enhanced breathing control. *American Journal of Physical Anthropology, 109*, 341–363.

Maurel, N., Lavaste, F., & Skalli, W. (1997). A three-dimensional parameterized finite element model of the lower cervical spine. Study of the influence of the posterior articular facets. *Journal of Biomechanics, 30*, 921–931.

McHenry, H. M. (1992). Body size and proportions in early hominids. *American Journal of Physical Anthropology, 87*, 407–431.

Mercer, S. R., & Bogduk, N. (2003). Clinical anatomy of ligamentum nuchae. *Clinical Anatomy, 16*, 484–493.

Meyer, M. R. (2003). Vertebrae and language ability in early hominids. *PaleoAnthropology, 1*, 20–21.

Meyer, M. R. (2005). *Functional Anatomy of the Homo erectus axial Skeleton from Dmanisi, Georgia.* (Doctoral Dissertation), University of Pennsylvania, Philadelphia.

Meyer, M. R. (2008). Skeletal indications for distance locomotion in early *Homo erectus*. *American Journal of Physical Anthropology, 135*, 155.

Meyer, M. R. (2012). Functional anatomy of the thoracic vertebrae in early *Homo*. *American Journal of Physical Anthrolopogy, Supplement*, 214.

Morgan, E. (2002). Was man more aquatic in the past. What happens when you change the paradigm? *Nutrition and Health, 16*, 23–24.

Morgan, J., Ljunggren, G., & Read, R. (1967). Spondylosis deformans (vertebral osteophytosis) in the dog. *Journal of Small Animal Practices, 8*, 57.

Mosimann, J. E. (1970). Size allometry: Size and shape variables with characterizations of the lognormal and generalized gamma distributions. *Journal of the American Statistical Association, 65*, 930–945.

Moskovich, R. (2001). Biomechanics of the cervical spine. In M. Nordin & V. H. Frankel (Eds.), *Basic biomechanics of the musculoskeletal system* (3rd ed., pp. 286–317). Philadelphia: Lippincott Williams & Wilkins.

Nalley, T. K. (2013). Suspensory behaviors and the neck: A comparative analysis of the cervical vertebrae of extant primates. *American Journal of Physical Anthrolopogy, 150*, 205.

Napier, J. R. (1959). Fossil metacarpals from Swartkrans. *Fossil Mammals of Africa, 17*, 1–18.

Nathan, H. (1962). Osteophytes of the vertebral column. *Journal of Bone and Joint Surgery, 44-A*, 243–268.

Nathan, H. (1968). Compression of the sympathetic trunk by osteophytes of the vertebral column in the abdomen: An anatomical study with pathological and clinical considerations. *Surgery, 63*, 609–625.

Nevell, L., & Wood, B. (2008). Cranial base evolution within the hominin clade. *Journal of Anatomy, 212*, 455–468.

Nichols, K. A., & Zilhlman, A. L. (2002). Skeletal and dental evidence of aging in captive western lowland gorillas: A preliminary report. In J.M. Erwin & P.R. Hof (Eds.), *Aging in nonhuman primates* (Vol. 31, pp. 22–31). Basel: Karger.

Niemitz, C. (2002). A theory on the evolution of the habitual orthograde human bipedalism—the "Amphibische Generalistentheorie". *Anthropologischer Anzeiger, 60*, 3–66.

Nordin, M., & Weiner, S. S. (2001). Biomechanics of the lumbar spine. In M. Nordin & V. H. Frankel (Eds.), *Basic biomechanics of the musculoskeletal system* (3rd ed., pp. 256–285). Philadelphia: Lippincott Williams & Wilkins.

Nordquist, L. (1964). The sagittal diameter of the spinal cord and subaracnoid space in different age groups. *Acta Radiologica, 227*, 1–96.

O'Higgins, P., Johnson, D. R., & Paxton, S. K. (1989). The relationship between age, size and shape of mouse thoracic vertebrae: A scanning electron microscopic study. *Journal of Anatomy, 163*, 57–66.

O'Higgins, P., Milne, N., Johnson, D. R., Runnion, C. K., & Oxnard, C. E. (1997). Adaptation in the vertebral column: A comparative study of patterns of metameric variation in mice and men. *Journal of Anatomy, 190*, 105–113.

Ogden, J. A., & Ganey, T. M. (2003). Development and maturation of the spine and spinal cord. In V. Arlet, A. L. Carl, & R. L. DeWald (Eds.), *Spinal deformities: The comprehensive text* (pp. 54–70). New York: Thieme.

Onan, O. A., Heggeness, M. H., & Hipp, J. A. (1998). A motion analysis of the cervical facet joint. *Spine, 23*, 430–439.

Ostrofsky, K. R., & Churchill, S. E. (2013). Sex determination by discriminant function analysis of lumbar vertebrae. *American Journal of Physical Anthrolopogy, 150*, 212.

Palmqvist, P., Arribas, A., & Martínez-Navarro, B. (1999). Ecomorphological study of large canids from the lower Pleistocene of southeastern Spain. *Lethaia, 32*, 75–88.

Patla, A. E., Adkin, A., & Ballard, T. (1999). Online steering: Coordination and control of body center of mass, head and body reorientation. *Experimental Brain Research, 129*, 629–634.

Penning, L., & Wilmink, J. T. (1987). Rotation of the cervical spine. A CT study in normal subjects. *Spine, 12*, 732–738.

Perez, P. J., Gracia, A., Martinez, I., & Arsuaga, J. L. (1997). Paleopathological evidence of the cranial remains from the Sima de los Huesos Middle Pleistocene site (Sierra de Atapuerca, Spain). Description and preliminary inferences. *Journal of Human Evolution, 33*, 409–421.

Pontzer, H. (2012). Ecological energetics in early homo. *Current Anthropology, 53*, S346–S358.

Pope, M. H. (2001). Cervical spine biomechanics. In M. Szpalski & R. Gunzburg (Eds.), *The degenerative cervical spine* (pp. 29–37). Philadelphia: Lippincott Williams & Wilkins.

Robinson, J. T. (1972). *Early hominid posture and locomotion.* Chicago: University of Chicago Press.

Romanes, G. J. (1964). Motor pools of the spinal cord. In J. C. Eccles & J. P. Schade (Eds.), *Organization of the spinal cord* (pp. 93–119). Elsevier.

Rose, M. D. (1984). A hominine hip bone, KNM-ER 3228, from East Lake Turkana, Kenya. *American Journal of Physical Anthropology, 63*, 371–378.

Rummel, J., Blum, Y., & Seyfarth, A. (2010). Robust and efficient walking with spring-like legs. *Bioinspiration and Biomimetics, 5*, 046004.

Rydevik, B., Brisby, H., & Olmarker, K. (2001). Where does the pain come from? In M. Szpalski & R. Gunzburg (Eds.), *The degenerative cervical spine* (pp. 39–43). Philadelphia: Lippincott Williams & Wilkins.

Sanders, W. J. (1990). Weight transmission through the lumbar vertebrae and sacrum in australopithecines. *American Journal of Physical Anthropology, 81*, 289.

Sanders, W. J. (1995). *Function, Allometry, and Evolution of the Australopithecine Lower Precaudal Spine.* (PhD Dissertation), New York University.

Sanders, W. J. (1996). Implications of structure and scaling of catarrhine lumbosacral vertebrae for australopithecine positional behavior. *American Journal of Physical Anthropology, 22*, 206–207.

Sanders, W. J. (1998). Comparative morphometric study of the australopithecine vertebral series Stw-H8/H41. *Journal of Human Evolution, 34*, 249–302.

Scheuermann, H. W. (1920). Kyphosis dorsalis juvenilis. *Ugeskrift for Læger, Copenhagen, 82*, 385–393.

Scheuermann, H. W. (1921). Kyphosis dorsalis juvenilis. *Zeitschrift fur Orthopädische und Chiropraktik, 41*, 4–11.

Schiess, R., & Haeusler, M. (2013). No skeletal dysplasia in the Nariokotome boy KNM-WT 15000 (homo erectus)-A reassessment of congenital pathologies of the vertebral column. *American Journal of Physical Anthropology, 150*, 365–374.

Schmitt, D. (2003). Insights into the evolution of human bipedalism from experimental studies of humans and other primates. *Journal of Experimental Biology, 206*, 1437–1448. doi:10.1242/jeb.00279.

Schultz, A. H. (1930). The skeleton of the trunk and limbs of higher primates. *Human Biology, 3*, 303.

Schultz, A. H. (1955). The position of the occipital condyles and of the face relative to the skull base in primates. *American Journal of Physical Anthropology, 13*, 97–120.

Sellers, W. I., Dennis, L. A., W, J. W., & Crompton, R. H. (2004). Evaluating alternative gait strategies using evolutionary robotics. *Journal of Anatomy, 204*, 343–351.

Shapiro, I., & Frankel, V. H. (1989). Biomechanics of the cervical spine. In M. Nordin & V. H. Frankel (Eds.), *Basic biomechanics of the musculoskeletal system* (pp. 209–224). Philadelphia: Lea & Febiger.

Shea, B. T. (1985). Bivariate and multivariate growth allometry: Statistical and biological considerations. *Journal of Zoology London A, 206*, 367–390.

Sherman, J. L., Nassaux, P. Y., & Citrin, C. M. (1990). Measurements of the normal cervical spinal cord on MR imaging. *American Journal of Neuroradiology, 11*, 369–372.

Shi, C., Nishizawa, S., Adachi, K., & Endo, B. (1995). A comparative morphological analysis of vertebral bodies in humans and some other mammals. *Anthropological Science*, 467–484.

Shi, C., Nishizawa, S., Adachi, K., & T., K. (1999). Morphological comparison of the vertebrae in humans and some other mammals. *Anthropological Science*, 3–19.

Shimoda, Y., Nagaoka, T., Moromizato, K., Sunagawa, M., Hanihara, T., Yoneda, M., & Ishida, H. (2011). *Degenerative changes of the spine in people from prehistoric Okhotsk culture and two ancient human groups from Kanto and Okinawa*. Japan: Anthropological Science (Advance Publication). 21.

Sinclair, D. C. (1969). *Human growth after birth*. London: Oxford University Press.

Smith, C. J. (1960). Osteoarthritis: Degenerative joint disease. *Annals of Internal Medicine, 53*, 88–95.

Snodgrass, J. J., & Galloway, A. (2001). Lumbar vertebral dimensions as a function of age, sex, and ancestry. *Proceedings of the American Academy of Forensic Sciences, 7*, 275.

Solow, B., & Tallgren, A. (1976). Head posture and craniofacial morphology. *American Journal of Physical Anthropology, 44*, 417–435.

Standring, S. (2008). *Gray's anatomy. The anatomical basis of clinical practice*. Edinburgh: Churchill Livingstone Elsevier.

Stanley, J. H., Schabel, S. I., Frey, G. D., & Hungerford, G. D. (1986). Quantitative analysis of the cervical spinal canal by computed tomography. *Neuroradiology, 28*, 139–143.

Storvik, S. G., & Stemper, B. D. (2011). Axial head rotation increases facet joint capsular ligament strains in automotive rear impact. *Medical and Biological Engineering and Computing, 49*, 153–161.

Straus, W., & Wislocki, G. (1932). On certain similarities between sloths and slow lemurs. *Bulltin of the Museum of Comparative Zoology (Harvard), 74*, 45–56.

Straus, W. L., & Cave, A. J. E. (1957). Pathology and the posture of Neanderthal Man. *Quarterly Review of Biology, 32*, 348–363.

Tatarek, N. E. (2005). Variation in the human cervical neural canal. *Spine, 5*, 623–631.

Thompson, D. A. W. (1961). *On growth and form (An* (abridged ed.)). Cambridge: University Press.

Trinkaus, E. (1986). The Neandertals and modern human origins. *Annual Review of Anthropology, 15*, 193–218.

Trinkaus, E. (1987). Bodies, brawn, brains and noses: Human ancestors and human predation. In M. H. Nitecki & D. V. Nitecki (Eds.), *The evolution of human hunting* (pp. 107–145). New York: Plenum Press.

Vallois, H. V. (1937). La duree de la vie chez l'homme fossile. *Anthropologie, Paris, 47*, 499–532.

Van der Merwe, A. E., Iscan, M. Y., & L'Abbè, E. N. (2006). The pattern of vertebral osteophyte development in a South African population. *International Journal of Osteoarchaeology, 16*, 459–464.

van Roy, P., Barbaix, E., & Clarijs, J. P. (2001). Functional anatomy of the cervical spine. In M. Szpalski & R. Gunzburg (Eds.), *The degenerative cervical spine* (pp. 3–27). Philadelphia: Lippincott Williams & Wilkins.

Vidal, P. P., Graf, W., & Berthoz, A. (1986). The orientation of the cervical vertebral column in unrestrained awake animals. I. Resting position. *Experimental Brain Research, 61*, 549–559.

Walker, A. (1993). Perspectives on the Nariokotome discovery. In A. Walker & R. Leakey (Eds.), *The Nariokotome Homo erectus skeleton* (pp. 411–430). Cambridge, MA: Harvard University Press.

Walker, A., & Leakey, R. E. (1993). The postcranial bones. In A. Walker & R. E. Leakey (Eds.), *The Nariokotome Homo erectus Skeleton* (pp. 95–160). Cambridge, MA: Harvard University Press.

Ward, C. V. (1991). *Functional Anatomy of the Lower Back and Pelvis of the Miocene Hominoid Proconsul nyanzae from Mfangano Island*. Baltimore, MD, Kenya: Johns Hopkins University.

Ward, C. V. (2002). Interpreting the posture and locomotion of *Australopithecus afarensis*: Where do we stand? *American Journal of Physical Anthropology, Suppl, 35*, 185–215.

Weidenreich, F. (1943). The skull of *Sinanthropus pekinensis*. *Palaenotologica Sinica New Series, D10*, 1–484.

Wescott, D. J. (2000). Sex variation in the second cervical vertebra. *Journal of Forensic Science, 45*, 462–466.

Westenhöfer, M. (1942). *Der Eigenweg des Menschen*. Berlin: Mannstaedt & Co.

Wilke, H. J., Kettler, A., Wenger, K. H., & Claes, L. E. (1997). Anatomy of the sheep spine and its comparison to the human spine. *Anatomical Record, 247*, 542–555.

Williams, S. A. (2011). *Evolution of the Hominoid Vertebral Column*. (PhD. Doctoral Dissertation), University of Illinois at Urbana-Champaign, Urbana, Illinois.

Wilson, V. J., Schor, R. H., Suzuki, I., & Park, B. R. (1986). Spatial organization of neck and vestibular reflexes acting on the forelimbs of the decerebrate cat. *Journal of Neurophysiology, 55*, 514–526.

Xiang, Y., Yakushin, S. B., Kunin, M., Raphan, T., & Cohen, B. (2008). Head stabilization by vestibulocollic reflexes during quadrupedal locomotion in monkey. *Journal of Neurophysiology, 100*, 763–780.

Yamamoto, I., Panjabi, M. M., Crisco, T., & Oxland, T. (1989). Three-dimensional movements of the whole lumbar spine and lumbosacral joint. *Spine, 14*, 1256–1260.

Chapter 6
The Shoulder Girdle of KSD-VP-1/1

Stephanie M. Melillo

Abstract Extant humans and non-human apes differ markedly in shoulder girdle anatomy. Our understanding of the evolutionary history of this region was previously limited by poor fossil preservation, but over the past decade a number of impressively complete scapulae and clavicles have been described for the genus *Australopithecus*. However, independent analyses have reached different conclusions regarding the morphological affinity of each specimen and the degree of difference among specimens. This study provides a more detailed comparative description of the KSD-VP-1/1 scapula and clavicle, which constitute the oldest substantial evidence of hominin shoulder girdle anatomy currently known. The results suggest that the adult *Australopithecus afarensis* scapula is morphologically distinct, but more similar to that of modern humans than previously recognized. Some aspects of clavicle morphology are similar to non-human apes, but are also variably present in Pleistocene hominins. If comparable methodology is employed, no difference exists among *Australopithecus* specimens. When this morphology is considered with reference to a parsimony-based model of the chimpanzee–human last common ancestor, the adult *Australopithecus* shoulder girdle is derived toward morphology associated with emphasis on a manipulatory function of the pendant upper limb.

Keywords Scapula • Clavicle • *Australopithecus* • Geometric morphometrics

S.M. Melillo (✉)
Department of Human Evolution, Max Planck Institute for Evolutionary Anthropology, Deutscher Platz 6, 04103 Leipzig, Germany
e-mail: stephanie_melillo@eva.mpg.de

Introduction

The shoulder girdle (scapula and clavicle) is the skeletal link between the upper limb and thorax. It consequently plays a central role in locomotion for most primates. This is particularly true of hominoids, a group distinguished by shared adaptation to overhead use of the upper limb during arboreal locomotion (Mivart 1867; Frey 1923; Gregory 1928; Schultz 1930, 1936; Miller 1932; Ashton and Oxnard 1963; Roberts 1974; Etter 1984; Hunt 1991a, b; Larson 1993; Gebo 1996; Ward 2007). However, humans are unique among hominoids in lacking a number of these features (Table 6.1).

The shoulder girdle ceased functioning in a propulsive and weight-bearing manner with the evolution of obligate bipedalism in the hominin lineage. In modern humans, it functions instead to position the upper limb for myriad non-locomotor tasks (Dempster 1965; Itoi et al. 2009) and it is generally recognized that the emphasis is on use of the arms and hands below shoulder level (Inman et al. 1944; Roberts 1974). Thus, the conventional understanding is that overhead adaptation was lost over the course of human evolution as the function of the upper limb became purely manipulatory (Aiello and Dean 1990). Such functions are central to multiple hypotheses of the evolution of bipedalism (Darwin 1871; Lovejoy 1981). Unique features of the human shoulder girdle have also figured into proposals of adaptation to endurance running and high-speed throwing in *Homo erectus* (Bramble and Lieberman 2004; Roach et al. 2013).

Given the functional and morphological differences between humans and non-human apes, it is not surprising that shoulder girdle morphology has been a focal point in the debate over the locomotion and positional behavior of *Australopithecus afarensis* (Ward 2002, 2013; Larson 2013). Although epigenetically responsive and uniquely derived features of the lower limb and pelvis provide strong evidence that terrestrial bipedalism was a major component of *Au. afarensis* positional behavior (Parsons 1914; LeGros Clark 1947; Tardieu 1981; Shefelbine et al. 2002; Lovejoy 2005a, b),

Yohannes Haile-Selassie and Denise F. Su (eds.), *The Postcranial Anatomy of Australopithecus afarensis: New Insights from KSD-VP-1/1*, Vertebrate Paleobiology and Paleoanthropology, DOI 10.1007/978-94-017-7429-1_6

Table 6.1 Features of the primate shoulder girdle thought to reflect adaptation to overhead use of the upper limb, and their expression in extant hominoids and quadrupedal cercopithecids

	Homo sapiens	*Pan*	*Gorilla*	*Pongo*	*Hylobates*	Quadrupedal cercopithecids
Girdle position	Low	High	High	High	High	Low
Shoulder breadth	Wide	Intermediate	Wide	Intermediate	Intermediate	Narrow
Scapula position	Dorsal	Dorsal	Dorsal	Dorsal	Dorsal	Lateral
Orientation of scapular spine[a]	Transverse	Cranial	Cranial	Intermediate	Cranial	Transverse
Extent of muscle attachment along scapular spine	Medial	Lateral	Lateral	Variable	Lateral	Medial
Size and projection of the acromion process	Large and projecting	Large and projecting	Large and projecting	Intermediate and projecting	Intermediate and Projecting	Small and not projecting
Orientation of glenoid[a]	Lateral	Cranial	Cranial	Cranial	Cranial	Lateral
Glenoid shape	Ovate to round[b]	Round	Round	Ovate	Round	Narrow
Infraspinous fossa shape[c]	Broad	Narrow	Narrow	Narrow	Very narrow	Very narrow
Clavicle length[d]	Long	Intermediate[e]	Intermediate	Long	Long	Short
Clavicle transverse curvature[f]	Internal and external	Internal and external	Primarily external	Variable[g]	Primarily internal	Primarily external
Clavicle coronal curvature[h]	Variable[i]	Superior and inferior	Variable[j]	Superior and inferior	Primarily superior	Primarily inferior

Extant non-human apes share a number of features that are derived relative to the hypothesized primitive condition, commonly modeled after quadrupedal cercopithecids. Humans lack a number of these features

[a]Relative to the axillary border

[b]Glenoid shape varies among modern human populations. Indigenous populations from sub-Saharan Africa exhibit a more ovate glenoid on average than populations from Europe (Vallois 1932). More than half of the humans in this sample come from the archaeological collection curated by the Iziko Museum in Cape Town, South Africa

[c]The craniocaudal dimension of the fossa relative to size. Broca (1878), Frey (1923), Schultz (1930) and others define the craniocaudal dimension as breadth and the ML dimension as length. This is intuitive with reference to the scapula of quadrupedal monkeys, because the ML dimension is the larger of the two. Although the opposite is true of the hominoid scapula, the terminology was maintained to avoid confusion about which dimensions are anatomical homologous in comparative studies including monkeys and apes. This convention has persisted in more recent studies considering only hominoid morphology (e.g. Green and Alemseged 2012). The use of the terms "broad" and "narrow" are retained to refer to craniocaudal dimensions of the blade, for the sake of consistency

[d]Relative to trunk length

[e]The clavicle of *Pan paniscus* is a considerably shorter clavicle than that of *Pan troglodytes*

[f]The curvature of the clavicle in the transverse plane, visible in superior or inferior view. Among primates, the medial portion of the shaft varies from straight to internally curved (i.e., curving toward the body) whereas the lateral portion varies from straight to externally curved (i.e., curving away from the body)

[g]*Pongo* clavicles are variable in transverse curvature. Some specimens closely resemble hylobatids whereas others exhibit both internal and external curves, similar to *Pan*

[h]The curvature of the clavicle in the coronal plane, best seen in posterior view. The medial portion of the shaft varies from inferiorly to superiorly curved, whereas the lateral portion varies from straight to inferiorly curved

[i]Most modern human clavicles lack pronounced coronal curvature, but there is considerable variation. Olivier (1951c, 1956) created a qualitative classification: Type I clavicles exhibit one long inferior curve, Type II clavicles exhibit both superior and inferior curves, and Type III clavicles exhibit a superior curve laterally

[j]*Gorilla* clavicles may exhibit a superior curve medially whereas others are straight. The inferior curve is consistently present

no consensus exists with reference to the importance of arboreal locomotion. Many authors have called attention to the presence of primitive features of the *Au. afarensis* upper limb (Senut 1980; Stern and Susman 1983; Susman et al. 1984; Stern 2000; Green and Alemseged 2012; Churchill et al. 2013). Ward (2002) suggests that the existence of morphological stasis over long periods is the strongest support for the hypothesis that primitive features were retained in *Au. afarensis* due to continued selection for effective arboreal locomotion. In a recent review of the evolutionary history of the shoulder, Larson (2013) suggests that, "Of all the regions of the early hominin upper limb, the shoulder perhaps displays the largest number of primitive features" (p. 257).

Prior to 2006, the anatomy of the early hominin shoulder girdle was not well known. Our understanding was based on a small number of fragmentary specimens, most notably A.L. 288-1 l, the scapula of the "Lucy" partial skeleton, and Sts 7, a similarly preserved specimen attributed to *Au. africanus* (Broom and Robinson 1950; Campbell 1966; Oxnard 1968a; Ciochon and Corruccini 1976; Vallois 1977; Vrba 1979; Johanson et al. 1982; Stern and Susman 1983; Stern 2000). These specimens are sufficiently preserved to measure the dimensions of the glenoid fossa and the angle between the glenoid and "ventral bar,[1]" or bar-glenoid angle. Stern and Susman (1983) reported that both specimens lack the laterally facing glenoid that characterizes modern humans, but Mensforth et al. (1990) and Inouye and Shea (1997) suggested that allometry may account for the cranial orientation of the A.L. 288-1 l glenoid.

A fairly complete *Au. afarensis* clavicle (A.L. 333x-6/9) was described by Lovejoy et al. (1982), although the specimen is primarily known from a qualitative comparison conducted by Ohman (1986) rather than quantitative analysis (but see Barker 2006). Ohman (1986) suggests that the lack of pronounced shaft curvature in the coronal plane (visible in anterior or posterior view) in conjunction with the anterior position of the deltoid muscle origin indicates a low position of the shoulder girdle in *Au. afarensis*. In contrast, Larson (2007, 2013) draws on the comparative study of Voisin (2006a) and concludes that A.L. 333x-6/9 retains primitive coronal curvature and thus a high shoulder girdle position. None of these studies presents quantitative data on shaft curvature of *Australopithecus* clavicles.

Three relatively new discoveries preserve formerly unknown anatomical regions of the hominin shoulder and expand the fossil sample both chronologically and ontogenetically. The *Kadanuumuu* partial skeleton (KSD-VP-1/1) from Woranso-Mille, Ethiopia is attributed to *Au. afarensis*. Dated to 3.56-3.54 Ma (Saylor et al. 2016), the clavicle (KSD-VP-1/1f) and scapula (KSD-VP-1/1g) provide the oldest substantial[2] evidence of hominin shoulder girdle anatomy (Haile-Selassie et al. 2010). The *Selam* partial skeleton (DIK-1-1) is likewise attributed to *Au. afarensis* and is dated to 3.3 Ma (Alemseged et al. 2006). The scapulae of this young juvenile are nearly completely preserved, though they lack secondary centers of ossification. The *Au. sediba* partial skeleton MH2 includes the most complete adult *Australopithecus* scapula and clavicle currently known, dated to 1.98 Ma (Berger et al. 2010; Pickering et al. 2011).

The comparative analyses of each of these specimens have come to somewhat different conclusions. Alemseged et al. (2006) concluded that the juvenile DIK-1-1 scapula exhibits the greatest morphological affinity to *Gorilla*. Green and Alemseged (2012) confirmed the preliminary study and further proposed that the *Au. afarensis* scapula retained a growth pattern similar to *Gorilla* or *Pan*. The MH2 scapula is described as most similar to *Pongo* in terms of overall morphology and the clavicle is suggested to indicate a high position of the shoulder girdle (Churchill et al. 2013).

Haile-Selassie et al. (2010) reported that the adult KSD-VP-1/1 scapula shows more similarity to those of modern humans than to *Pan* or *Gorilla*. They also noted similarity to *Homo* with regard to the relative length of the clavicle. Overall, the KSD-VP-1/1 shoulder girdle specimens are described as morphologically and functionally unique (Haile-Selassie et al. 2010). This preliminary study reported only a few angular metrics of the scapula, clavicle length, and midshaft circumference. Other features that are thought to be functionally or phylogenetically informative (e.g., glenoid shape, infraspinous fossa size and shape, clavicle curvature, and various soft tissue features) were not discussed. Further, the initial comparative analysis included only one fossil specimen (A.L. 288-1 l) and no Asian apes. Although comparisons of the KSD-VP-1/1, DIK-1-1, and MH2 shoulder girdle remains have been conducted (Green and Alemseged 2012; Churchill et al. 2013), the results are affected by notable differences in the manner that osteometrics were defined and collected for each fossil specimen and by the choice of variables included in analyses.

[1]This feature is referred to as the ventro-axillary crest in the present study, following the terminology for the axillary border described by Vallois (1932).

[2]A small clavicle fragment from Saitune Dora, Middle Awash, Ethiopia (STD-VP-2/893) is attributed to *Ardipithecus kadabba*, but little can be inferred about early hominin shoulder anatomy based on this specimen.

This study presents a more detailed anatomical description and comparative analysis of the KSD-VP-1/1 scapula and clavicle, including comparisons with DIK-1-1 and MH2 conducted using equivalent methodology. It further considers how the morphology preserved by these specimens compares to a parsimony-based model of the chimpanzee–human last common ancestor (CLCA). Specifically, the hypothesis that the adult *Au. afarensis* shoulder girdle retains primitive morphology is tested. This evolutionary framework is used to make adaptive interpretations.

Materials

Table 6.2 details the composition of the extant comparative sample. Specimens are curated by the following institutions: the Museum of Comparative Zoology (Cambridge), the American Museum of Natural History (New York), the National Museum of Natural History (Washington, D.C.), the Iziko Museum (Cape Town), and the Royal Museum of Central Africa (Tervuren). The sample consists primarily of adult and fully fused specimens, but some *Pongo* and hylobatid specimens with incompletely fused epiphyses were included to increase sample size.[3] In cases where the inferior angle of the scapula was incompletely fused, the secondary center was fully ossified and glued onto the primary center or the cartilaginous model was partially ossified, with the landmark located on the ossified portion. Incomplete fusion of the vertebral border of the scapula and the sternal end of the clavicle is less of a concern because the addition of the epiphyses has a negligible impact on the size and shape of these bones. Wild caught individuals comprise the majority of the non-human sample. The human sample consists of both archaeological specimens (primarily individuals of southern African origin) and present-day specimens (including individuals of Asian, African, and European descent).

The fossil sample is listed in Table 6.3. Data come from firsthand examination of original specimens or casts, or are taken from the literature where indicated. A cast of the reconstructed MH2 scapula was provided by the Malapa Casting Project (University of Witwatersrand).

[3]Elements of the shoulder girdle often remain unfused into adulthood in orangutans and gibbons (Schultz 1941, 1944). Some of the specimens included in this study were classified as adult in museum records but had incompletely fused vertebral borders and/or inferior angles of the scapula and sternal ends of the clavicle. This does not significantly impact the overall shape of the bone.

Table 6.2 Size and composition of the extant comparative sample

Taxonomic group	*N*	Species	Sex		
			M	F	U
Homo	59	*H. sapiens*, archaeological	20	11	10
		H. sapiens, present	10	8	0
Pan	64	*P. troglodytes*	18	15	12
		P. paniscus	10	7	0
		Pan sp.	0	1	1
Gorilla	39	*G. gorilla*	4	7	0
		G. beringei	17	9	2
Pongo	30	*P. pygmaeus*	7	12	1
		P. abelii	2	7	0
		Pongo sp.	0	1	0
Hylobatidae	8	*H. lar*	2	3	0
		H. muelleri	0	1	0
		Symphalangus syndactylus	0	2	0
Total	200		90	84	26

Methods

General anatomical terminology follows *Gray's Anatomy* (Johnston et al. 1947), while more specialized descriptors are taken from the work of Vallois (1928, 1929, 1932, 1946) and Eickstedt (1925) for the scapula and Olivier (1951a, b, c, d; 1953, 1954a, b; 1956) for the clavicle. Anatomical axes are abbreviated as follows: AP (anteroposterior/ly), ML (mediolateral/ly), and SI (superoinferior/ly).

Reconstruction of Damaged Fossil Specimens

The KSD-VP-1/1 scapula preserves most of the infraspinous fossa, but is slightly damaged at the inferior angle and in the region where the spine meets the vertebral border. These regions were manually reconstructed by adding sculpting material to a plaster cast, following the lines and curves of preserved bone.

The original length of the KSD-VP-1/1 clavicle was estimated using ordinary least squares regression of length on midshaft size ($\sqrt{\text{midshaft minimum dimension} \times \text{perpendicular dimension}}$). Initially, two estimates were considered: one derived from a reference sample including chimpanzees, gorillas, and modern humans and the second from a reference sample including all extant apes. A slightly longer estimate results when the reference sample includes *Pongo* and hylobatids, but this regression may produce less accurate results for hominins. When both regressions were applied to nearly complete fossil clavicles (e.g., MH2 and KNM-WT 15000), the length predicted by the African ape and human

Table 6.3 Composition of the fossil sample

Taxon	Site	Specimen(s)	Developmental stage	Data source	References
Scapulae					
Australopithecus afarensis	Woranso-Mille	KSD-VP-1/1g	Adult	Original fossil specimen	Haile-Selassie et al. (2010)
	Hadar	A.L. 288-1 1	Adult	Original fossil specimen and literature	Johanson et al. (1982) Stern and Susman (1983)
	Dikika	DIK-1-1	Young juvenile	Literature	Alemseged et al. (2006) Green and Alemseged (2012)
Australopithecus africanus	Sterkfontein	Sts 7	Adult	Original fossil specimen and literature	Broom and Robinson (1950) Ciochon and Corruccini (1976) Vallois (1977) Vrba (1979)
Australopithecus sp.	Sterkfontein	Stw 431	Adult	Original fossil specimen and literature	Toussaint et al. (2003) Berger (1994)
Australopithecus sediba	Malapa	MH2	Adult	Fossil cast and literature	Berger et al. (2010) Churchill et al. (2013)
Homo erectus	West Turkana	KNM-WT 15000	Late juvenile	Original fossil specimen and literature	Walker and Leakey (1993)
Homo heidelbergensis	Sima de los Huesos	AT-320, AT-343, AT-794	Adult	Literature	Carretero et al. (1997)
		AT-801	Late juvenile		
Clavicles					
Australopithecus afarensis	Woranso-Mille	KSD-VP-1/1f	Adult	Original fossil specimen	Haile-Selassie et al. (2010)
	Hadar	A.L. 333x-6/9	Adult	Original fossil specimen and literature	Ohman (1986)
Australopithecus sp.	Sterkfontein	Stw 431	Adult	Original fossil specimen and literature	Toussaint et al. (2003) Berger (1994)
Australopithecus sediba	Malapa	MH2	Adult	Fossil cast and literature	Berger et al. (2010) Churchill et al. (2013)
Homo habilis	Olduvai Gorge	O.H. 48	Adult	Literature	Napier (1965) Oxnard (1968b) Day and Scheuer (1989)
Homo erectus	Koobi Fora	KNM-ER 1808	Adult	Original fossil specimen	Leakey (1973)
	West Turkana	KNM-WT 15000	Late juvenile	Original fossil specimen and literature	Walker and Leakey (1993)
Homo antecessor	Gran Dolina	ATD6-50	Adult	Literature	Carretero et al. (1999)
Neanderthals	Multiple sites	Multiple specimens	Adult	Literature	Trinkaus (1983) Vandermeersch and Trinkaus (1995) Carretero et al. (1999) Voisin (2006b)
Homo sapiens	Omo Kibish	Omo 1	Adult	Original fossil specimen and literature	Voisin (2008)

regression closely approximates the observed length, whereas the all-ape regression overestimates length. Therefore, only estimates derived from the African ape and human regression are considered.

Geometric Morphometric Analysis

Geometric morphometrics (GM) is a landmark-based method of shape comparison (Bookstein 1998; Zelditch et al. 2004; Slice 2005; Mitteroecker and Gunz 2009). GM offers several advantages over traditional morphometric methods and some of these are particularly pertinent to the scapula and clavicle. Quantitative comparison of scapula morphology has traditionally employed a number of angular metrics. Although these methods have produced a sizeable body of knowledge, they have some drawbacks. For instance, it is problematic to isolate the relative influence that each leg of an osteometric angle has on the final measurement. The axillary border–glenoid angle is commonly discussed in describing differences in glenoid orientation among primates. However, this metric also contains information about the inclination of the axillary border, which has the potential to vary independent of glenoid orientation. Thus, there is the potential for two taxa to have similar axillary border–glenoid angles that result from different morphological patterns: one with a more cranially oriented glenoid and the other with a more medially inclined axillary border. Procrustes superimposition is one possible solution to this problem. It is a best-fit method that registers specimens according to the overall geometry of the shape, with equal contribution from all landmarks. This enables simultaneous comparison of the orientation of each component of the scapula (i.e., axillary border, glenoid, spine, etc.).[4]

[4]An alternative solution is to register specimens according to a biologically or functionally meaningful criterion, such as the *in vivo* position of a bone in the skeleton. This approach requires the identification of a baseline that has a consistent geometric relationship with the rest of the skeleton, across all taxa considered. Secure identification of such a baseline is problematic in this case, because data on the resting position of the scapula and clavicle among non-human primates are scarce. It has been suggested that *in vivo* orientation can be achieved by using the line running through the superior-most point of the superior angle and the inferior-most point of the inferior angle as the registration baseline (Churchill et al. 2013). Although the clinical study cited in support of this assumption (Sobush et al. 1996) demonstrates that such an approach would be appropriate for comparing a subsample of modern humans, the same may not be true of cross-species comparisons. Vallois (1928) provides a cursory qualitative description of inter-taxon differences in resting orientation of the scapula among primates. It suggests that the line intersecting the superior and inferior angles is not parallel to the vertebral column in all apes (*Pongo* in particular). CT scans of preserved museum specimens also support a difference in resting orientation across genera (personal observation).

One aspect of clavicle morphology that has long been of interest in paleoanthropology is shaft curvature. One method of measuring clavicle curvature, originally described by Olivier (1951d) but employed more recently by Voisin (2006a, b, 2008), divides the "s" shape of the clavicle into medial and lateral components and then quantifies the curvature of each component as a ratio of depth to length. This approach can be problematic because it is difficult to identify the boundaries of the medial and lateral components when the degree of curvature is very slight. Perhaps even more problematic is the fact that the presence of an "s" shape is not consistent within or across ape species. Semilandmark methods (Bookstein 1997; Mitteroecker and Gunz 2009) offer a solution to this problem because data collection is not affected by curvature pronouncement and curves of different forms (e.g., "s" and "j" shaped curves) can be included in the same analysis. Further, semilandmark methods can distinguish differences in curvature that are conflated when using ratios of linear measurements. Two clavicles may exhibit curves with similar ratios but different shapes. For instance, one curve may be of average length but especially deep and the other is of average depth but especially short. Alternatively, the depth and length measurements may be identical but the point of maximum depth may be located at a different place along the curve, such that one curve is symmetric while the other is not.

Another advantage of GM is that researchers do not need to select the dimensions that will be considered before conducting the comparative analysis. This is likely another issue that has contributed to the different results of the KSD-VP-1/1, DIK-1-1, and MH2 analyses. All landmarks contribute equally to GM ordination analyses. The result of such analyses is the identification of features that account for the primary axes of variation in a dataset without considering group membership (in the case of a principal components analysis) or that best distinguish predefined groups (in the case of a discriminant analysis). This permits researchers to avoid the problems associated with analyzing datasets that are composed of many strongly dependent metrics or datasets that lack potentially discriminating metrics. However, despite the advantages of GM methods, traditional metric and qualitative methods remain indispensable for comparing morphology of highly fragmentary fossil specimens and for considering the presence or absence of anatomical structures.

A Microscribe digitizing arm (Immersion Corporation; San Jose, CA) and Microsoft Excel were used to collect 3D landmarks on the scapula (Fig. 6.1). Scapula landmarks are similar to those described by Young (2006, 2008), with the addition of one landmark at the acromial angle and two more on the acromioclavicular articular surface. Scapula landmarks were superimposed using a full Procrustes fit in MorphoJ (Klingenberg 2011). Centroid size (CS) is a scalar value generated during Procrustes superimposition that represents

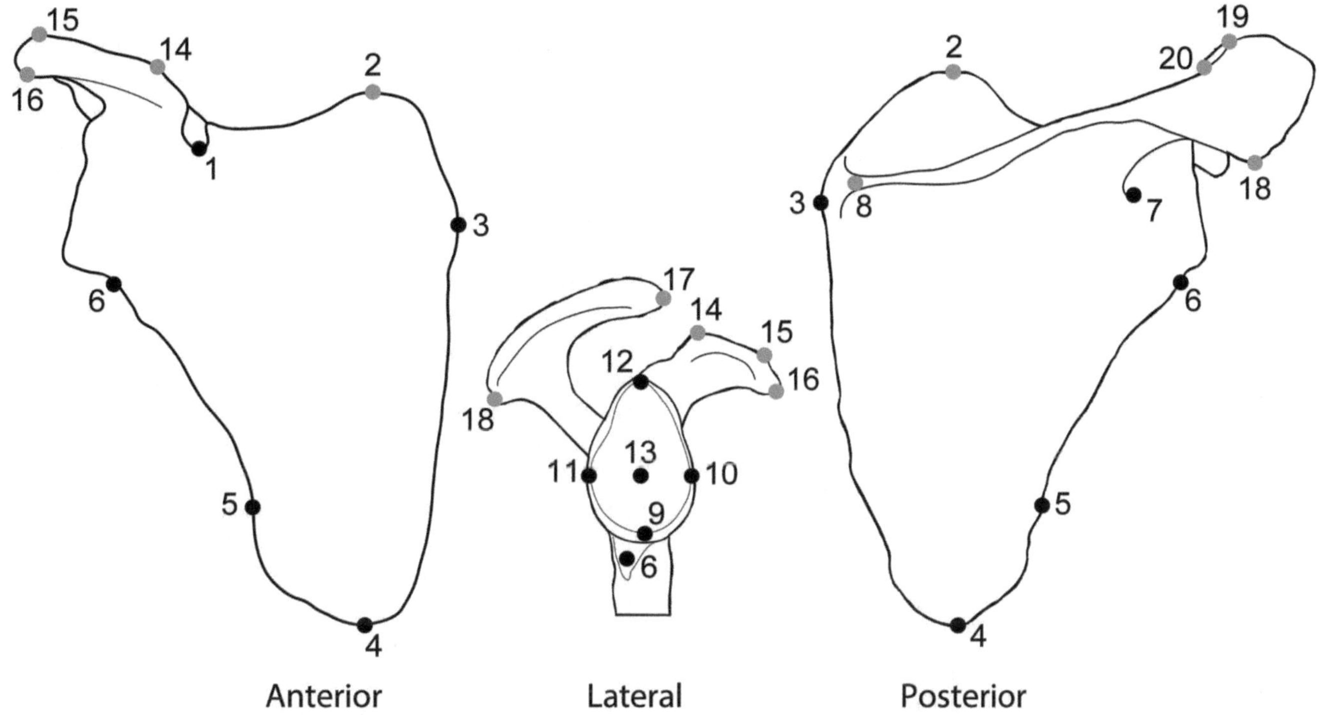

Fig. 6.1 Location of 3D scapula landmarks: *1* The deepest point in the suprascapular notch (where a notch is not present the homologous landmark is the point where the medial limit of the coracoid meets the superior border), *2* the superior-most point on the superior angle, *3* the point where a line following the axis of the base of the spine meets the medial border, *4* the inferior-most point on the inferior angle, *5* the centroid of the region where the teres major and teres minor origins overlap on the axillary border, *6* the centroid of the infraglenoid tubercle, *7* the deepest point in the spinoglenoid notch, *8* the medial extent of muscle attachment along the crest of the spine, *9* the inferior-most point of the glenoid articular surface, *10* the anterior-most point of the glenoid articular surface, *11* the posterior-most point of the glenoid articular surface, *12* the superior-most point of the glenoid articular surface, *13* the deepest point of the glenoid fossa, *14* the superolateral extent of the conoid ligament attachment surface, *15* the superolateral extent of the coracoid process, *16* the inferolateral extent of the coracoid process, *17* the distal-most point of the acromion, *18* the point of maximum curvature of the acromial angle, *19* the lateral extent of the clavicular articular surface, and *20* the medial extent of the clavicular articular surface. Landmarks 1–20 comprise the complete landmark dataset. The *Kadanuumuu* landmark dataset consists of the black landmarks, whereas the gray landmarks are not preserved

the overall size of each specimen. Comparisons of scapula CS are conducted using traditional quantitative methods.

It is difficult to collect 3D landmarks on the clavicle with a Microscribe because there is no well-defined anatomical feature that follows shaft curvature and is present in all specimens. Instead, the midline curve of the clavicle shaft is quantified in 2D (Fig. 6.2). Photographs in superior and posterior views were used to characterize shaft curvature in the transverse and coronal planes, respectively. The position of each specimen was standardized with the AP axis of the acromial end oriented either parallel (superior view) or perpendicular (posterior view) to the horizontal. The midline curve of the clavicle in each digital image was established by making a series of AP (superior view) or SI (posterior view) thickness measurements across the length of the bone and marking the midline point at each measurement location in ImageJ (Rasband 2012). Next, a curve connecting the midline markers was drawn and resampled with 50 equidistant points. The medial- and lateral-most points constrained the remaining semilandmarks, which were permitted to slide along their curves according to the criterion of minimizing bending energy (Bookstein 1997). Data collection and superimposition of clavicle curvatures was carried out using the TPS software series (Rohlf 1998).

Because GM analyses require all specimens to have the same number of coordinates, incomplete fossil specimens cannot be compared against complete extant specimens. Accordingly, two versions of the GM datasets are discussed. The complete datasets include all landmarks and semilandmarks, but only extant specimens. The *Kadanuumuu* datasets include fossil and extant specimens, but only the subset of landmarks and semilandmarks preserved by the KSD-VP-1/1 scapula (after minor reconstruction of the inferior angle and vertebral border as described above) and clavicle (without reconstruction). In other words, the comparative specimens in the *Kadanuumuu* dataset are broken virtually to imitate fossil preservation. Contrasting the results from the complete and *Kadanuumuu* datasets permits an assessment of how drastically incomplete fossil preservation affects the comparative analysis.

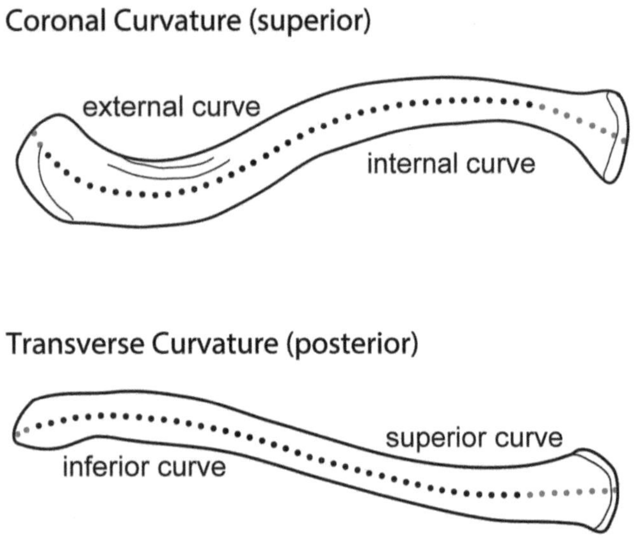

Fig. 6.2 2D clavicle semilandmarks sampling midshaft curvature in the coronal and transverse planes. The complete dataset consists of 50 semilandmarks. There is some uncertainty in determining the number of semilandmarks missing from KSD-VP-1/1f due to incomplete preservation. The best estimate is illustrated here, in which the *Kadanuumuu* dataset consists of 40 semilandmarks (black) with 10 deleted (gray) from the comparative sample to imitate fossil preservation. The minimum estimate posits that 5 semilandmarks are missing from KSD-VP-1/1f (4 deleted at the sternal end and 1 at the acromial end) and a maximum estimate posits that 15 semilandmarks are missing (12 deleted at the sternal end and 3 at the acromial end)

There is some uncertainty in determining the number of semilandmarks to drop from the *Kadanuumuu* clavicle dataset to imitate the breakage of KSD-VP-1/1f. As a starting point, specimens from the comparative sample that approximately matched KSD-VP-1/1f in terms of shaft thickness and curvature (including *Homo*, *Pan,* and fossil hominin specimens) were identified. Images of these comparative specimens were overlaid on images of KSD-VP-1/1f, then scaled and rotated in image editing software and used as templates to draw portions of the shaft missing from the fossil. The midline curves of these qualitative reconstructions were then drawn and resampled as described above to determine the number of semilandmarks located on reconstructed portions of the fossil versus preserved portions. The goal of this process was only to approximate the number of landmarks to be dropped from the comparative sample, and the qualitative reconstructions of the fossil were not used in comparative analysis. The consensus best estimate generated from this process was that 8 semilandmarks are missing at the sternal end and 2 at the acromial end (Fig. 6.2). To bracket the uncertainty inherent to this estimate, two additional *Kadanuumuu* datasets for the clavicle were created: a minimum estimate (missing 4 semilandmarks at the sternal end and 1 at the acromial end) and a maximum estimate (missing 12 semilandmarks at the sternal end and 3 at the acromial end).

A principal component analysis (PCA) was conducted for each GM dataset to explore the pattern of morphological variation. In PCA plots, each point represents a single scapula or clavicle. Gray wireframes represent the consensus shape of the entire hominoid sample, while the overlying black wireframes represent the landmark and semilandmark displacements implied by movement along each axis.

A Procrustes ANOVA accompanied by post hoc pairwise comparisons was used to test for differences among groups. Tests were conducted in the R statistical platform using the "procD.lm" and "pairwiseD.test" functions of the geomorph package (Adams et al. 2013). This nonparametric approach uses Procrustes distances to construct a test statistic analogous to the F-ratio of a univariate ANOVA. Significance is established using a permutation procedure (1,000 iterations) that randomly reshuffles group membership to determine the distribution of test statistics possible under the null hypothesis that groups do not differ. The *p*-value of an observation is calculated as the frequency that a permutated test statistic exceeds an observed test statistic (Anderson 2001).

Landmark data for the scapula and the clavicle were collected twice for a sample of 16 extant specimens. These data were added to the complete datasets and intra-observer error was assessed in two ways. First, PCA analyses were conducted and the first two components plotted to visualize the difference between trials relative to the total variation in the datasets. Second, the Procrustes distance between the first and second trials was calculated and considered relative to the average pairwise distance among individuals within a species. In all PCA analyses, the first and second trials fell very close in principal component space. The Procrustes distance between trials ranged from 25 to 30% of the average pairwise distance among individuals of the same species. Thus, intra-observer error was judged to be low relative to variation among taxa as well as within a species.

Reconstruction of Ancestral Morphology

A working model of the primitive condition is required to assess the hypothesis of morphological stasis in the *Au. afarensis* shoulder. To establish this CLCA model, the complete landmark datasets were mapped onto a molecular phylogenetic tree comprising extant hominoids using squared-change parsimony (Maddison 1991; Rohlf 2001; Klingenberg and Gidaszewski 2010). A consensus phylogenetic tree was downloaded from *10kTrees* (Arnold et al. 2010). It is constructed from mitochondrial and autosomal genetic data using a Bayesian inference method, with branch lengths proportional to time. A tree including all extant species and subspecies in the comparative sample was downloaded and then terminal taxa were merged to the species level for all great apes and to the family level for the hylobatids. The parsimony analysis treats each landmark dataset as a single multidimensional character. The cost of moving from one shape to another is the squared Procrustes distance between the two shapes. Given the consensus shapes of the terminal taxa, each node is reconstructed to minimize cost over the entire tree (Klingenberg and Gidaszewski 2010). The analysis was conducted in MorphoJ software (Klingenberg 2011). Although this method has a number of limitations (Losos 2011), it is simply the quantitative formalization of the traditional comparative approach. Like all phylogenetic analyses, the results have the potential to change depending on the taxa included (Strait and Grine 2004).

Traditional Comparative Analysis

Traditional metrics for the scapula are defined in Table 6.4. Glenoid size ($\sqrt{\text{height} \times \text{breadth}}$) is used to standardize size variation, in keeping with Green and Alemseged (2012) and Churchill et al. (2013). Clavicle midshaft size ($\sqrt{\text{midshaft minimum dimension} \times \text{perpendicular dimension}}$) is used where glenoid size is not available (i.e., for comparisons with Stw 431). Most of the scapula osteometrics are calculated from 3D landmarks (see below), in an effort to present data for KSD-VP-1/1 that are comparable to metrics reported for the juvenile DIK-1-1 scapulae (Green and Alemseged 2012). As a result, these values differ from those that were published previously (Haile-Selassie et al. 2010).

Table 6.4 Definitions of the traditional scapula metrics considered here

Metric (Abbreviation)	Definition
Axillary border–spine angle (ASA)	The angle between the line running through landmarks 3 and 7 and the line running though landmarks 9 and 4
Axillary border–glenoid angle (AGA)	The angle between the line running through landmarks 12 and 9 and the line running though landmarks 9 and 4
Spine-glenoid angle (SGA)	The angle between the line running through landmarks 12 and 9 and the line running though landmarks 3 and 7
Infraspinous vertebral border–axillary border angle (IVAA)	The angle between the line running through landmarks 9 and 4 and the line running though landmarks 3 and 4
Bar-glenoid angle (BGA)	The angle between a line running through landmarks 12 and 9 and a line running along the ventral bar (ventro-axillary crest), measured directly with a goniometer
Glenoid version (VER)	The angle between a line running through landmarks 10 and 11 and a line parallel to the base of the spine (measured directly with a goniometer in superior view with the spine oriented parallel to the horizontal)[a]
Glenoid size (GS)	The square root of the product of the distance between landmarks 12 and 9 and the distance between landmarks 10 and 11
Glenoid index (GI)	The distance between landmarks 10 and 11 divided by the distance between landmarks 12 and 9, and multiplied by 100
Relative scapula length (RSL)	The distance between landmarks 3 and 13, divided by glenoid size
Relative infraspinous fossa breadth (RIFB)	The distance between landmarks 3 and 4, divided by glenoid size
Infraspinous fossa index (IFI)	The distance between landmarks 3 and 4, divided by the distance between landmarks 3 and 13, and multiplied by 100
Relative spine thickness (RST)	The thickness of the crest of the spine (measured with calipers at the midpoint of the base of the spine), divided by glenoid size and multiplied by 100
Relative axillary border length (RAL)	The distance between landmarks 6 and 4 divided by clavicle midshaft size[b]

Landmarks are illustrated in Fig. 6.1

[a]Negative values indicate retroversion and positive values indicate anteversion

[b]Clavicle midshaft size is defined as the square root of the product of the minimum dimension of the clavicle at midshaft and its perpendicular dimension

Table 6.5 Definitions of the traditional clavicle metrics considered here

Metric (Abbreviation)	Definition
Clavicle length (CL)	The maximum length of the clavicle, measured directly with an osteometric board
Relative clavicle length (RCL)	Clavicle length (CL) divided by the articular breadth of the distal humerus
Clavicle midshaft size (MS)	The square root of the product of the minimum dimension of the clavicle at midshaft and its perpendicular dimension
Clavicle midshaft index (MI)	The minimum dimension of the clavicle at midshaft divided by its perpendicular dimension, and multiplied by 100

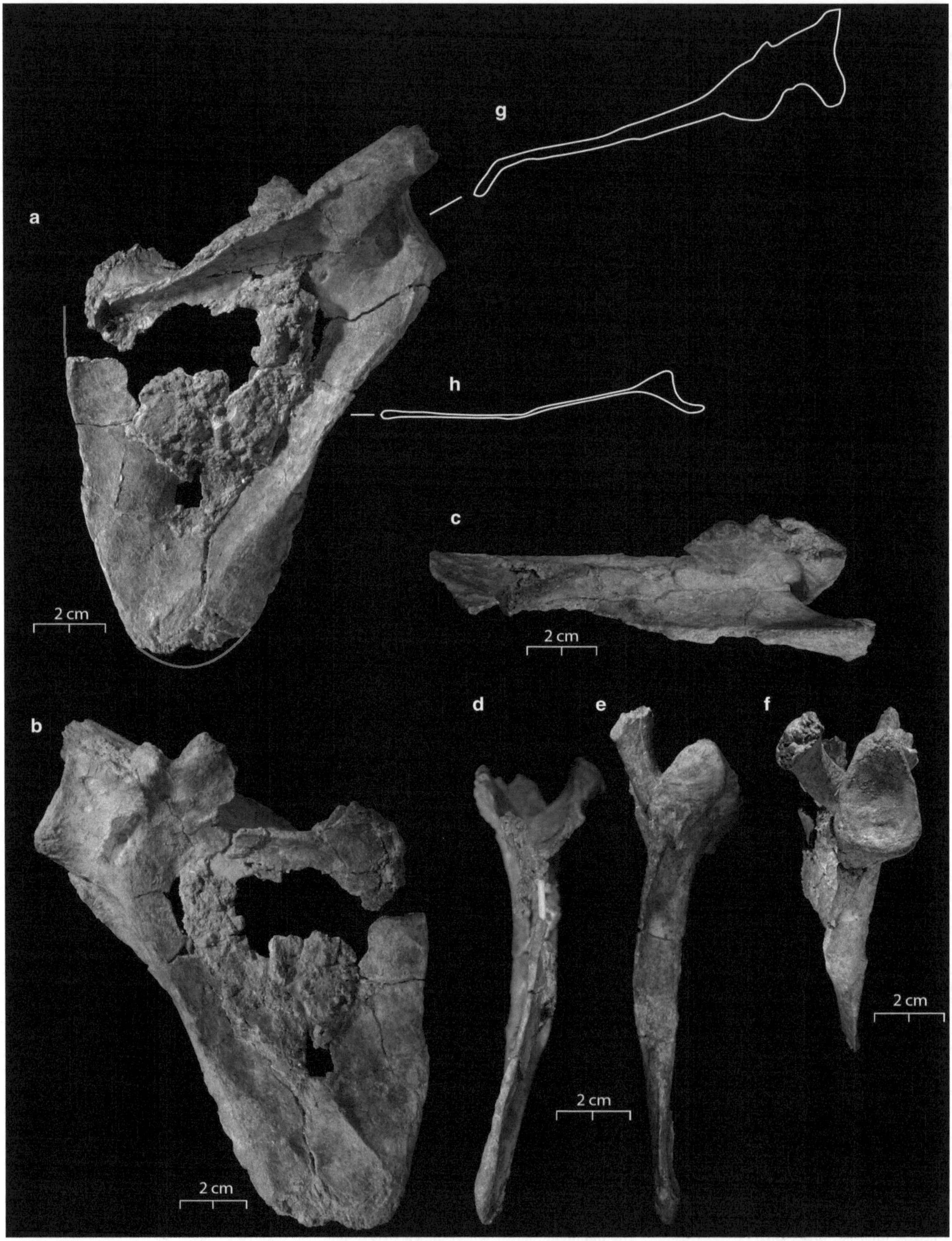

Fig. 6.3 KSD-VP-1/1g in posterior (**a**), anterior (**b**), superior (**c**), medial (**d**), lateral (**e**) views and parallel to the glenoid plane (**f**). **g** Shows a cross-section of the blade and glenoid, taken from a plane passing through the anterior- and poster-most points on the glenoid cavity and the medial extent of the base of the spine and (**h**) shows a cross-section of the vertebral border. Gray lines (**a**) show reconstruction of the vertebral border and inferior angle

Traditional metrics for the clavicle are defined in Table 6.5. A number of authors (e.g., Schultz 1930; Larson 2007) have used clavicohumeral index to make cross-taxon comparisons of clavicle length because humerus length is reported to scale nearly isometrically with body mass among apes (Jungers 1994). Here, difference in body size is standardized using humerus distal articular surface width, because it is the best-preserved measurement on the heavily exfoliated KSD-VP-1/1 humerus and it is also preserved in other hominin partial skeletons like Stw 431. Jungers and Susman (1984) report that the articular width of the distal humerus also scales nearly isometrically in African apes. It is perhaps even preferable to use an articular surface dimension. Ruff (1990) argues that articular dimensions have a more direct mechanical link to body mass than long bone lengths and that such dimensions tend to be less affected by differences in body proportions that are specific to certain taxonomic groups.

The R statistical platform and PAST software (Hammer et al. 2001) were used to calculate angles and distances from landmarks, and for further analysis. The angle between two vectors was calculated using the "angle" function described by Claude (2008). Angular osteometrics were generally normally distributed within genera, thus no further adjustment was required before statistical comparison. Means, ranges, and 95% confidence intervals are presented to place single fossil observations in a comparative context.

Descriptions

Scapula (KSD-VP-1/1g)

Preservation

The specimen is a right scapula preserving much of the blade, spine, and glenoid (Fig. 6.3). The coracoid was sheared off at its base. Only the root of the acromion process is present, measuring about 40 mm from the spinoglenoid notch to its preserved lateral extent. The majority of the supraspinous fossa is missing, broken medial to the suprascapular notch. The maximum dimension of the specimen is 210 mm, as measured from the superolateral extent of the root of the acromion to the inferior extent of the inferior angle.

The center of the blade is highly fractured and some pieces are missing. The matrix covering this area could not be removed without further damage. Despite fracturing, only minor distortion exists. It is most noticeable at the medial extent of the spine, which is displaced slightly anteriorly. The glenoid cavity is well preserved. Its articular surface is undamaged except for a small patch of erosion at its apex. The dorsal half of the supraglenoid tubercle is present. As preserved, it measures 11.6 mm wide by 13.6 mm long. The ventral half of the supraglenoid tubercle was sheared off along with the coracoid. The damaged surface from which the coracoid would have arisen begins at the glenoid rim and has a maximum length of 33.3 mm. This serves as a maximum length for the base of the coracoid, although is it likely a considerable overestimate. On the costal surface of the neck of the scapula, there is a rectangular surface where cortical bone has been flaked away. The axillary border is in good apposition and undamaged with the exception of minor fossilization cracks. The inferior-most portion of the inferior angle is missing. The vertebral border is preserved beginning just below the base of the spine and its contour is intact. The base of the spine is preserved for its entire length. A minor (1–3 mm) crack runs along the junction of the spine and the blade. There is a small amount of deformation due to crushing in the region of the spinoglenoid notch. The crest of the spine and its attendant muscle attachment surfaces are preserved along the root of the acromion and just medial to the midpoint of the spine.

Anatomy

The glenoid is moderately cranially oriented and noticeably retroverted (oriented posteriorly with reference to the blade, see Stewart 1962; Fig. 6.3g). The glenoid fossa is strongly curved both craniocaudally and dorsoventrally. The concavity is more than 5 mm deep at its deepest point, which is located a few millimeters above, and just anterior to, the geometric center of the articular surface. There is a small secondary depression, ovoid in shape, present on the articular surface inferiorly. The outline of the glenoid recalls a rounded triangle (Fig. 6.3f). The inferior edge of the rim is nearly straight, while the anterior and posterior edges curve gently toward the apex. The glenoid does not exhibit a notch. The articular surface proper (i.e., only subchondral bone) is roughly symmetric and measures 35.4 mm SI by 24.6 mm AP. The maximum dimensions of the glenoid including the rim are 39.0 mm SI by 29.3 mm AP.

The articular surface reaches the rim anteriorly, but is withdrawn from the rim posteriorly and at the anteroinferior corner. In these locations the bone is roughened, marking the attachment surface for the labrum, joint capsule, and glenohumeral ligaments. The roughening varies in thickness (2–5 mm) and is most marked at the posteroinferior corner, which blends into the infraglenoid tubercle. The infraglenoid tubercle is a linear (as opposed to mound-like) rugosity situated at the inferoposterior corner of the glenoid. Anterior to the tubercle, the surface is markedly rugose and preserves two large foramina. In posterior view, the tubercle is bounded medially by the dorso-axillary crest.

The axillary border measures 160 mm from the inferior-most point on the glenoid rim to its inferior-most preserved point. With minor reconstruction at the inferior angle, the original length of the axillary border is estimated to be about 165 mm. When viewed posteriorly (Fig. 6.3a), the superior half of the border is slightly concave. When viewed laterally (Fig. 6.3e), the superior three-fourths of the border is rectangular, but tapers to a thin ridge at the teres major apophysis. Just below the origin of the long head of the triceps muscle, the AP thickness of the border is 17.4 mm. The axillary border exhibits the ventral sulcus pattern (sulcus axillaris subscapularis of Eickstedt 1925, Fig. 6.3h). The medio-axillary crest (the structure narrowly defined as the axillary border) is well developed but fades away just inferior to the point where the ridge of the teres major apophysis begins. A V-shaped depression is created by the overlap of these linear features, which separate the attachment surface of subscapularis (the ventro-axillary surface) from those of teres minor and teres major (dorso-axillary surface). The ventro-axillary crest is sharp and exhibits some rugosity. The rectangular ventro-axillary surface is faintly concave superiorly and flat inferiorly. It faces laterally and has a maximum width of about 16 mm. The maximum breadth of the dorso-axillary surface is about 4 mm. A poorly developed dorso-axillary crest is visible in posterior view. The medio- and dorso-axillary crests are continuous with the infraglenoid tubercle.

When the vertebral border is aligned perpendicular to the horizontal, the base of the spine is nearly parallel to the horizontal. In a typical human scapula, the inferior face of the base of the spine curves inferiorly where it meets the vertebral border, creating a rounded corner at the superomedial limit of the infraspinous fossa. This feature is clearly present on KSD-VP-1/1g, indicating that the base of the spine is preserved in its entirety. It extends the full length of the blade and measures 99 mm long from the deepest point of the spinoglenoid notch to its medial-most preserved point (which lacks a few millimeters of the vertebral border). The original ML dimension of the spine (measured from spinoglenoid notch to the medial limit of the vertebral border) is estimated to be about 106 mm. The original scapula morphological length (sensu Schultz 1930) is estimated to be about 116 mm. In lateral view, the spine meets the blade at an angle of about 40°. The root of the acromion has an ovoid cross-section where it is broken, but it becomes more triangular near the spinoglenoid notch due to an inferior thickening. A faint linear ridge runs along the thickening, similar to the "spinal buttressing" described by other authors (Walker and Pickford 1983; Senut et al. 2004). The root of the acromion is robust, measuring 20.1 mm SI at its lateral extent and about 16 mm where the spine meets the blade, in the spinoglenoid notch. Two muscle attachment surfaces can be differentiated on the dorsal surface: the insertion of the intermediate fibers of the trapezius muscle superiorly and the origin of the posterior fibers of the deltoid muscle inferiorly. The superior surface is fairly smooth and curves superiorly at its lateral extent. The inferior surface is the larger of the two and is more rugose. Following a small zone of damage, the crest of the spine is preserved near its midpoint. Here, the surface of the crest is rugose and SI expanded relative to the base, indicating muscle attachment well past the midpoint of the spine. The thickness of the crest at the midpoint of the base of the spine is 7.2 mm.

The spinoglenoid notch is small, circular, and roughly symmetric (Fig. 6.3c). Its anterior border is formed by the dorsal surface of the scapular neck, which is deeply excavated. The notch is about 17.2 mm deep, as measured in superior view (with the spine held parallel to the horizontal) from a line tangent to the glenoid surface to the deepest point in the notch. The outlet of the notch is extremely narrow. The glenoid rim is separated from the anterior surface of the acromion root by only a few millimeters. A nutrient foramen is preserved in the medial third of the superior face of the base of the spine.

KSD-VP-1/1g preserves a suprascapular notch (Fig. 6.3b). The notch is asymmetric with a blunted V-shape. It compares well with the human Type II suprascapular notch described by Rengachary et al. (1979). There are two vascular depressions associated with the suprascapular notch. The larger one meets the suprascapular notch on its medial edge while the smaller depression passes through the suprascapular notch at its deepest point. It is not possible to establish a precise estimate of the height or area of the supraspinous fossa, but the trend of the medial portion of the notch suggests it would have continued superiorly resulting in a well-defined superior angle.

The infraspinous fossa is nearly completely preserved (Fig. 6.3a). The original breadth (traditionally defined as the craniocaudal dimension) is estimated to be 135 mm, as measured from the point where the base of the spine would meet the vertebral border to the inferior-most point on the reconstructed inferior angle. There is a depression just medial to the axillary border that stretches along the entire length of the infraspinous fossa. This depression is bounded laterally by the dorso-axillary crest and medially by the gentle convexity of the center of the fossa. The teres major attachment area is absorbed into the inferior angle, rather than projecting laterally past the axillary border. It is bounded superiorly and medially by the inferior extension of the medio-axillary crest, which continues all the way to the vertebral border. At its inferior extent, the vertebral border thickens and begins to curve laterally. Thus, it seems that the inferomedial extent of the inferior angle is fully preserved. Less than 5 mm and no more than 10 mm of the inferior angle is missing at its apex.

The majority of the costal surface (subscapular fossa) is concave. The point of maximum concavity is located at the superolateral portion of the subscapular fossa. A nutrient foramen and associated vascular depression is present where the neck of the scapula meets the subscapular fossa.

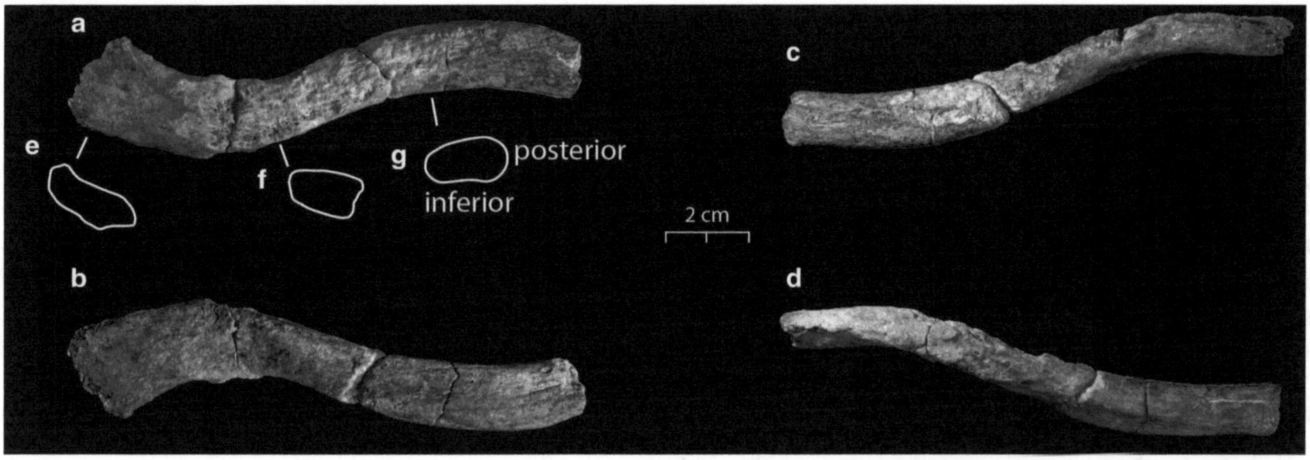

Fig. 6.4 KSD-VP-1/1f in superior (**a**), inferior (**b**), anterior (**c**) and posterior (**d**) views. **e–g** show shaft cross-sections at the indicated locations

A transverse depression runs across the superior third of the subscapular fossa that corresponds to the base of the spine. Near the midpoint of this depression, a single nutrient foramen is preserved. A number of subscapular ridges are present, trending (with varying obliquity) from superolateral to inferomedial.

The vertebral border is preserved for 107 mm, which is roughly two-thirds of its original length. The border is in excellent apposition. The shape of the border (in anterior or posterior view) is straight, rather than concave or convex (*sensu* Graves 1921). Its margin is roughened for the insertion of the serratus anterior muscle. On the costal surface the roughening is thin and confined to the margin superiorly but widens gradually where the vertebral border meets the inferior angle. The attachment surface on the dorsal surface of the vertebral border is most rugose inferiorly, in the region where the rhomboid major, teres major, and latissimus dorsi muscles attach. When viewed medially (Fig. 6.3d), the preserved portion of the supraspinous fossa bends anteriorly, forming an angle with the infraspinous fossa (the subscapular angle *sensu* Johnston et al. 1947) of about 140°.

Clavicle (KSD-VP-1/1f)

Preservation

The specimen is a left clavicle shaft missing the acromial epiphysis and a considerable portion of the sternal extremity (Fig. 6.4). The maximum dimension, as preserved, is 125.6 mm. At the medial point of breakage, small pits are present on the superior and inferior surfaces.

The majority of the upper surface suffers from exfoliation, severe enough to reveal cancellous bone in some places. Original bone surface is present only at the medial and lateral ¼ of this surface. The specimen was broken near midshaft and the adjoining surfaces of the break were weathered, with loss of cortical bone anteriorly and posteriorly. It appears that no length was lost as a result of the weathering, but apposition is imperfect. The contour of the anterior border is slightly offset at the break, although the contour of the posterior border is not disrupted. Two additional fossilization cracks are present, each approximately at the maxima of the internal and external curves. On the lower surface, there are two bone flakes in the lateral crack that are displaced slightly superiorly. None of these fossilization artifacts significantly alter the dimensions of the specimen.

Anatomy

The shaft exhibits pronounced internal and external curves in the transverse plane (Fig. 6.4a, b). Curvature in the coronal plane (Fig. 6.4c, d) is less conspicuous, but both a superior curve and an inferior curve are present. If the acromial end is positioned parallel to the horizontal, the lateral end of the bone sits about 3 cm higher than the medial end (as preserved).

The shaft is SI compressed throughout its length and presents two major surfaces (upper and lower) separated by two borders (anterior and posterior). The anterior border is bluntly rounded with the exception of the lateral-most 10 mm, where it is somewhat flattened. The posterior border is bluntly rounded from the medial end until about midshaft.

At the center of the internal curve, the shaft is markedly SI compressed, with an oval cross-section (Fig. 6.4g). The maximum and minimum dimensions at this point are 18.9 mm and 10.1 mm, respectively. Severe exfoliation to the upper surface prevents precise measurement of cross-sectional dimensions at midshaft, but an estimate of the original maximum and minimum dimensions are 17 mm by 10 mm. Just lateral to midshaft, the cross-section comprises a rounded anterior border, slightly convex upper and lower surfaces, and a flattened posterior surface. This flat posterior surface extends about 30 mm ML. It is anteroinferiorly inclined and meets the upper and lower surfaces at well-defined angles (Fig. 6.4f). It is bounded superiorly and inferiorly by rugose crests. The superior crest is partially preserved. The inferior crest is fully preserved and fades out roughly at midshaft, just lateral to a large nutrient foramen. A small conoid tubercle is present at the lateral extent of the posterior surface.

The acromial end is AP expanded relative to the rest of the shaft with maximum dimensions (at the lateral point of breakage) of 24.5 mm by 9.4 mm. The upper surface of the acromial end is slightly concave and the lower surface is slightly convex, producing a semilunar cross-section (Fig. 6.4e). Similar morphology is present on the Omo 1 clavicle, which is followed laterally by a lipping of the superior surface where it meets the acromial articular surface. The fact that the lipping is present on KSD-VP-1/1f suggests that the diaphysis is very close to fully preserved at the acromial end.

The specimen exhibits a considerable amount of torsion, which is further pronounced due to the SI compression of the medial portion of the shaft. When the AP long axis of the acromial end is positioned parallel to the horizontal, the AP long axis of the medial portion of the shaft twists about 30° inferiorly.

The upper surface of KSD-VP-1/1f is flat medially with no discernible roughening for the attachment of pectoralis major. The origin for the anterior fibers of the deltoid is located at the anterior edge of the upper surface. The scar is partially exfoliated, but its medial-most extent is preserved and ends only a few millimeters medial to the center of the external curve. Laterally, it takes the form of two crests bounding a linear depression.

The lower surface exhibits a shallow subclavian groove just medial to midshaft. There is no indication of a costal tuberosity. The acromial end preserves a roughened patch that appears to be the attachment surface for the trapezoid ligament, although it is damaged by acid etching. There is a faint vascular depression on the posterior border medially. The acromial portion of the posterior border is sharp and presents a slightly lipped inferior surface.

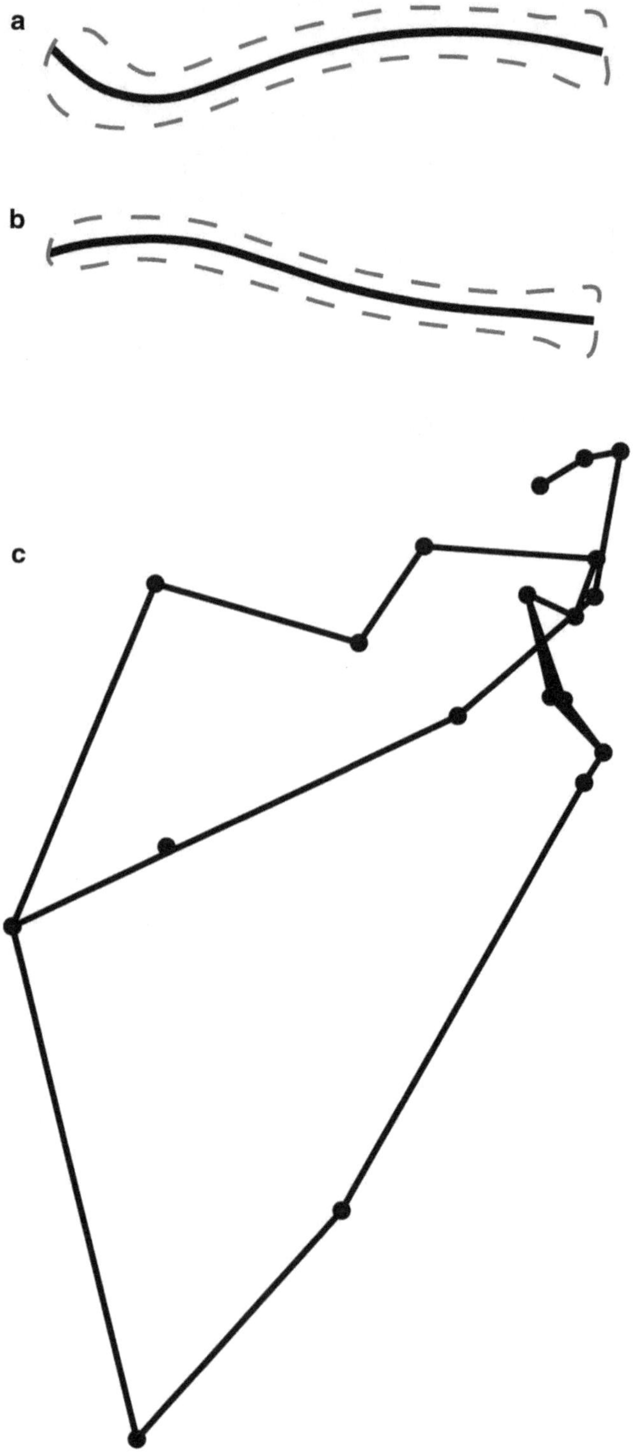

Fig. 6.5 Model of clavicle and scapula morphology in the chimpanzee–human last common ancestor (CLCA) generated from parsimony analysis. Transverse curvature is shown in superior view (**a**) and coronal curvature is shown in posterior view (**b**) for a left clavicle. A right scapula is shown in posterior view (**c**). See Figs. 6.6 and 6.7 for comparison with extant taxa

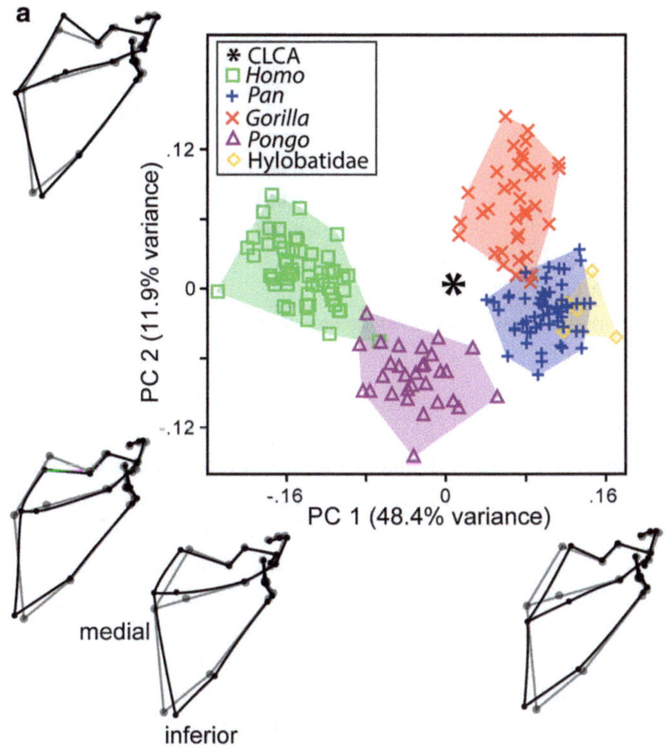

Fig. 6.6 Results of the PCA of the complete landmark set (**a**) and the *Kadanuumuu* landmark set (**b**). The reduction of landmarks associated with incomplete fossil preservation does not change the overall clustering pattern nor shape changes associated with the principal component axes. The first PC is primarily driven by variation in the obliquity of the scapula. The KSD-VP-1/1 scapula is significantly less oblique than either *Pan* or *Gorilla* scapulae, falling between *Pongo* and *Homo* along PC 1 and near the center of the *Homo* range along PC 2. The MH2 scapula falls near the center of the *Homo* cluster. The location of the CLCA model is overlaid in (**a**) for comparison, but did not contribute to the analysis

Results

Reconstruction of Ancestral Morphology

The CLCA model generated from the parsimony analysis is depicted in Fig. 6.5. The scapula (Fig. 6.5c) exhibits a cranially oriented glenoid (most similar to the *Pongo* consensus), a spine that is intermediate in orientation between *Pongo* on one hand and *Gorilla* and *Pan* on the other, a craniocaudally narrow infraspinous fossa, and a trapezius attachment that terminates well before the medial border. The internal and external curvature of the clavicle in the transverse plane is pronounced, most similar to the *Pan troglodytes* consensus (Fig. 6.5a). Curvature in the coronal plane is slight but both a superior and an inferior curve are present (Fig. 6.5b). This result is generally consistent with the CLCA described by Larson (2007, 2009, 2013), although this model does not include data on clavicle length or shoulder girdle elevation. The problems associated with predicting girdle elevation from clavicle coronal curvature are addressed below (see "Discussion").

Comparative Analysis of the Scapula (KSD-VP-1/1g)

Geometric morphometrics

The primary axes of variation in scapula shape among extant apes concern the obliquity of the scapula,[5] and the proportions of the blade. The results from the PCAs of the complete and *Kadanuumuu* landmark sets are very similar, because most of the landmarks that exhibit drastic displacement in the analysis of the complete landmark set (Fig. 6.6a) are also represented in the *Kadanuumuu* set (Fig. 6.6b). In the PCA

[5] The term "obliquity" is used to describe an aspect of scapula shape that incorporates a number of strongly correlated morphological features, including the orientation of the spine, the craniocaudal dimension of the infraspinous fossa and the orientation of the glenoid. The following osteometrics quantify these morphological features: axillary border–spine angle (ASA), infraspinous fossa breadth (RIFB), and axillary border–glenoid angle (AGA). Correlation coefficients between these metrics range from 0.92 to 0.75 among extant hominoids. This term was used in an equivalent sense by Miller (1932).

Table 6.6 Results of pairwise comparisons of scapula shape among extant hominoids and fossil hominins, as represented by the *Kadanuumuu* landmark set

	MH2	KSD-VP-1/1g	*Homo*	*Pan*	*Gorilla*	*Pongo*	Hylobatidae
MH2	–	0.73	0.87	**0.01**	0.08	0.34	**<0.001**
KSD-VP-1/1g	0.12	–	0.53	0.12	0.20	0.73	**0.01**
Homo	0.08	0.12	–	**<0.001**	**<0.001**	**<0.001**	**<0.001**
Pan	0.24	0.18	0.25	–	**<0.01**	**<0.001**	**<0.001**
Gorilla	0.20	0.16	0.22	0.09	–	**<0.001**	**<0.01**
Pongo	0.15	0.10	0.15	0.12	0.12	–	**<0.001**
Hylobatidae	0.32	0.26	0.34	0.14	0.15	0.22	–

Procrustes distances are reported below the diagonal and *p*-values (probability that a given distance will be observed under the null hypothesis of similarity) are reported above the diagonal (significant differences in boldface)

plots, genera fall out in discrete clusters with a small amount of overlap, indicating that variation is taxonomically structured. Thus, KSD-VP-1/1g preserves the anatomical regions that have the highest discriminatory potential.

With regard to the *Kadanuumuu* analysis (Fig. 6.6b), the landmarks showing the greatest displacements along PC 1 determine the orientation of the spine, followed by the relative dimensions and orientations of the axillary and vertebral borders (which together determine the position of the apex of the inferior angle), and lastly, the orientation of the glenoid. These features account for the majority (68.7%) of the variance in the dataset. The wireframes show that differences in spine orientation among taxa result from variation in the position of landmark 3 (marking the medial extent of the spine), whereas the position of landmark 7 (marking the lateral extent of the spine) is consistent among taxa.

Hylobatids occupy the positive extreme of PC 1. The hylobatid scapula has a very cranially oriented scapular spine and glenoid. The infraspinous fossa is craniocaudally narrow and the axillary border is relatively long, which results in the apex of the inferior angle being medially positioned. The scapulae of chimpanzees and gorillas score similarly along PC 1. These taxa also exhibit a cranially oriented spine and glenoid, a craniocaudally narrow infraspinous fossa, and a long axillary border, but the degree of scapular obliquity is less pronounced in *Pan* and *Gorilla* than it is in hylobatids. The negative extreme of PC 1 is occupied by modern human specimens, which exhibit a transversely oriented spine and a laterally directed glenoid. The infraspinous fossa is craniocaudally broad and the axillary border is short, which results in a laterally positioned inferior angle. Orangutans are intermediate between these extremes (the *Pongo* mean PC 1 score is approximately zero), primarily due to the fact that the spine of the orangutan scapula is typically less cranially oriented than is the case among the other non-human apes.

The KSD-VP-1/1 scapula exhibits less obliquity than non-human apes, falling between *Pongo* and *Homo* along PC 1. The MH2 scapula falls near the center of the *Homo* cluster.[6] Traditional metrics representing spine and glenoid orientation likewise show that both adult *Australopithecus* specimens fall above the 95% confidence interval of non-human apes and below the interval of humans (Table 6.7: ASA and AGA). The infraspinous fossa of KSD-VP-1/1g is broad, which is unique to *Homo*, but the axillary border is longer than is typical of *Homo* (Table 6.8: RIFB and RAL). This results in the inferior angle apex being positioned roughly at the ML center of the blade.

When the complete landmark set is considered (Fig. 6.6a), the second PC further distinguishes *Gorilla* from *Pongo* according to differences in the pronouncement of the superior angle and the mediolateral dimension of the blade (See footnote 6). The elimination of the landmark at the superior angle (landmark 2) results in much less discrimination along this axis (Fig. 6.6b). *Homo*, *Pan*, and the adult *Australopithecus* specimens occupy the center of this axis.

A Procrustes ANOVA demonstrates significant differences in scapula shape among hominoids ($p < 0.001$). The results of the pairwise comparisons using the *Kadanuumuu* landmark set are presented in Table 6.6. Scapula morphology is highly distinctive among extant apes, and all genera in the comparative sample differ significantly from each other. The KSD-VP-1/1 and MH2 scapulae are least similar to hylobatids, followed by *Pan* and *Gorilla*. The nearest neighbor of KSD-VP-1/1g is *Pongo*, followed by MH2 and *Homo* (which are roughly equidistant), whereas the nearest neighbor of MH2 is *Homo*, followed by KSD-VP-1/1g.

The size of the KSD-VP-1/1 scapula, as represented by centroid size of the *Kadanuumuu* landmark set (Table 6.8: CS), is larger than any modern human in the comparative sample. It falls at the upper end of the *Pan* and *Pongo* ranges

[6] The MH2 scapula is nearly complete, permitting an analysis that includes more landmarks than those comprising the *Kadanuumuu* set. If all landmarks preserved in MH2 are considered, the specimen remains within the modern human distribution and distinct from non-human apes. MH2 lacks the extreme expansion of the supraspinous fossa present in *Gorilla* but has a defined superior angle, which distinguishes it from *Pongo*.

Table 6.7 Summary of scapula angular metrics

	ASA	AGA	BGA	SGA	IVAA	VER
AT-320[a]	–	140	–	–	–	–
AT-801[a, b]	55–60	144	–	89.0	–	–
KNM-WT 15000 E[b, c]	69.6	137.8	137.6	111.8[d]	–	–
MH2	44.7	125.2	133	98.4	48.4	−1.5
Sts 7	–	–	133	–	–	0
DIK-1-1[b, e]	37.8	121.0	128.9	96.9[d]	–	–
A.L. 288-1 1	–	–	132	–	–	–
KSD-VP-1/1g	43.8	123.5	134	99.4	46.9	−19
Homo	50.2	131.9	143.0	97.6	49.6	−6.4
	(48.7–51.6)	(130.6–133.2)	(141.5–144.4)	(96.1–99.1)	(48.8–50.3)	(−4.8 to 7.9)
	[39.2–62.3]	[121.3–143.4]	[130–159]	[84.5–108.3]	[43.8–55.4]	[7.5–21]
Pan	20.1	112.6	128.3	87.1	40.1	−8.2
	(19.2–21.0)	(111.3–113.8)	(126.9–129.7)	(85.7–88.4)	(39.5–40.8)	(−7 to 9.4)
	[13.2–28.8]	[101.8–121.9]	[117–139.5]	[73.9–97.9]	[33.2–46.7]	[2–17.3]
Gorilla	25.5	118.0	131.6	87.0	51.2	−4.4
	(24.2–26.8)	(116.4–119.7)	(129.8–133.3)	(85.4–88.6)	(50.1–52.3)	(−2.9 to 5.9)
	[17.1–31.5]	[109.9–127.6]	[121–142]	[76.3–95.0]	[45.4–58.4]	[3.5–14]
Pongo	35.9	117.0	131.6	98.7	44.3	−5.8
	(33.8–38.0)	(114.8–119.1)	(129.7–133.5)	(96.2–101.2)	(43.0–45.6)	(−3.8 to 7.9)
	[23.9–48.0]	[105.0–135.4]	[122.3–145]	[83.0–112.1]	[38.2–50.5]	[4.2–16]
Hylobatidae	13.1	103.8	116.1	88.1	55.2	−14.6
	(9.6–16.6)	(96.4–111.3)	(111.2–120.8)	(80.1–96.2)	(48.6–61.9)	(−11.1 to 18.2)
	[8.4–15.7]	[99.4–114.3]	[110–126.5]	[79.5–94.3]	[51.6–63.7]	[− 7 to 18]

Data presented for living taxa are: means (95% confidence intervals) and [ranges]. *ASA* axillary border–spine angle, *AGA* axillary border–glenoid angle, *BGA* bar–glenoid angle, *SGA* spine–glenoid angle, *IVAA* infraspinous vertebral border–axillary border angle, *VER* glenoid version; see Table 6.4 for metric definitions

[a]Carretero et al. (1997)

[b]Juvenile individual with unfused glenoid fossa

[c]Green and Alemseged (2012)

[d]Complement of angle reported in Green and Alemseged (2012)

[e]Averages of the left and right sides, as reported by Green and Alemseged (2012)

Table 6.8 Summary of scapula linear metrics and scapula size

	GS	GI	RSL	RIFB	IFI	RST	RAL	CS
AT-320[a, b]	28.8	63.8	–	–	–	–	–	–
AT-343[a, b]	29.2	62.2	–	–	–	–	–	–
AT-794[a, b]	32.7	67.0	–	–	–	–	–	–
KNM-WT 15000 E	–	–	–	4.8	–	–	–	–
MH2 (UW88-56)	25.0	68.6	3.2	3.8	119.0	23.6	13.1	142.8
Stw 431	–	–	–	–	–	–	12.8	–
Sts 7[c]	26.9	66.4	–	–	–	–	–	–
DIK-1-1[d]	–	–	3.7[e]	3.0	–	–	–	–
A.L. 288-1 l	18.3	65.9[f]	–	–	–	–	–	–
KSD-VP-1/1g	29.0	67.1	4.0	4.7	116.6	24.8	12.6	209.7
Homo	25.5	68.6	3.6	4.4	121.5	28.0	11.3	165.4
	(24.7–26.2)	(67.4–69.7)	(3.6–3.7)	(4.3–4.5)	(118.5–124.6)	(25.8–30.2)	(10.9–11.7)	(161.6–169.2)
	[20.0–32.3]	[59.2–77.6]	[3.0–4.1]	[3.5–5.4]	[97.4–148.2]	[13.4–46.7]	[7.7–15.8]	[135.6–198.5]
Pan	24.7	73.2	4.1	3.2	78.8	16.1	15.0	183.2
	(24.2–25.1)	(71.8–74.6)	(4.0–4.1)	(3.1–3.3)	(76.5–81.1)	(14.7–17.5)	(14.6–15.4)	(179.7–186.6)
	[21.2–28.8]	[62.6–86.8]	[3.4–4.7]	[2.4–3.7]	[56.5–81.1]	[6.5–26.0]	[12.4–18.4]	[160.5–217.1]
Gorilla	35.6	71.8	4.4	3.2	73.3	8.5	14.6	269.7
	(34.8–38.3)	(70.2–73.4)	(4.2–4.5)	(3.1–3.3)	(70.8–75.8)	(7.3–9.7)	(14.1–15.2)	(257.0–282.5)
	[28.2–47.7]	[60.5–80.6]	[3.8–5.3]	[2.4–4.1]	[50.9–87.2]	[3.4–19.6]	[11.7–17.7]	[204.0–320.6]
Pongo	27.6	66.4	3.6	3.7	103.6	14.0	11.9	188.1
	(26.4–28.8)	(64.6–68.2)	(3.5–3.7)	(3.6–3.8)	(98.2–108.9)	(12.0–16.1)	(11.4–12.3)	(179.4–196.9)
	[23.1–34.5]	[55.8–75.2]	[3.1–4.7]	[3.1–4.3]	[76.0–136.8]	[5.2–26.4]	[10.6–13.5]	[155–234.0]
Hylobatidae	11.4	76.8	5.6	2.4	43.4	6.9	13.5	95.8
	(10.4–12.4)	(73.5–80.2)	(5.2–6.1)	(2.2–2.6)	(37.4–49.3)	(4.3–9.6)	(11.5–15.6)	(83.8–107.7)
	[10.2–11.9]	[73.7–79.5]	[5.4–6.3]	[2.3–2.6]	[36.2–47.8]	[5.2–9.1]	[12.4–15.4]	[85.0–111.2]

Data presented for living taxa are: means (95% confidence intervals) and [ranges]. *GS* glenoid size, *GI* glenoid index, *RSL* relative scapula length, *RIFB* relative infraspinous fossa breadth, IFI infraspinous fossa index, *RST* relative spine thickness, *RAL* relative axillary border length, *CS* centroid size of the *Kadanuumuu* landmark set; see Table 6.4 for metric definitions

[a]Carretero et al. (1997)

[b]Glenoid dimensions include rim and labrum attachment surface

[c]Sts 7 was reconstructed to correct for a large fossilization crack that runs through the center of the glenoid

[d]Averages of the left and right sides, as reported by Green and Alemseged (2012)

[e]Green and Alemseged (2012) measured scapula length from the point where the spine meets the vertebral border to the superior-most point on the glenoid fossa

[f]Johanson et al. (1982) report maximum dimensions of the glenoid that include rim and labrum attachment surface that result in a GI of 70

and among the smallest *Gorilla* individuals. The size of the glenoid (GS), however, is within the human range and *Pongo* 95% confidence interval. In contrast, the A.L. 288-1l glenoid is smaller than any extant great ape, including humans. To further investigate the pattern of size among the different developmental components of the scapula, centroid size was calculated for the subset of landmarks representing the glenoid (9–13) and those preserved on the KSD-VP-1/1g blade (1, 3–7). With reference to the relationship between blade size and glenoid size, KSD-VP-1/1g is most similar to *Gorilla* females and some *Pongo* specimens. This underscores the fact that the blade is relatively large.

Traditional metrics and qualitative features

Examination of traditional metrics permits comparison in cases where landmark data is not available or a sufficient number of landmarks are not preserved. Despite the fact that AGA is a more discriminatory metric than BGA, BGA remains useful for understanding the relationship between size and morphological variation in the available fossil sample. The size of the KSD-VP-1/1 glenoid is at the upper end of the *Homo* range, but BGA is similar to the much smaller specimens MH2, A.L. 288-1 l and Sts 7 (Table 6.7). Thus, observations from this sample do not support an allometric explanation of glenoid orientation in *Au. afarensis* (Mensforth et al. 1990; Inouye and Shea 1997).

The axillary border of the *Kadanuumuu* scapula is similar to humans in lacking strong medial inclination, but is absolutely and relatively longer. Axillary border length can also be measured on the fragmentary specimen Stw 431. When size-adjusted, the axillary border length of this South African *Australopithecus* specimen likewise falls above the *Homo* and *Pongo* confidence intervals but below the *Pan* and *Gorilla* intervals (Table 6.8: RAL). The Stw 431 axillary border is similar to that of KSD-VP-1/1g in exhibiting a well-defined ventro-axillary crest, an anteroposteriorly wide ventro-axillary surface, and a large, linear infraglenoid tubercle. MH2 also exhibits this morphology, albeit in a more gracile form. The existence of a clear ventral sulcus pattern in all available *Australopithecus* specimens is consistent with the proposal that it is the plesiomorphic condition (Trinkaus 2008). This is interesting in light of the fact that the dorsal sulcus pattern is dominant among Neanderthals, the bisulcate pattern is dominant among Upper Paleolithic *Homo sapiens,* and the ventral sulcus pattern is dominant among most modern human populations (Vallois 1932; Trinkaus 2008; Voisin 2012).

The relative breadth of the KSD-VP-1/1g infraspinous fossa falls above the range of all non-human apes, and is even notably broad compared to modern humans (Table 6.8: RIFB). The KNM-WT 15000 scapula exhibits a similarly broad fossa, even without correcting for the unfused inferior angle. The MH2 infraspinous fossa is relatively narrower, but the difference between KSD-VP-1/1g and MH2 does not exceed the range of variation in extant *Homo*.

A few additional features that are not well represented in the landmark dataset deserve comparative comment. Variation in glenoid shape among apes is small relative to variation in the other shape attributes discussed above, thus it does not heavily influence the PCA. However, significant differences in glenoid shape exist. The glenoid fossae of hylobatids, *Pan* and *Gorilla* tend to be rounder than those of *Pongo* or *Homo* (Table 6.8: GI). The GI of the adult *Australopithecus* sample is most comparable to *Pongo*. A similarly ovate glenoid persists in early and middle Pleistocene hominins and in many Neanderthals (Churchill and Trinkaus 1990; Carretero et al. 1997; Di Vincenzo et al. 2012).

The scapular spine of KSD-VP-1/1g differs from *Pan* and *Gorilla* in more than orientation alone. The *Kadanuumuu* dataset does not include information about the degree to which the spine stretches across the blade and the extent of the associated attachment surfaces of the trapezius and deltoid muscles.[7] In many *Pan* and especially *Gorilla* scapulae, the spine blends into blade well before it reaches the vertebral border and the muscle attachment surface on the crest of the spine thins medially, terminating around the midpoint of the spine's base. In contrast, the human scapular spine projects strongly across the entire blade and the crest of the spine is rugose and superoinferiorly expanded across its length. About half of the *Pongo* individuals in the comparative sample exhibit morphology similar to *Pan* or *Gorilla* and the others tend toward a pattern somewhat similar to humans, but with much less pronounced superoinferior expansion of the crest.

The base of the spine stretches across the entire blade on the KSD-VP-1/1 scapula and the preserved portions of the crest suggest that a rugose muscle attachment surface stretched well past the midpoint of the spine's base. The crest of the spine is fully preserved in the MH2 scapula. On this specimen, the spine also stretches across the entire blade and muscle attachment surface is present throughout its length. In addition, the specimen exhibits a conspicuously human-like deltoid tubercle. The features described above are adequately quantified by a measurement of the thickness of the crest of the spine at its midpoint (Table 6.8: RST). Both fossils are most similar to *Homo* in this respect.

Modern humans exhibit a large amount of variation in the shape of the suprascapular notch, an indentation of the superior border associated with the suprascapular nerve, artery, and ligament. Rengachary et al. (1979) provide a typology for classifying notch variation: type I "notches" are wide and occupy nearly the entire superior border

[7]Among extant hominoids, the trapezius attachment surface tends to extend further medially than the deltoid attachment surface; however, this is not always the case.

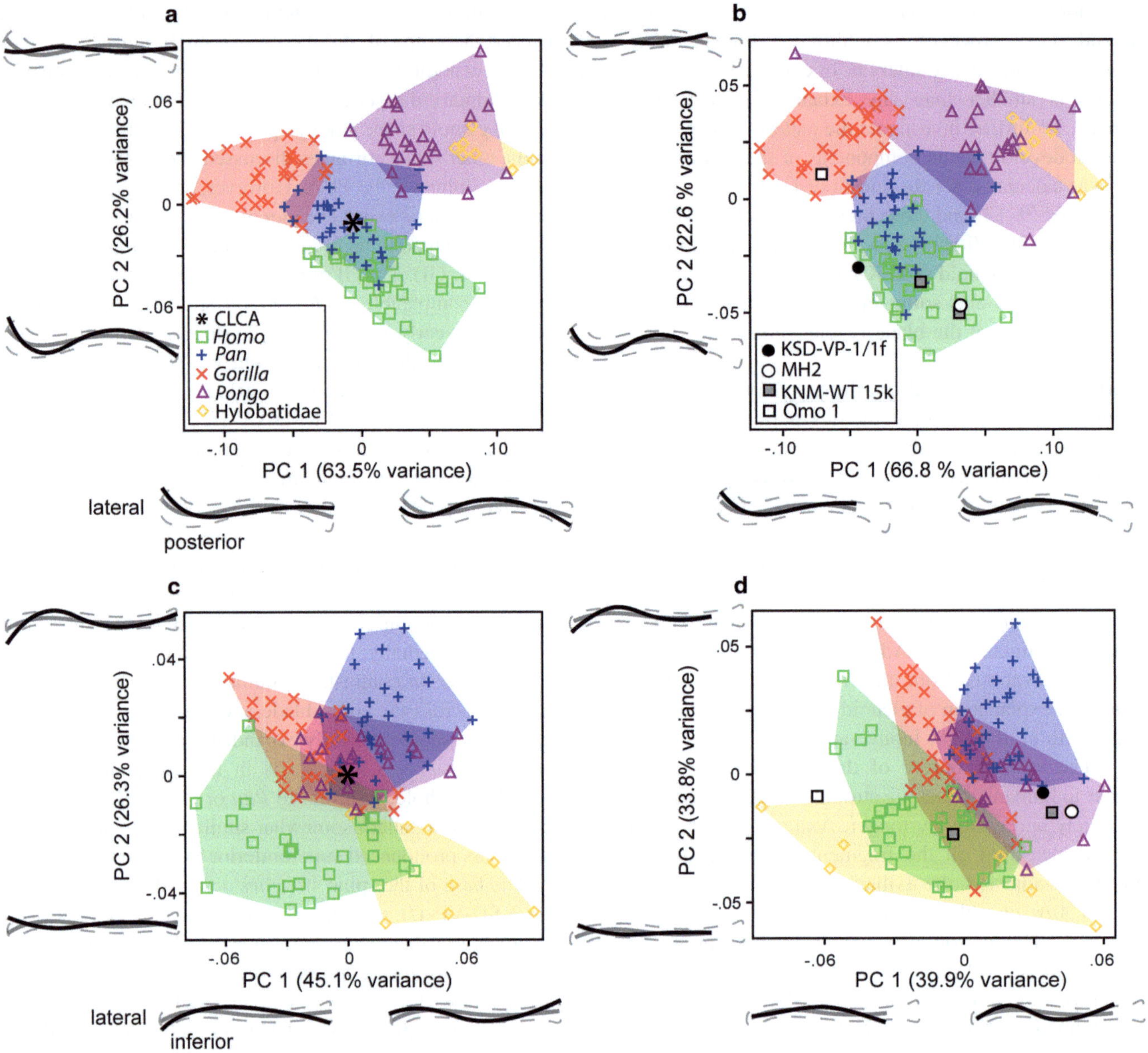

Fig. 6.7 Results of the PCA of the complete and *Kadanuumuu* datasets for transverse clavicle curvature (**a** and **b**, respectively) and the complete and *Kadanuumuu* datasets for coronal curvature (**c** and **d**, respectively). All *Australopithecus* specimens fall within the *Homo* cluster for transverse curvature but KSD-VP-1/1f, MH2 and the left clavicle of KNM-WT 15000 exhibit more superior curvature than *Homo*. The best estimate *Kadanuumuu* datasets are shown here, but results are consistent for the minimum and maximum estimates. The location of the CLCA model is overlaid in (**a**) and (**c**) for comparison, but did not contribute to the analysis

between the coracoid base and the superior angle. Types II through VI are increasingly narrow, constricted, and more readily recognized as a notch. About 80% of human scapulae exhibit a notch of either type II or III. Although some *Gorilla* and *Pan* scapulae exhibit a wide indentation that occupies much of the superior border (thus fitting the definition of the type I notch), the type II notch is present only in the *Homo* comparative sample, KSD-VP-1/1g and MH2. The Sts 7 scapula does not exhibit a type II notch, although a slight depression is present medial to the coracoid process.

Clavicle (KSD-VP-1/1f)

Geometric morphometrics

Results of the GM analyses of clavicle curvature are shown in Fig. 6.7. For both transverse and coronal curvature, the complete datasets (Fig. 6.7a, c) and the *Kadanuumuu* datasets (Fig. 6.7b, d) describe highly similar shape changes along the PC axes and taxa generally retain their positions relative to one another, despite the loss of some discrimination. With regard to the different estimates of the number of semilandmarks missing from KSD-VP-1/1f, results from the dataset in which the fewest landmarks are assumed missing and the best estimate are indistinguishable, while those from the dataset in which the most landmarks are assumed missing showed greater overlap among taxa and changes in the amount of variance accounted for by each axis. The results described below are consistent regardless of which *Kadanuumuu* clavicle dataset is considered.

In the analysis of transverse curvature, specimens falling at the positive extreme of PC 1 exhibit an internal curve followed by an external curve that is short relative to overall clavicle length. Those falling at the negative extreme lack an obvious internal curve but exhibit a relatively long external curve. This axis primarily distinguishes *Gorilla* from hylobatids. *Homo* is distinguished from Asian apes along PC 2, which captures variation in the overall pronouncement of transverse curvature. *Homo* and *Pan* overlap considerably along PC 2, especially when considering the *Kadanuumuu* landmark sets. All fossil specimens fall within *Homo* range, but KNM-WT 15000 D (right side) and KSD-VP-1/1f fall within or close to the region of overlap with *Pan.*[8] A pronounced S-shaped curve is also reconstructed at the CLCA node and thus appears to be a synapomorphy of the (*Pan*, *Homo*) clade, with stronger phylogenetic than functional implications.

Taxa differ less with regard to coronal curvature, even when the bone is complete. The first PC distinguishes specimens according to the direction and amount of curvature of the medial shaft. At the negative extreme are specimens that curve inferiorly both medially and laterally, such that the entire shaft is occupied by a single inferior curvature. These correspond to the "Type I" human clavicles described by Olivier (1951c, 1956). At the positive extreme of PC 1 are specimens that exhibit a superior curve medially, which occupies the majority of the length of the shaft. Although this axis primarily distinguishes *Homo* from hylobatids, most specimens are intermediate along this axis with a large amount of overlap between humans and non-human great apes. Specimens are further distinguished along PC 2 by the pronouncement of the inferior curve. *Homo* and hylobatids differ from the other apes in lacking a pronounced inferior curve at the lateral end of the shaft. However, this difference is obscured in the *Kadanuumuu* dataset (compare Fig. 6.7c, d). KSD-VP-1/1f, MH2 and KNM-WT 15000 C (left side) fall outside the range of the extant *Homo* sample. These specimens exhibit a superior curve that is similar to *Pongo* or *Pan* but lack the pronounced inferior curve that is typically present in *Pan.*[9] Voisin (2006b, 2008) has previously reported that Pleistocene hominin clavicles also tend to exhibit more superior curvature than those of extant humans, with a number of Neanderthals exhibiting coronal curvature similar to KNM-WT 15000 C.

The GM datasets for the clavicle in superior and posterior views were combined for the purpose of comparing overall clavicle shape among taxonomic groups. A Procrustes ANOVA demonstrates significant differences among hominoids ($p < 0.001$) and Table 6.9 reports the results of the pairwise comparisons. All comparative taxa differ significantly from each other. Procrustes distance indicates KSD-VP-1/1f is least similar to hylobatids and *Pongo* and most similar to KNM-WT 15000 and *Pan*. MH2 also shows close similarity with KNM-WT 15000.

Traditional metrics and qualitative features

Table 6.10 reports data on clavicle length. The length estimate for KSD-VP-1/1f reported here comes from a regression of length on midshaft size (see Table 6.10 footnotes), and it is in accordance with the estimate published previously by Haile-Selassie et al. (2010) that used midshaft circumference as the independent variable. The raw length estimates for KSD-VP-1/1f, A.L. 333x-6/9 and Stw 431 fall within the overlapping ranges of all great apes. In contrast, the MH2 clavicle is absolutely shorter than that of most great apes, except the bonobo (mean CL = 105 mm; 95% C. I. = 103–107 mm). After size normalization, the relative clavicle length (RCL) of KSD-VP-1/1f falls at the mean of modern humans and outside the 95% confidence intervals of all non-human apes (Table 6.10). The RCL values of Stw 431, MH2, and KNM-WT 15000 are similar to KSD-VP-1/1f and also exceed the *Pan* and *Gorilla* confidence intervals. Humans differ from *Pan, Gorilla,* and *Pongo* in the location and nature of the clavicular deltoid origin. In non-human apes the attachment surface tends to be located on the superior surface of the clavicle, whereas in humans it tends to be located further anteriorly, at the junction of the superior surface and anterior border. Although there is variation in this feature, all early hominin clavicles tend to exhibit a more anterior position. Differences in the medial extent of the deltoid attachment surface have been less discussed. The enthesis is laterally restricted in

[8]A.L. 333x-6/9 was not included in the GM analysis because it is less complete than KSD-VP-1/1f and thus would have required a further reduction in semilandmarks. Qualitatively, it exhibits a fairly pronounced internal curve and a relatively long external curve, but the S-shape is less pronounced than is the case for KSD-VP-1/1f. It would likely fall in the region where *Homo* and *Pan* overlap as well.

[9]A.L. 333x-6/9 and O.H. 48 are too incomplete to be included in the analysis. Qualitatively, both specimens exhibit a superior curve that is slightly less marked than in KSD-VP-1/1f. They lack a pronounced inferior curve, like all hominins considered here. Accordingly, they would likely fall near the other *Australopithecus* specimens and KNM-WT 15000 C. The KNM-WT 15000 specimens are more complete than KSD-VP-1/1f. When the full set of semilandmarks preserved in these specimens is considered, KNM-WT 15000 C remains outside the human range. KNM-WT 15000 D remains inside the human range along PC 1, but falls at the positive extreme of the extant *Homo* scatter (5th highest score).

Table 6.9 Results of pairwise comparisons of clavicle shape among extant hominoids and fossil hominins

	Omo 1	KNM-WT 15000 D	KNM-WT 15000 C	MH2	KSD-VP-1/1f	*Homo*	*Pan*	*Gorilla*	*Pongo*	Hylobatidae
Omo 1	–	0.25	0.12	0.12	0.47	**0.04**	**0.01**	**0.05**	**<0.001**	**<0.001**
KNM-WT 15000 D	0.13	–	0.90	0.83	0.81	1.0	0.61	0.18	0.19	0.10
KNM-WT 15000 C	0.15	0.06	–	1.0	0.60	0.43	0.35	**0.04**	0.19	0.13
MH2	0.15	0.07	0.02	–	0.56	0.35	0.29	**0.02**	0.17	0.15
KSD-VP-1/1f	0.10	0.07	0.09	0.09	–	0.30	0.50	0.20	**0.04**	**<0.01**
Homo	0.13	0.03	0.08	0.08	0.08	–	**<0.001**	**<0.001**	**<0.001**	**<0.001**
Pan	0.14	0.06	0.08	0.08	0.07	0.06	–	**<0.001**	**<0.001**	**<0.001**
Gorilla	0.13	0.10	0.13	0.14	0.09	0.09	0.07	–	**<0.001**	**<0.001**
Pongo	0.17	0.10	0.10	0.10	0.13	0.10	0.08	0.12	–	**0.01**
Hylobatidae	0.18	0.12	0.11	0.11	0.16	0.13	0.14	0.17	0.07	–

To quantify overall shape, the *Kadanuumuu* landmark datasets of transverse and coronal curvature are combined. Procrustes distances are reported below the diagonal and *p*-values are reported above the diagonal (significant differences in boldface). The values presented here derive from the best estimate of the number of semilandmarks missing from KSD-VP-1/1f but results are similar using the minimum and maximum estimates (see text for details)

Table 6.10 Summary of clavicle metrics

	CL	RCL	MS	MI
Omo 1	157.5	–	12.3	61.1
Neanderthals	156.1 ± 15.1[a] [155.0–178.5]	–	–	68.4[b] (86.6–50.2)
Sima de los Huesos	–	–	–	66.7 ± 8.1[b]
ATD6-50	161.5	–	–	75.5[b]
KNM-WT 15000 D	130.5[c]	3.2	10.0	59.7
KNM-WT 15000 C	130.4[c]	3.2	9.7	58.7
KNM-ER 1808	142.8[d]	–	11.0	84.9
O.H. 48	141–155[e]	–	–	70.3[b]
MH2	107.5	3.2	8.0	73.8
Stw 431	139.2[d]	3.3	10.8	66.7
A.L. 333x-6/9	134.0[d]	–	10.2	71.1
KSD-VP-1/1f	157.2[d]	3.3	12.9	53.9
Homo	139.5 (136.5–142.5) [113.0–165.0]	3.3 (3.3–3.4) [3.0–3.9]	10.1 (9.7–10.5) [7.2–16.0]	71.9 (69.9–74.0) [57.7–87.6]
Pan	121.9 (118.3–125.5) [100.0–169.0]	2.7 (2.6–2.8) [2.3–3.2]	9.7 (9.4–9.9) [7.7–12.4]	70.0 (68.0–72.1) [49.8–95.1]
Gorilla	164.0 (157.0–171.0) [129.0–201.0]	2.5 (2.4–2.6) [1.9–3.2]	13.5 (12.9–14.2) [10.3–17.4]	69.7 (67.1–72.2) [56.8–95.2]
Pongo	165.0 (157.3–172.7) [140.0–212.5]	3.6 (3.5–3.7) [3.2–3.9]	11.7 (11.0–12.3) [8.8–15.5]	82.4 (78.2–86.6) [65.3–99.0]
Hylobatidae	97.6 (87.3–108.0) [81.0–112.0]	4.6 (4.2–5.0) [4.0–5.4]	5.7 (5.1–6.4) [4.9–6.8]	60.7 (51.4–70.1) [48.7–79.7]

Data presented for living taxa are: means (95% confidence intervals) and [ranges]
CL clavicle length, *RCL* relative clavicle length, *MS* Midshaft size, *MI* midshaft index; see Table 6.5 for metric definitions
[a]Vandermeersch and Trinkaus (1995); mean length ± 1 SD
[b]Carretero et al. (1997, 1999); slight difference in measurement definition of midshaft index
[c]Walker and Leakey (1993) report preserved lengths. These specimens are not only unfused but also eroded at their acromial ends, so these measurements are slight underestimates of complete length, regardless of growth-related changes
[d]Clavicle length (CL) is estimated from regression of length on midshaft size (MS). The reported estimate derives from a reference sample of African apes and humans: $CL = 8.6 \times MS + 46.3$ ($F = 309.2$, $r^2 = 0.68$, $p < 0.001$). A longer estimate results if the reference sample includes all hominoids: $CL = 8.9 \times MS + 44.9$ ($F = 379.1$, $r^2 = 0.69$, $p < 0.001$)
[e]Day and Scheuer (1989); length is an estimated range

humans. In contrast, it can extend well past the external curve in non-human apes, occasionally near midshaft. KSD-VP-1/1f, A.L. 333x-6/9 and MH2 exhibit medial restricted deltoid attachments, similar to *Homo*. The deltoid attachment scar of KSD-VP-1/1f also qualitatively resembles that of *Homo*. The crest–depression–crest pattern described above (see "Descriptions") is fairly common in human clavicles, but not in those of non-human apes.

The subclavius muscle is present in both humans and non-human apes, but the associated subclavian groove occurs much more commonly among humans (Voisin 2012). KSD-VP-1/1f preserves a subclavian groove along the inferior face of the internal curve.

The KSD-VP-1/1f shaft is strongly SI compressed or "platicleidic" (Fig. 6.4g). The midshaft index is most similar to KNM-WT 15000 and differs from all extant hominoids (including extant humans). In the KSD-VP-1/1 and KNM-WT 15000 clavicles, the strong SI compression continues medially in the region of the pectoralis major origin. "Platicleidic" clavicles are present in high frequencies among Neanderthals and in the sample from Gran Dolina and Sima de los Huesos (Carretero et al. 1997, 1999), though the morphology is not ubiquitous among early hominins. For instance, the midshaft indices of A.L. 333x-6/9, MH2 and O.H. 48 do not differ significantly from extant *Homo*.

Partridge et al. (2003) commented on variation in shaft cross-section in the region of the conoid tubercle, noting that the cross-section of the Stw 606 clavicle is semilunar due to the presence of a dorsally projecting and flange-like conoid tubercle, whereas other *Australopithecus* clavicles from

Hadar and Sterkfontein exhibit an approximately trapezoidal cross-section with a well-defined superoposterior corner. The KSD-VP-1/1 clavicle exhibits the latter pattern.

Discussion

How Different Are Australopithecus *Scapulae?*

When data are collected from fossil specimens using comparable methods, all *Australopithecus* specimens are quite similar. The recently described specimens (DIK1-1, KSD-VP-1/1, and MH2) reinforce the previous observations on Sts 7 and A.L. 288-1 l, that the *Australopithecus* glenoid is more cranially oriented than modern humans. This is true regardless of variation in size. Yet these more complete specimens importantly permit the comparison of more discriminatory metrics. Spine orientation was likely previously underestimated for Sts 7 and A.L. 288-1 l (Larson 1995; Vrba 1979) due to insufficient preservation. As noted above (see "Comparative analysis of the scapula: Geometric morphometrics"), an accurate estimate of spine orientation relies on preservation of the medial portion of the spine. However, Sts 7 and A.L. 288-1 l preserve only about 20 mm of the spine at its lateral extent, where there are no systematic differences among taxa.

The KSD-VP-1/1, DIK-1-1, and MH2 scapulae were compared by Churchill et al. (2013) and Green and Alemseged (2012), but these studies use fossil data that differ in the manner that metrics were defined and collected. Green and Alemseged (2012) report a significant difference in scapula shape between KSD-VP-1/1g and the right DIK-1-1 scapula (the difference with left scapula approaches significance). This finding is heavily influenced by the difference in axillary border–spine angle, because (as shown above) spine orientation is the primary discriminator of ape scapula shape. When the axillary border–spine angle of KSD-VP-1/1g is measured using the methodology described by Green and Alemseged (2012), the degree of difference with DIK-1-1 is only about 6°, as opposed to the previously reported 12°. Given the standard deviations reported, this amount of difference can be explained by inter-individual variation. In comparing the available metrics (AGA, ASA, SGA, RSL, and RIFB), the Euclidean distance between DIK-1-1 and KSD-VP-1/1 is easily observed between two adults of the same species. Further, the landmark-based analysis of scapula shape does not support morphological affinity between the adult *Australopithecus* scapula and that of *Gorilla*. These results are not consistent with the proposal of *Gorilla*-like juvenile morphology and an African ape-like growth trajectory (Green and Alemseged 2012). More research is needed to better understand the ontogeny of the *Au. afarensis* scapula.

Churchill et al. (2013) report that the MH2 scapula differs significantly from *Homo* and is most similar to *Pongo*. In contrast, the results of the present GM analyses suggest that the MH2 scapula is morphologically most similar to extant *Homo* and to KSD-VP-1/1g. Similar results are obtained using the metric definitions described by Green and Alemseged (2012). These conflicting results are not easily explained, although the manner in which metrics are defined is clearly an important factor. The results reported here are, however, in accordance with the observations of Churchill and colleagues in identifying overall shape similarity between *Australopithecus* and *Pongo*. The most pronounced differences between MH2 and KSD-VP-1/1g are in infraspinous fossa shape and overall size. However, the magnitude of morphological difference between these two specimens (as judged by Procrustes distance or Euclidean distance) is comparable to what is seen among individuals of the same species. Thus, available evidence does not support the idea that either of these specimens is an extreme outlier.

Although the overall shape of KSD-VP-1/1g is slightly more similar to *Pongo* than to *Homo*, there are a number of features that clearly distinguish this specimen from a *Pongo* scapula. Most notably, it is less oblique (ASA and AGA fall outside the *Pongo* 95% confidence intervals) and the infraspinous fossa is broader (RIFB falls outside the observed *Pongo* range, comparable only to *Homo*). In addition, KSD-VP-1/1g lacks the narrow cranial extension of the glenoid fossa that characterizes *Pongo*. Although the superior angle and acromial angle are not preserved in KSD-VP-1/1g, the scapulae of the MH2 and DIK-1-1 partial skeletons suggest that a well-defined superior angle and acromial angle characterize the genus *Australopithecus*, whereas these features are poorly developed in Asian apes. In addition, overall clavicle curvature differs from *Pongo*.

Functional Morphology

Variation in spine and glenoid orientation has long been associated with differences in locomotion across primates, with more suspensory taxa exhibiting more oblique scapulae (Miller 1932; Ashton and Oxnard 1964; Young 2008). However, recent work shows that this form–function relationship does not necessarily hold within apes. Gorillas exhibit suspensory morphology although they are largely terrestrial (Remis 1998; Larson and Stern 2013) and orangutans are highly suspensory yet their scapula has quadrupedal affinities (Young 2008). In addition, Larson and

Stern (2013) recently reported that despite differences in scapula shape, chimpanzees and orangutans show similar recruitment patterns in rotator cuff muscles during suspension, climbing, and walking. These results prompted Larson and Stern to advise caution in making functional interpretations from scapula shape alone.

The human shoulder, on the other hand, cannot be understood in the same functional framework as other primates (i.e., the quadrupedal to suspensory gradient), because it does not play a central role in propulsion or support of the body. A habitually pendant position and purely manipulatory function of the human upper limb is thought to be indicated by a laterally facing glenoid, transverse spine, and broad infraspinous fossa (Inman et al. 1944; Roberts 1974; Larson 2013).

The preceding analyses show that the KSD-VP-1/1 scapula exhibits spine and glenoid orientation intermediate between humans and non-human apes, in conjunction with a broad infraspinous fossa. The intermediate morphology of the fossil specimen thus raises the question of which functional framework is more appropriate. Basic morphological comparisons cannot address this issue alone, but an understanding of evolutionary history can contribute to functional interpretation by considering trait polarity (Simpson 1953; Bock 1980; Harvey and Pagel 1991; Ward 2002, 2013). Given the CLCA model specified by the parsimony analysis, the *Australopithecus* scapula appears to be derived in the direction of *Homo*, whereas the clavicle is less derived. This is consistent with the identification of derived morphology in other regions of the upper limb (Drapeau et al. 2005; Drapeau and Ward 2007) and an extensively derived lower limb (reviewed in Lovejoy 2005a, b; Ward et al. 2012). Alternatively, if the CLCA model based on extant taxa is inaccurate and CLCA scapula and clavicle morphology more closely approximated that of *Australopithecus*, then *Au. afarensis* shoulder girdle morphology may indeed be primitive. A CLCA that lacks the pronounced scapular obliquity shared between *Pan* and *Gorilla* should be seriously considered, given the CLCA morphology suggested by *Ardipithecus ramidus* (Lovejoy et al. 2009).

Similarity in shoulder girdle function between humans and *Australopithecus* is also suggested by the location of muscle attachment sites, the trapezius in particular. The kinematic process of arm elevation differs between humans and non-human apes. The trapezius muscle inserts along the complete length of the scapular spine in humans and contributes to a force–couple mechanism that produces rotation of the scapula on the thorax (Inman et al. 1944; Lucas 1973; Itoi et al. 2009). In contrast, the trapezius attachment tends to terminate laterally in chimpanzee, orangutan, and especially gorilla scapulae. Electromyographic studies have shown that this muscle is inactive during arm elevation in non-human apes, indicating a difference in the mechanism of scapular rotation (Tuttle and Basmajian 1977; Larson et al. 1991). The KSD-VP-1/1 and MH2 scapulae exhibit the human-like pattern in this regard: the base of the spine reaches the medial border and muscle insertion extends medially, positioning the trapezius to contribute to scapular rotation in *Australopithecus*.

Shoulder Girdle Configuration

A number of questions about hominin shoulder evolution relate to the configuration of the girdle in articulation, including the elevation of the girdle and shoulder breadth. Identifying change in shoulder girdle configuration over the course of human evolution requires the ability to reconstruct these parameters from fossil remains.

Coronal curvature of the clavicle and the location of the deltoid origin are thought to reflect shoulder girdle height (Ohman 1986; Voisin 2006a; Ward 2013). Observations on coronal curvature have contributed to the proposal that the shoulder girdle was elevated in Neanderthals (Voisin 2006b, 2010), *Au. sediba* (Churchill et al. 2013), *Au. afarensis* and the CLCA (Larson 2007, 2009, 2013; but see Ohman (1986) for the opposite interpretation). KSD-VP-1/1f, MH2, and a number of other hominin specimens exhibit more superior curvature than is typical of extant humans, yet the location of the deltoid origin is human-like. Unfortunately, a predictive relationship between the above-mentioned features and girdle height has not yet been demonstrated.

The results of this study suggest that a relationship between coronal curvature and girdle height, if present, would not produce very precise predictions across genera due to the large amount of variation within species and overlap among genera (Fig. 6.7c). All non-human apes have an elevated girdle but do not share in common a pattern of curvature. Although a superior curve is present in hylobatids and pronounced in many *Pan* and *Pongo* specimens, the same is not true of many *Gorilla* specimens. Further, a fair number of *Homo* specimens exhibit some superior curvature, as indicated by the overlap between non-human apes in PC 1 score (Fig. 6.7c). Thus, it is possible to select extant *Homo* clavicles that are quite similar to early hominins or that differ markedly. In as far as coronal curvature is taken as an indication of shoulder girdle height, the observation of similarities between a number of Pliocene and Pleistocene specimens does not indicate a drastic change in girdle position with the evolution of *Homo erectus*, as previously suggested (Larson 2007, 2009, 2013; Bramble and Lieberman 2004; Roach et al. 2013). Meyer (2016) suggests that morphology of the KSD-VP-1/1 cervical vertebrae reflect some independence in rotation of the head and thorax.

Clavicle length is also discussed in relation to shoulder girdle configuration. The KSD-VP-1/1 clavicle is similar to extant *Homo* with regard to absolute and relative length. This is consistent with the suggestion that the upper thorax was broad in KSD-VP-1/1 (Latimer et al. 2016), without requiring the scapulae to be positioned laterally on the thorax (as described by Larson 2007, 2009). The MH2 clavicle is absolutely shorter, but not markedly different in relative length.

Conclusions

This study presents a detailed comparative analysis of the KSD-VP-1/1shoulder girdle. The scapula exhibits spine and glenoid orientation that is intermediate between human and non-human apes and an expanded infraspinous fossa, which is unique to humans. The clavicle exhibits a superior curve in the coronal plane that is distinct from modern humans but also present in some Pleistocene hominins. Muscle attachment locations show notable similarity to humans. The same features are present in other clavicle and scapula fossils attributed to *Australopithecus*. A number of features that are common in later Pleistocene hominins, but not in modern humans, are also present in KSD-VP-1/1. These include a superoinferiorly compressed and strongly twisted clavicle shaft, a relatively narrow glenoid, and notable glenoid retroversion.

When shoulder girdle morphology is compared to a CLCA model generated from extant hominoids, *Australopithecus* appears to be derived toward morphology associated with an emphasis on a manipulatory function of the pendant upper limb. It is not clear to what extent the primitive morphology present in KSD-VP-1/1 and MH2 would have enhanced climbing or suspensory ability relative to living humans. Future research aimed at identifying epigenetically responsive features that distinguish shoulders involved in propulsion from those that are not would be of great value in addressing questions about the habitual behavior of individuals.

Acknowledgments I am grateful to Yohannes Haile-Selassie for the invitation to study this material and to all the members of the Woranso-Mille Paleontological Research Project for their many years of work at the KSD-VP-1 excavation. I thank Scott Simpson, Richard Klein, and anonymous reviewers for their comments on this manuscript. Many museum curators and staff helped in accessing collections, including Lyman Jellema, Judy Chupasko, Wim Wendelen, Linda Gordon, Brendon Billings, Bernhard Zipfel, Stephany Potze, and Alemu Ademassu. A special thank you to members of the Malapa project and to Steve Churchill for casts of MH2. Field and laboratory research at Woranso-Mille is authorized by the Ethiopian Authority for Research and Conservation of Cultural Heritage and the Afar Regional State. Financial support for this research was provided by the Leakey Foundation and Stanford University.

References

Adams, D. C., Otárola-Castillo, E., & Paradis, E. (2013). Geomorph: an R package for the collection and analysis of geometric morphometric shape data. *Methods in Ecology and Evolution, 4*(4), 393–399.

Aiello, L., & Dean, C. (1990). *An introduction to human evolutionary anatomy*. London: Academic Press.

Alemseged, Z., Spoor, F., Kimbel, W. H., Bobe, R., Geraads, D., Reed, D., et al. (2006). A juvenile early hominin skeleton from Dikika, Ethiopia. *Nature, 443*(7109), 296–301.

Anderson, M. J. (2001). A new method for non-parametric multivariate analysis of variance. *Austral Ecology, 26*, 32–46.

Arnold, C., Matthews, L. J., & Nunn, C. L. (2010). The 10kTrees website: A new online resource for primate phylogeny. *Evolutionary Anthropology: Issues, News, and Reviews, 19*(3), 114–118.

Ashton, E. H., & Oxnard, C. E. (1963). The musculature of the primate shoulder. *The Transactions of the Zoological Society of London, 29*(7), 553–650.

Ashton, E. H., & Oxnard, C. E. (1964). Functional adaptations in the primate shoulder girdle. *Proceedings of the Zoological Society of London, 142*(1), 49–66.

Barker, K. B. (2006). Functional morphology of the hominoid clavicle. Master Thesis. University of Missouri-Columbia

Berger, L. R. (1994). Functional morphology of the hominoid shoulder, past and present. Ph.D. Dissertation. University of Witwatersrand, Johannesburg.

Berger, L. R., de Ruiter, D. J., Churchill, S. E., Schmid, P., Carlson, K. J., Dirks, P. H., et al. (2010). *Australopithecus sediba*: A new species of *Homo*-like australopith from South Africa. *Science, 328*(5975), 195–204.

Bock, W. J. (1980). The definition and recognition of biological adaptation. *American Zoologist, 20*(1), 217–227.

Bookstein, F. L. (1997). Landmark methods for forms without landmarks: Morphometrics of group differences in outline shape. *Medical Image Analysis, 1*(3), 225–243.

Bookstein, F. L. (1998). A hundred years of morphometrics. *Acta Zoologica Academiae Scientiarum Hungaricae, 44*(1–2), 7–59.

Bramble, D. M., & Lieberman, D. E. (2004). Endurance running and the evolution of *Homo. Nature, 432*(7015), 345–352.

Broca, P. (1878). Sur les indices de largeur de l'omoplate chez l'homme, les singes et dans la série des mammifères. *Bulletin et Memoires de la Societe d'Anthropologie de Paris, 1*, 66.

Broom, R., & Robinson, J. T. (1950). Further evidence of the structure of the Sterkfontein ape-man *Plesianthropus*. (Vol. 4, Transvaal Museum Memoir). Pretoria: Transvaal Musuem.

Campbell, B. G. (1966). *Human evolution: An introduction to man's adaptations*. Chicago: Aldine Publishing Company.

Carretero, J. M., Arsuaga, J. L., & Lorenzo, C. (1997). Clavicles, scapulae and humeri from the Sima de los Huesos site (Sierra de Atapuerca, Spain). *Journal of Human Evolution, 33*(2), 357–408.

Carretero, J. M., Lorenzo, C., & Arsuaga, J. L. (1999). Axial and appendicular skeleton of *Homo antecessor*. *Journal of Human Evolution, 37*(3), 459–499.

Churchill, S. E., Holliday, T. W., Carlson, K. J., Jashashvili, T., Macias, M. E., Mathews, S., et al. (2013). The upper limb of *Australopithecus sediba. Science, 340*(6129), 1233477.

Churchill, S. E., & Trinkaus, E. (1990). Neandertal scapular glenoid morphology. *American Journal of Physical Anthropology, 83*(2), 147–160.

Ciochon, R. L., & Corruccini, R. S. (1976). Shoulder joint of Sterkfontein *Australopithecus*. *South African Journal of Science, 72*, 80–82.

Claude, J. (2008). *Morphometrics with R* (Use R!), New York: Springer.

Darwin, C. (1871). *The descent of man*. London: J. Murray.

Day, M. H., & Scheuer, J. L. (1989). Olduvai hominid 48: A clavicle. *Human Biology Budapest, 19*, 9–14.

Dempster, W. T. (1965). Mechanisms of shoulder movement. *Archives of Physical Medicine and Rehabilitation, 46*, 49–70.

Di Vincenzo, F., Churchill, S. E., & Manzi, G. (2012). The Vindija Neanderthal scapular glenoid fossa: Comparative shape analysis suggests evo-devo changes among Neanderthals. *Journal of Human Evolution, 62*(2), 274–285.

Drapeau, M. S., & Ward, C. V. (2007). Forelimb segment length proportions in extant hominoids and *Australopithecus afarensis*. *American Journal of Physical Anthropology, 132*(3), 327–343.

Drapeau, M. S., Ward, C. V., Kimbel, W. H., Johanson, D. C., & Rak, Y. (2005). Associated cranial and forelimb remains attributed to *Australopithecus afarensis* from Hadar, Ethiopia. *Journal of Human Evolution, 48*(6), 593–642.

Eickstedt, E. F. (1925). Variationen am Axillarrand der Scapula: Sulcus axillaris teretis und Sulcus axillaris subscapularis. *Anthropologischer Anzeiger, 2*(4), 217–228.

Etter, H. F. (1984). L'omoplate des primates superieurs: la relation entre sa forme et sa fonction. *Archives suisses d'anthropologie generale, 48*(1), 31–51.

Frey, H. (1923). Untersuchungen über die Scapula, speziell über ihre äußere Form und deren Abhängigkeit von der Funktion. *Anatomy and Embryology (Berl), 68*(1), 277–324.

Gebo, D. L. (1996). Climbing, brachiation, and terrestrial quadrupedalism: historical precursors of hominid bipedalism. *American Journal of Physical Anthropology, 101*(1), 55–92.

Graves, W. W. (1921). The types of scapulae. A comparative study of some correlated characters in human scapulae. *American Journal of Physical Anthropology, 4*(2), 111–128.

Green, D. J., & Alemseged, Z. (2012). *Australopithecus afarensis* scapular ontogeny, function, and the role of climbing in human evolution. *Science, 338*(6106), 514–517.

Gregory, W. K. (1928). Were the ancestors of man primitive brachiators? *Proceedings of the American Philosophical Society, 67*(2), 129–150.

Haile-Selassie, Y., Latimer, B. M., Alene, M., Deino, A. L., Gibert, L., Melillo, S. M., et al. (2010). An early *Australopithecus afarensis* postcranium from Woranso-Mille, Ethiopia. *Proceedings of the National Academy of Sciences of the United States of America, 107* (27), 12121–12126.

Hammer, Ø., Harper, D. A. T., & Ryan, P. D. (2001). PAST: Paleontological Statistics Software Package for education and data analysis. *Palaeontolia Electronica 4*.

Harvey, P. H., & Pagel, M. (1991). *Comparative method in evolutionary biology (Oxford series in ecology and evolution)*. Oxford: Oxford University Press.

Hunt, K. D. (1991a). Mechanical implications of chimpanzee positional behavior. *American Journal of Physical Anthropology, 86*(4), 521–536.

Hunt, K. D. (1991b). Positional behavior in the Hominoidea. *International Journal of Primatology, 12*(2), 95–118.

Inman, V. T., Saunders, M., & Abbott, L. C. (1944). Observations on the function of the shoulder joint. *The Journal of Bone and Joint Surgery, 26*(1), 1–30.

Inouye, S. E., & Shea, B. T. (1997). What's your angle? size correction and bar-glenoid orientation in "Lucy" (AL 288-1). *International Journal of Primatology, 18*(4), 629–650.

Itoi, E., Morrey, B. F., & An, K. N. (2009). Biomechanics of the shoulder. In C. A. Rockwood, F. A. Matsen, S. B. Lippitt, & M. A. Wirth (Eds.), *The Shoulder* (pp. 213–265). Philadelphia: Elsevier Health Sciences.

Johanson, D. C., Lovejoy, C. O., Kimbel, W. H., White, T. D., Ward, S. C., Bush, M. E., et al. (1982). Morphology of the Pliocene partial hominid skeleton (AL 288-1) from the Hadar formation, Ethiopia. *American Journal of Physical Anthropology, 57*(4), 403–451.

Johnston, T., Whillis, J., & Gray, H. (1947). *Gray's Anatomy: descriptive and applied* (29th ed.). London: Longmans, Green and Co.

Jungers, W. L. (1994). Ape and hominid limb length. *Nature, 369*, 194.

Jungers, W. L., & Susman, R. L. (1984). Body size and skeletal allometry in African apes. In R. L. Susman (Ed.), *The pygmy chimpanzee* (pp. 131–177). New York: Plenum Press.

Klingenberg, C. P. (2011). MorphoJ: An integrated software package for geometric morphometrics. *Molecular Ecology Resources, 11*(2), 353–357.

Klingenberg, C. P., & Gidaszewski, N. A. (2010). Testing and quantifying phylogenetic signals and homoplasy in morphometric data. *Systematic Biology, 59*(3), 245–261.

Larson, S. G. (1993). Functional morphology of the shoulder in primates. In D. Gebo (Ed.), *postcranial adaptation in nonhuman primates* (pp. 45–69). DeKalb: Northern Illinois University Press.

Larson, S. G. (1995). New characters for the functional interpretation of primate scapulae and proximal humeri. *American Journal of Physical Anthropology, 98*(1), 13–35.

Larson, S. G. (2007). Evolutionary transformation of the hominin shoulder. *Evolutionary Anthropology, 16*(5), 172–187.

Larson, S. G. (2009). Evolution of the hominin shoulder: Early Homo. In: The first humans—origin and early evolution of the Genus Homo (pp. 65–75). Netherlands: Springer.

Larson, S. G. (2013). Shoulder morphology in early hominin evolution. In K. E. Reed, J. G. Fleagle, & R. E. Leakey (Eds.), *The paleobiology of Australopithecus* (pp. 247–261). Netherlands: Springer.

Larson, S. G., & Stern, J. T. (2013). Rotator cuff muscle function and its relation to scapular morphology in apes. *Journal of Human Evolution*

Latimer et al. (2016) The Thoracic Cage of KSD-VP-1/1. In *The Thoracic Cage of KSD-VP-1/1*

Larson, S. G., Stern, J. T., & Jungers, W. L. (1991). EMG of serratus anterior and trapezius in the chimpanzee: Scapular rotators revisited. *American Journal of Physical Anthropology, 85*(1), 71–84.

Leakey, R. E. (1973). Further evidence of lower Pleistocene hominids from east Rudolf, North Kenya, 1973. *Nature, 248*, 653–656.

LeGros Clark, W. E. (1947). Observations on the anatomy of the fossil Australopithecinae. *Journal of Anatomy, 81*(3), 300–333.

Losos, J. B. (2011). Seeing the forest for the trees: The limitations of phylogenies in comparative biology. *The American Naturalist, 177*(6), 709–727.

Lovejoy, C. O. (1981). The origin of man. *Science, 211*(4480), 341–350.

Lovejoy, C. O. (2005a). The natural history of human gait and posture. Part 1. Spine and pelvis. *Gait and Posture, 21*(1), 95–112.

Lovejoy, C. O. (2005b). The natural history of human gait and posture. Part 2. Hip and thigh. *Gait and Posture, 21*(1), 113–124.

Lovejoy, C. O., Johanson, D. C., & Coppens, Y. (1982). Hominid upper limb bones recovered from the hadar formation: 1974–1977 collections. *American Journal of Physical Anthropology, 57*(4), 637–649.

Lovejoy, C. O., Suwa, G., Simpson, S. W., Matternes, J. H., & White, T. D. (2009). The great divides: *Ardipithecus ramidus* reveals the postcrania of our last common ancestors with African Apes. *Science, 326*(5949), 73–73, 100–106.

Lucas, D. B. (1973). Biomechanics of the shoulder joint. *Archives of Surgery, 107*(3), 425–432.

Maddison, W. P. (1991). Squared-change parsimony reconstructions of ancestral states for continuous-valued characters on a phylogenetic tree. *Systematic Biology, 40*(3), 304–314.

Mensforth, R. P., Latimer, B., & Senturia, S. (1990). A review of the functional significance of the AL-288 axilloglenoid angle. *American Journal of Physical Anthropology, 81*(2), 267–268.

Meyer, M. R. (2016). Chapter 5: The cervical vertebrae of KSD-VP-1/1. In: Y. Haile-Selassie & D. F. Su (Eds.), *The postcranial anatomy of* Australopithecus afarensis: *New insights from KSD-VP-1/1* (pp. 63–111). Dordrecht: Springer.

Miller, R. A. (1932). Evolution of the pectoral girdle and fore limb in the Primates. *American Journal of Physical Anthropology, 17*(1), 1–56.

Mitteroecker, P., & Gunz, P. (2009). Advances in geometric morphometrics. *Evolutionary Biology, 36*(2), 235–247.

Mivart, G. (1867). On the appendicular skeleton of the primates. *Philosophical Transactions of the Royal Society of London, 157*, 299–429.

Napier, J. R. (1965). Comment on New discoveries in Tanganyika: Their bearing on hominid evolution by P.V. Tobias. *Current Anthropology, 6*(4), 402–403.

Ohman, J. C. (1986). The first rib of hominoids. *American Journal of Physical Anthropology, 70*(2), 209–229.

Olivier, G. (1951a). Anthropologie de la clavicule. I: La clavicule de l'Australien. *Bulletins et Mémoires de la Société d'anthropologie de Paris, 2*(1), 67–85.

Olivier, G. (1951b). Anthropologie de la clavicule. II: La clavicule du Vietnamien. *Bulletins et Mémoires de la Société d'anthropologie de Paris, 2*(1), 86–99.

Olivier, G. (1951c). Anthropologie de la clavicule. III: La clavicule du francais. *Bulletins et Mémoires de la Société d'anthropologie de Paris, 2*(4), 121–157.

Olivier, G. (1951d). Techniques de mesure des courbures de la clavicule. *C. R. Associated Anatomy, 69*, 753–764.

Olivier, G. (1953). Anthropologie de la clavicule. V: La clavicule des Mélanésiens. *Bulletins et Mémoires de la Société d'anthropologie de Paris, 4*(5), 553–561.

Olivier, G. (1954a). Anthropologie de la clavicule. VII: La clavicule des Japonais. *Bulletins et Mémoires de la Société d'anthropologie de Paris, 5*(1), 47–56.

Olivier, G. (1954b). Anthropologie de la clavicule. VIII: La clavicule des Amérindiens. *Bulletins et Mémoires de la Société d'anthropologie de Paris, 5*(3), 144–153.

Olivier, G. (1956). Anthropologie de la clavicule. XIII: Conclusions générales. *Bulletins et Mémoires de la Société d'anthropologie de Paris, 7*(5), 404–447.

Oxnard, C. E. (1968a). A note on the fragmentary Sterkfontein Scapula. *American Journal of Physical Anthropology, 28*(2), 213–217.

Oxnard, C. E. (1968b). A note on the Olduvai clavicular fragment. *American Journal of Physical Anthropology, 29*(3), 429–431.

Parsons, F. G. (1914). The characters of the English thigh-bone. *Journal of Anatomy and Physiology, 48*(3), 238–267.

Partridge, T. C., Granger, D. E., Caffee, M. W., & Clarke, R. J. (2003). lower pliocene hominid remains from Sterkfontein. *Science, 300* (5619), 607–612.

Pickering, R., Dirks, P. H., Jinnah, Z., de Ruiter, D. J., Churchil, S. E., Herries, A. I., et al. (2011). *Australopithecus sediba* at 1.977 Ma and implications for the origins of the genus *Homo*. *Science, 333* (6048), 1421–1423.

Rasband, W. S. (2012). *ImageJ*. Bethesda, Maryland, USA: U.S. National Institutes of Health.

Remis, M. J. (1998). The Gorilla Paradox. In E. Strasser, J. G. Fleagle, A. L. Rosenberger, & H. M. McHenry (Eds.), *Primate locomotion: Recent advances* (pp. 95–106). New York: Springer.

Rengachary, S. S., Burr, D., Lucas, S., Hassanein, K. M., Mohn, M. P., & Matzke, H. (1979). Suprascapular entrapment neuropathy: A clinical, anatomical, and comparative study. Part 2: Anatomical study. *Neurosurgery, 5*(4), 447–451.

Roach, N. T., Venkadesan, M., Rainbow, M. J., & Lieberman, D. E. (2013). Elastic energy storage in the shoulder and the evolution of high-speed throwing in *Homo*. *Nature, 498*(7455), 483–486.

Roberts, D. (1974). structure and function of the Primate Scapula. In F. A. Jenkins (Ed.), *Primate locomotion* (pp. 171–200). New York: Academic Press.

Rohlf, F. J. (1998). *Tps software series. Ecology & Evolution*. New York: SUNY, Stony Brook.

Rohlf, F. J. (2001). Comparative methods for the analysis of continuous variables: geometric interpretations. *Evolution, 55*(11), 2143–2160.

Ruff, C. (1990). Body mass and hindlimb bone cross-sectional and articular dimensions in anthropoid primates. In J. Damuth & B. J. MacFadden (Eds.), *Body size in Mammalian Paleobiology* (pp. 119–150). Cambridge: Cambridge University Press.

Saylor et al. (2016) The Geologic Context of Korsi Dora and the Partial3 Skeleton KSD-VP-1/1. In *The Postcranial Anatomy of* Australopithecus afarensis.

Schultz, A. H. (1930). The skeleton of the trunk and limbs of higher primates. *Human Biology, 2*(3), 303–438.

Schultz, A. H. (1936). Characters common to higher primates and characters specific for man. *The Quarterly Review of Biology, 11*(3), 259–283 and *11*(4), 425–455.

Schultz, A. H. (1941). Growth and development of the orangutan. *Contributions to Embryology, 182*, 57–110.

Schultz, A. H. (1944). Age changes and variability in gibbons. A morphological study on a population sample of a man-like ape. *American Journal of Physical Anthropology, 2*(1), 1–129.

Senut, B. (1980). New data on the humerus and its joints in Plio-Pleistocene hominids. *Collegium Anthropologicum, 4*(1), 87–93.

Senut, B., Nakatsukasa, M., Kunimatsu, Y., Nakano, Y., Takano, T., Tsujikawa, H., et al. (2004). Preliminary analysis of *Nacholapithecus* scapula and clavicle from Nachola, Kenya. *Primates, 45*(2), 97–104.

Shefelbine, S. J., Tardieu, C., & Carter, D. R. (2002). Development of the femoral bicondylar angle in hominid bipedalism. *Bone, 30*(5), 765–770.

Simpson, G. G. (1953). *The major features of evolution*. New York: Columbia University Press.

Slice, D. E. (Ed.). (2005). *Modern morphometrics in physical anthropology*. New York: Kluwer Press.

Sobush, D. C., Simoneau, G. G., Dietz, K. E., Levene, J. A., Grossman, R. E., & Smith, W. B. (1996). The lennie test for measuring scapular position in healthy young adult females: A reliability and validity study. *The Journal of Orthopaedic and Sports Physical Therapy, 23*(1), 39–50.

Stern, J. T. (2000). Climbing to the top: A personal memoir of *Australopithecus afarensis*. *Evolutionary Anthropology, 9*(3), 113–133.

Stern, J. T., & Susman, R. L. (1983). The locomotor anatomy of *Australopithecus afarensis*. *American Journal of Physical Anthropology, 60*(3), 279–317.

Stewart, T. D. (1962). Neanderthal Scapulae with special attention to the Shanidar Neanderthals from Iraq. *Anthropos, 57*(3/6), 779–800.

Strait, D. S., & Grine, F. E. (2004). Inferring hominoid and early hominid phylogeny using craniodental characters: The role of fossil taxa. *Journal of Human Evolution, 47*(6), 399–452.

Susman, R. L., Stern, J. T., & Jungers, W. L. (1984). Arboreality and bipedality in the Hadar hominids. *Folia Primatologica, 43*(2–3), 113–156.

Tardieu, C. (1981). Morphofunctional analysis of the articular surfaces of the knee joint in primates. In A. B. Chiarelli & R. S. Corruccini (Eds.), *Primate evolutionary biology* (pp. 68–80). Berlin: Springer Verlag.

Toussaint, M., Macho, G. A., Tobias, P. V., Partridge, T. C., & Hughes, A. R. (2003). The third partial skeleton of a late Pliocene hominin

(Stw 431) from Sterkfontein, South Africa. *South African Journal of Science, 99*, 215–223.

Trinkaus, E. (1983). *The Shanidar Neandertals*. New York, NY: Academic Press.

Trinkaus, E. (2008). Kiik-Koba 2 and Neandertal axillary border ontogeny. *Anthropological Science, 116*(3), 231–236.

Tuttle, R. H., & Basmajian, J. V. (1977). Electromyography of pongid shoulder muscles and hominoid evolution. I. Retractors of the humerus and rotators of the scapula. *Yearbook of Physical Anthropology, 20*, 491–497.

Vallois, H. V. (1928). L'omoplate humaine, étude anatomique et anthropologique. I. *Bulletins et Mémoires de la Société d'anthropologie de Paris, 9*(4), 129–168.

Vallois, H. V. (1929). L'omoplate humaine, étude anatomique et anthropologique. III. *Bulletins et Mémoires de la Société d'anthropologie de Paris, 10*(1), 110–191.

Vallois, H. V. (1932). L'omoplate humaine, étude anatomique et anthropologique. VI. *Bulletins et Mémoires de la Société d'anthropologie de Paris, 3*(1), 3–153.

Vallois, H. V. (1946). L'omoplate humaine, étude anatomique et anthropologique. XI. *Bulletins et Mémoires de la Société d'anthropologie de Paris, 7*(1), 16–100.

Vallois, H. V. (1977). Interpretation of the Scapula of *Plesianthropus transvaalensis*. *Journal of Human Evolution, 6*(8), 675–679.

Vandermeersch, B., & Trinkaus, E. (1995). The postcranial remains of the Régourdou 1 Neanderthal: The shoulder and arm remains. *Journal of Human Evolution, 28*(5), 439–476.

Voisin, J.-L. (2006a). Clavicle, a neglected bone: Morphology and relation to arm movements and shoulder architecture in primates. *The Anatomical Record. Part A, Discoveries in Molecular, Cellular, and Evolutionary Biology, 288*(9), 944–953.

Voisin, J.-L. (2006b). Krapina and other Neanderthal clavicles: A peculiar morphology? *Periodicum Biologorum, 108*(3), 331–339.

Voisin, J.-L. (2008). The Omo 1 hominin clavicle: Archaic or modern? *Journal of Human Evolution, 55*(3), 438–443.

Voisin, J.-L. (2012). Les caractères discrets des membres supérieurs: Un essai de synthèse des données. *Bulletins et Mémoires de la Société d'anthropologie de Paris, 24*(3–4), 107–130.

Voisin, J.-L. (2010). L'architecture de l'épaule au sein du genre Homo: nouvelles interprétations.*L'Anthropologie, 114*(3), 354–369.

Vrba, E. S. (1979). A new study of the scapula of *Australopithecus africanus* from Sterkfontein. *American Journal of Physical Anthropology, 51*(1), 117–129.

Walker, A. C., & Leakey, R. (1993). The postcranial bones. In A. Walker & R. Leakey (Eds.), *The Nariokotome Homo erectus skeleton* (pp. 95–160). Cambridge: Harvard University Press.

Walker, A. C., & Pickford, M. (1983). New postcranial fossils of *Proconsul africanus* and *Proconsul nyanzae*. In R. L. Ciochon & R. S. Corruccini (Eds.), *New interpretations of ape and human ancestry* (pp. 325–351). New York: Plenum Press.

Ward, C. V. (2002). Interpreting the posture and locomotion of *Australopithecus afarensis*: Where do we stand? *Yearbook of Physical Anthropology, 45*, 185–215.

Ward, C. V. (2007). Postcranial and locomotor adaptations of hominoids. In W. Henke, & I. Tattersall (Eds.), *Handbook of Paleoanthropology, Volume II: Primate Evolution and Human Origins* (pp. 1011–1030). Berlin Heidelberg: Springer.

Ward, C. V. (2013). Postural and locomotor adaptations of Australopithecus species. In K. E. Reed, J. G. Fleagle, & R. E. Leakey (Eds.), *The paleobiology of Australopithecus* (pp. 235–245). Netherlands: Springer.

Ward, C. V., Kimbel, W. H., Harmon, E. H., & Johanson, D. C. (2012). New postcranial fossils of *Australopithecus afarensis* from Hadar, Ethiopia (1990–2007). *Journal of Human Evolution, 63* (1), 1–51.

Young, N. M. (2006). Function, ontogeny and canalization of shape variance in the primate scapula. *Journal of Anatomy, 209*(5), 623–636.

Young, N. M. (2008). A comparison of the ontogeny of shape variation in the anthropoid scapula: Functional and phylogenetic signal. *The American Journal of Physical Anthropology, 136*(3), 247–264.

Zelditch, M. L., Swiderski, D. L., Sheets, H. D., & Fink, W. L. (2004). *Geometric morphometrics for biologists: A primer*. New York: Academic Press.

Chapter 7
The Thoracic Cage of KSD-VP-1/1

Bruce M. Latimer, C. Owen Lovejoy, Linda Spurlock, and Yohannes Haile-Selassie

Abstract Ribs are naturally fragile and, as a consequence, are rarely preserved in the fossil record. The costal elements recovered from the KSD-VP-1/1 partial skeleton are important evidence allowing reconstruction of the hominin thoracic cage. The ribs of KSD-VP-1/1 are examined with respect to their implications for the evolution of *Australopithecus afarensis* thoracic morphology. Angulation and torsion along the rib corpus and rib declination indicate a broad upper thorax and a deeply invaginated thoracic vertebral column. Implications for the early hominin thoracic *bauplan* are discussed.

Keywords *Australopithecus* • Thorax • Ribs

Introduction

The discovery of the 3.6-million-year old *Australopithecus afarensis* (KSD-VP-1/1) partial skeleton (Haile-Selassie et al. 2010a, b) from Woranso-Mille, Ethiopia provides rare evidence regarding the evolutionary development of the human and great ape thoracic cages. Differences in thoracic skeletal anatomies between modern African apes and humans have been noted earlier (Keith 1923; Schultz 1961; Latimer and Ward 1993; Jellema et al. 1993; Lovejoy 2005; Gomez-Olivencia et al. 2009; Ward et al. 2012; Bastir et al. 2013). However, owing to their natural fragility, there has been a lack of sufficient fossil rib material that could allow reconstruction of the chronology of these changes. Indeed, debate continues regarding the anatomy of the primitive hominin thoracic shape with some researchers suggesting that the chimpanzee/human last common ancestor (CLCA) had an African ape-like, inverted, "funnel-shaped" thorax (Schmid 1991; Schmid et al. 2013), while others reconstruct the CLCA as having a human-like "barrel-shaped" thorax (Lovejoy 2005; Haile-Selassie et al. 2010b). This simplistic dichotomy between "funnel-shaped" and "barrel-shaped" may too greatly simplify a more complex problem, a possibility that is discussed below.

The six ribs included in the KSD-VP-1/1 partial skeleton (Fig. 7.1), even while fragmentary, shed light on this issue, and allow the reconstruction of the thoracic shape in *Au. afarensis*. Because the features that characteristically distinguish hominin ribs are functionally interrelated (declination, torsion, and neck/shaft angle), even fragmentary ribs can be highly informative allowing reliable discrimination between hominoid thoracic shapes. They also provide evidence regarding the associated functions of the shoulder, the pelvis, and the lumbar and thoracic spines. As many of the anatomical features that we discuss below are unique to humans (and our bipedal ancestors), we also, where appropriate, include brief mention of certain pathological conditions that are similarly confined to humans. These conditions shed light on the selective pressures encountered by early hominins and are the direct result of our habitually upright posture.

Anatomical Descriptions

The following descriptions are intended for use by researchers utilizing cast replicas of the original specimens. Real anatomy is often difficult to discern from casts

B.M. Latimer (✉)
Department of Orthodontics School of Dental Medicine, Case Western Reserve University, Cleveland, OH 44106, USA
e-mail: bxlatimer@aol.com

B.M. Latimer · C.O. Lovejoy · L. Spurlock · Y. Haile-Selassie
Department of Physical Anthropology, Cleveland Museum of Natural History, 1 Wade Oval Drive, Cleveland, OH 44106, USA
e-mail: olovejoy@aol.com

Y. Haile-Selassie
e-mail: yhailese@cmnh.org

C.O. Lovejoy · L. Spurlock
Department of Anthropology and Division of Biomedical Sciences, Kent State University, Kent, OH 44242, USA
e-mail: lspurloc@kent.edu

Yohannes Haile-Selassie and Denise F. Su (eds.), *The Postcranial Anatomy of Australopithecus afarensis: New Insights from KSD-VP-1/1*, Vertebrate Paleobiology and Paleoanthropology, DOI 10.1007/978-94-017-7429-1_7

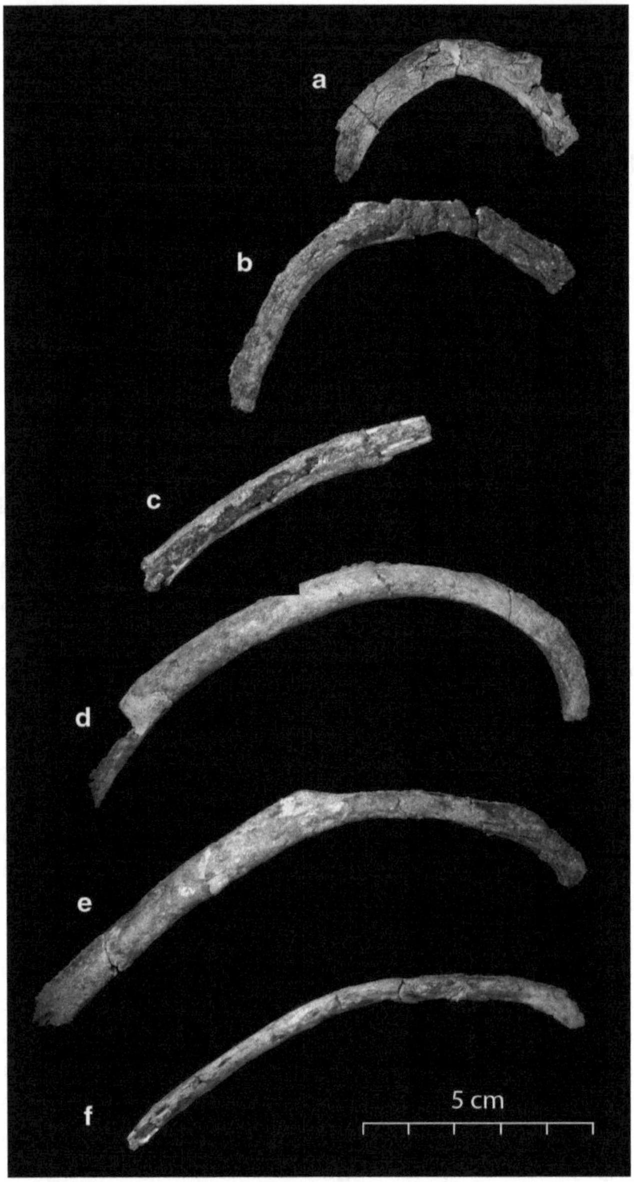

Fig. 7.1 Cranial views of the ribs of KSD-VP-1/1. **a** KSD-VP-1/1n, left second rib, reversed in image; **b** KSD-VP-1/1q, R. fifth, sixth or seventh rib fragment; **c** KSD-VP-1/1s, mid-thoracic rib fragment; **d** KSD-VP-1/1o, R. seventh or eighth rib fragment; **e** KSD-VP-1/1p, R. eighth or ninth rib fragment; **f** KSD-VP-1/1r, L. eleventh rib

particularly if there is significant postmortem damage, reconstruction, and/or supplementation of the original fossil prior to casting. Adherent matrix can also cause difficulties when using replicas. Following this, it should be noted that postmortem damage has removed many of the traditional anatomical landmarks and that caution should be practiced when reviewing the measurements presented here.

In addition, confusion can also result when using qualitative and relative comparative terminology (Franciscus and Churchill 2002) such as "highly curved," "roughened," or "robust." In the following descriptions, the reader should be aware that such qualitative terminology refers to ranges of variability normally encountered in human anatomy. When used, comparisons with other taxa will be explicitly noted in an attempt to reduce potential ambiguity. All measurements are presented in Table 7.1.

KSD-VP-1/1n (Figs. 7.1a, 7.2, and 7.3)

KSD-VP-1/1n is a fragmentary left second rib. Approximately two-thirds of the corpus is present from the head to the most ventral point, which is obliquely fractured such that the ventral third of the specimen is not present. Several postmortem transverse fractures occur along the body, most of which oppose well and do not alter the original anatomy. A large triangular fragment is absent in the vicinity of the posterior angle, the size of which required reinforcement of the rib body (see Figs. 7.1a, 7.2, and 7.3). This loss of bone has resulted in damage to the tubercle and required reconstruction that may have slightly altered the rib's original anatomy in the direction of increasing the curvature of the pleural margin (decreasing the radius of curvature). As a consequence, estimates of the original curvature and copular volume may be somewhat underestimated. Nevertheless, its curvature radius falls well within the human range and substantially below those of the African apes (see below and Haile-Selassie et al. 2010b). Despite the postmortem damage, it is nevertheless apparent that the curvature of the pleural margin indicates a human-like neck-corpus angle and the deep invagination of the vertebral column into the thorax. Several flakes of bone are missing from the cranial and caudal surfaces of the corpus but these do not alter overall anatomy. The crest for *M. scalene* posterior, although somewhat eroded, is visible along the dorsal margin of the corpus. There is a roughened tuberosity for the attachment of *M. serratus* anterior. The caudal surface is smooth and the costal groove, which begins immediately ventral to the tubercle, is shallow. The rib corpus is typically flat ventral to the tubercle.

KSD-VP-1/1q (Fig. 7.1b)

KSD-VP-1/1q is a fragmentary right fifth, sixth, or seventh rib. Postmortem damage does not allow for a more accurate numerical designation. The head and neck are eroded and no original subchondral surfaces remain. A triangular flake of bone is lost associated with a well-opposed transverse fracture across the costal angle. It is nevertheless apparent from the remaining portions that the costal angle is flexed indicating a deep invagination of the mid-thoracic vertebral

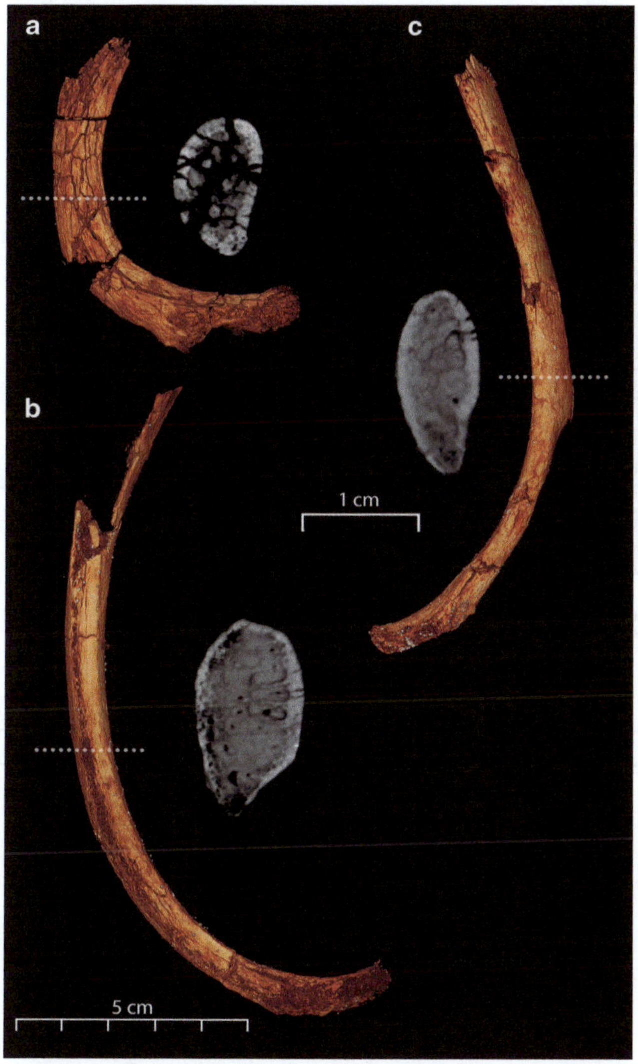

Fig. 7.2 CT-based views of KSD-VP-1/1 ribs (KSD-VP-1/1n, b, q). Postmortem damage is evident but the original curvature of the pleural margin remains largely undistorted. Dashed lines indicate areas of cross-sections. Figure from Ryan and Sukhdeo (2016)

column. In addition, despite the significant surface flaking and cortical exfoliation, the original curvature of the rib is apparent especially along the pleural margin. If oriented so that the undamaged pleural surface of the head and neck is in a vertical plane (see Jellema et al. 1993), it is clear that the remaining ventral portion of the rib corpus declines inferiorly and sigmoidally twists (caudal margin inferomedially, cranial edge superolaterally) along its long axis.

KSD-VP-1/1s (Fig. 7.1c)

KSD-VP-1/1s is a badly damaged middle segment of a mid-thoracic rib. Judging from its size and curvature, it is likely from a sixth, seventh, or eighth rib. Much of the original cortical surface is exfoliated, and no original subchondral surfaces remain intact.

KSD-VP-1/1o (Fig. 7.1d)

KSD-VP-1/1o is a fragment from the dorsal half of a right seventh or eighth rib. Much of the original cortical surface is damaged particularly around the head and neck, and no subchondral surfaces remain. Adhering matrix remains in several areas producing a roughened surface that should not be interpreted as anatomical. Despite the significant postmortem damage, the overall curvature and shaft dimensions are interpretable. When the pleural surface of the head and neck are oriented vertically (see Jellema et al. 1993), the corpus of the rib demonstrably declines and twists about its long axis. The costal neck axis is strongly flexed indicating posteriorly directed transverse vertebral processes and an invaginated thoracic vertebral column. The ventral third of the shaft is flattened and has a sharp inferior margin. Although erosion has damaged much of the pleural surface, a shallow costal groove is apparent.

KSD-VP-1/1p (Fig. 7.1e)

KSD-VP-1/1p is the dorsal half of a right eighth or ninth rib. Several transverse cracks occur along the shaft but these oppose well and do not distort the original curvature or torsion. The surface is eroded over much of the specimen although small areas of subchondral bone remain on the head and neck. The posterior angle is rugose and ventral to it the corpus flattens and presents a sharp caudal margin. The costal groove begins at the posterior angle and is deep. The shaft declines markedly such that when the cranial surface of the specimen is placed on a horizontal plane the ventral half of the corpus does not touch the surface. Associated with this declination is the obvious sigmoidal torsion along the rib corpus.

KSD-VP-1/1r (Fig. 7.1f)

KSD-VP-1/1r is the dorsal half of a left eleventh rib. Much of the surface is eroded, and several transverse cracks occur along the corpus but these oppose well and do not distort the original curvature and torsion. The surface around the head and neck is damaged and no original subchondral bone remains. The cranial margin of the shaft is sharp particularly along the ventral third of the

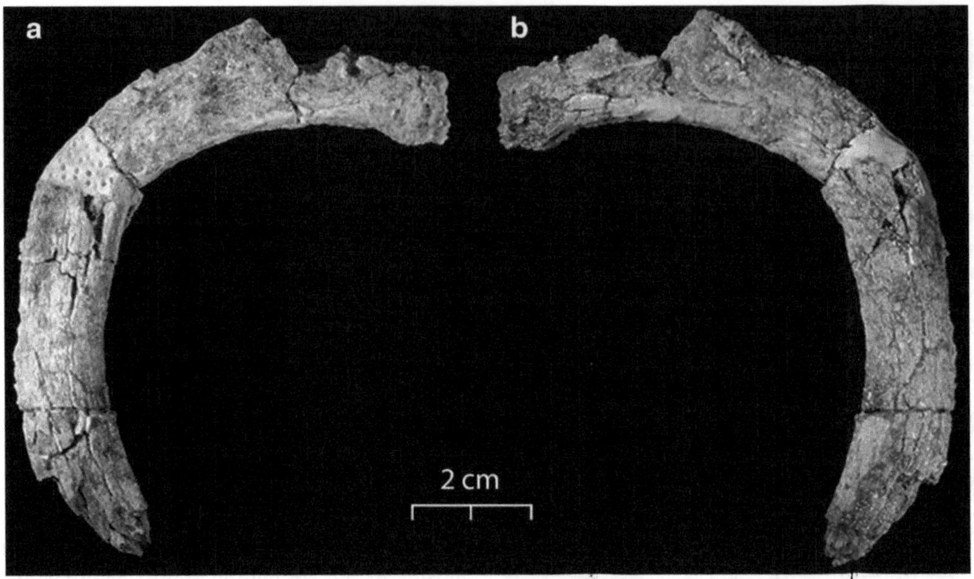

Fig. 7.3 KSD-VP-1/1n, second left rib. Cranial and caudal views, right and left, respectively

Table 7.1 Linear measurements of the preserved KSD-VP-1/1 ribs

Measurement	KSD-VP-1/1n	KSD-VP-1/1o	KSD-VP-1/1p	KSD-VP-1/1q	KSD-VP-1/1r	KSD-VP-1/1s
Maximum preserved chord length (MCL)	75.1	177.8	173.3	153.6	147.3	101.3
Maximum perpendicular from MCL to pleural margin	29.5	54	34.4	47.4	25.4	–
Transverse corpus dimension, estimated original mid-shaft – transverse	13.7	17.8	16.1	18.7	15.3	–
Craniocaudal corpus dimension, estimated original mid-shaft – cranial–caudal	8.0	9.5	8.1	10.0	6.7	–
Center of articular head to the center of tubercle	30.0	–	–	–	–	–

All measurements are in millimeters (mm)

fragment. A costal groove is apparent extending from the posterior angle ventrally. When the specimen is placed on a horizontal planar surface with its cranial edge facing down, the ventral half of the fragment does not touch the surface, an exercise that demonstrates rib declination. Associated with this declination is the sigmoidal torsion along the corpus.

Thoracic Shape and Function

In order to provide an appropriate anatomical context within which to view the KSD-VP-1/1 thoracic skeleton, it is necessary to compare and contrast human and African ape thoraces (see Fig. 7.4). Below we describe the anatomy and function of the thoraces of modern human and African apes and the implications for KSD-VP-1/1 and *Australopithecus* in general.

Modern Human

The human thoracic *bauplan* includes several linked, diagnostic features in the rib cage, which allow it to be reliably distinguished from that of the African apes. These include a relatively broad cupola or superior thoracic area, a ventrally invaginated vertebral column, and the declination and torsion of the rib bodies (Schultz 1961; Latimer and Ward 1993; Jellema et al. 1993; Franciscus and Churchill 2002; Haile-Selassie 2010b; Ward et al. 2012).

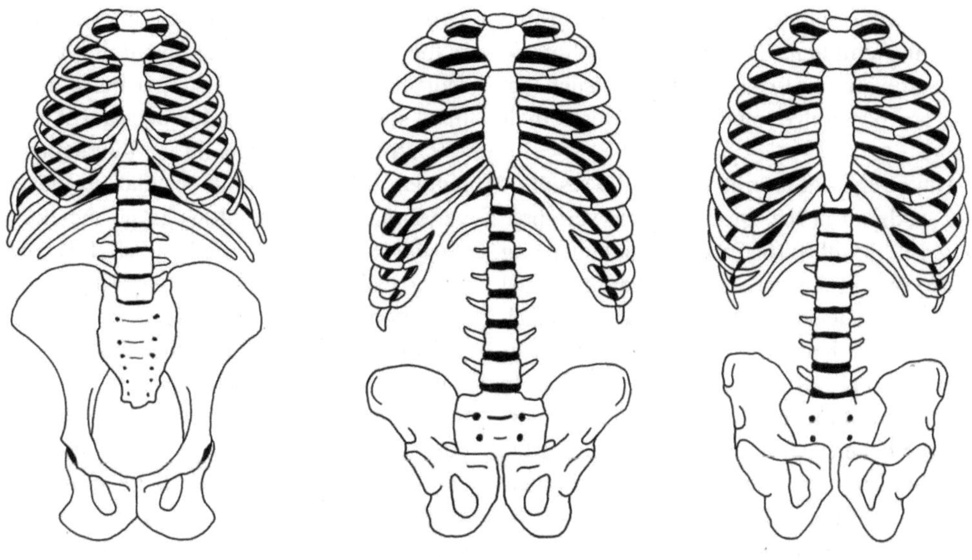

Fig. 7.4 A comparison of the thoracic cages and the pelvic girdles in a chimpanzee (left), KSD-VP-1/1 (center), and a human (right). The "barrel-shaped" human rib cage is transversely broad in its upper portion and narrowed at its bottom to conform to the sagittalized ilia. In contrast, the ape rib cage is narrowly constricted in its cranial segment and flares in its inferior segments to conform to the transversely broad false pelvis – forming the inverted "funnel-shaped" thorax. The KSD-VP-1/1 thorax is broad in its upper portion like the human but flares somewhat in its inferior portion to conform to the typical *Australopithecus* platypelloid pelvis – thus forming a "bell-shaped" thorax. Also note the long and flexible lumbar columns in the hominins in direct contrast to the ape's short and inflexible lumbar column

As noted, modern humans are distinguished from the African apes by having a "barrel-shaped" thorax in contrast to what is described as an inverted "funnel-shaped" thorax in chimpanzees and gorillas (Schultz 1961; Latimer and Ward 1993; Jellema et al. 1993). The "barrel-shaped" human condition is largely a product of the relatively broad upper thorax juxtaposed to the constricted and transversely narrow lower rib cage (see Fig. 7.4; Latimer and Ward 1993 and discussion below). The tapering or "waisting" of the caudal half of the human thorax is a reflection of the marked changes in the pelvis and iliocostal space of hominins and is the consequence of anatomical modifications required for habitual bipedality (Lovejoy 1974; Latimer and Ward 1993; Jellema et al. 1993; Lovejoy 2005; Lovejoy et al. 2016). The modifications of the hominin false pelvis are the result of the "sagittalization" of the iliac blades to provide a functioning abductor complex allowing stabilization of the torso during single limb stance.

In addition to the dramatic, easily recognized changes in the human pelvis are features associated with the long, flexible human lumbar spine. These vertebral characteristics are a necessary requirement for habitual bipedality and likely represent one of the earliest adaptations to this peculiar form of locomotion. As noted by Lovejoy and McCollum (2010), the early availability of at least a partial lumbar lordosis negated any requirement for the earliest bipeds to have *ever* utilized a, bent-hip-bent-knee (BHBK) gait. The unique series of alternating curvatures in the human spine enable the superjacent torso and head to balance over the supporting limbs, thus not requiring a BHBK gait. This contrasts markedly to that seen in the African apes, wherein the abbreviated lumbar spine is "entrapped" between the elongated iliac blades and is virtually immobile. The resulting inability to lordose the lumbar spine leads to the BHBK posture during bouts of facultative bipedality as this is the only mechanism available to allow balancing over the single supporting limb.

Because several of the features characterizing human ribs are related directly to the long, flexible thoracolumbar column, their ontogenetic development warrants some discussion. The development of the human thoracic cage is basically biphasic with the first major phase taking place during infancy and the second occurring during adolescence. During the first two years of life, human infants have a pyramidal-shaped thorax exhibiting a circular cross section with little or no rib declination and no rib torsion (Keith 1923; Bastir et al. 2013; Openshaw et al. 1984). As a consequence of this orientation of the ribs, the breathing of infants is largely diaphragmatic. Importantly, during this early period, infants also have nearly straight thoracolumbar vertebral columns with minimal spinal curvatures. Upon beginning to walk, children start to develop an incipient lumbar lordosis and a structural thoracic kyphosis. These two curves develop simultaneously as a balancing mechanism. During this early period, the developing thoracic kyphosis is relatively deeper than is the comparatively undeveloped lumbar lordosis, a condition that results in the somewhat rounded back profile characteristic of young children. The developing thoracic kyphosis causes the growing ribs to begin to decline such that their anterior ends are positioned caudally relative to their vertebral attachments.

Rib declination in young children is largely the result of the strong anterior wedging in the upper thoracic vertebrae (Latimer and Ward 1993) with further rib growth enhancing declination. It is also during this early period that the spinal column begins to invaginate into the thorax (Bastir et al. 2013) as part of the general hominoid adaptation that places the scapula dorsally for improved humeral circumduction. In humans, the invagination of the thoracic vertebral column is also a way to improve balance and improve leverage for the epaxial muscles. This anterior migration of the thoracic vertebral column also positions the column closer to the upper body's centroidal axis thereby reducing potentially damaging eccentric loading of the vertebral bodies and adjacent intervertebral discs (Latimer and Ward 1993; Ward et al. 2012). Prior to this invagination process, the most dorsal structures in an infant's back are the spinous processes of the thoracic vertebrae. In contrast, in older children and adults, the most posterior structures are the costal angles, confirming this anterior migration of the vertebral column into the thoracic cage.

The final adult configuration of the human thoracic cage occurs during adolescence, wherein further development of the lumbar lordosis and associated thoracic kyphosis cause the ribs to increasingly decline and twist along their corpora. Additional growth along the rib's anterior terminus results in additional rib declination and torsion. Another transformation that takes place during this period is the increased broadening of the upper thoracic cage (ribs 2–5), ultimately resulting in the adult human transversely broad upper torso (Jellema et al. 1993; Bastir et al. 2013).

During adolescence, several additional secondary structural modifications of the thorax take place leading to the eventual achievement of the adult thoracic shape. The lumbar lordosis increases, especially in females (Latimer and Ward 1993; Masharawi et al. 2010). These changes are, among extant hominoids, unique to humans, and as noted above, allow balancing of the upper body in a sagittal plane during bipedal walking and running. The transversely broad rib cage also increases the ability to control rotation and balance around the long, highly flexible lumbar region. This latter feature, the flexible lumbar region, is entirely lacking in extant African apes (Lovejoy 2005). In order to achieve the hominin lordotic curve, the articular facets of the lumbar vertebrae must become progressively more separated moving caudally down the column. This increasing interfacet distance permits hyperlordosis of the lumbar column and prevents impingement by the intervertebral facets of the intervening laminar region known as the *pars interarticularis*. Inadequate separation between the lumbar intervertebral facets can result in the condition known as spondylolysis and its pathological sequela spondylolisthesis, two unfortunate conditions confined solely to humans (Latimer and Ward 1993).

Ontogenetic changes in the thoracolumbar vertebrae also include differential craniocaudal growth along the anterior vertebral margins resulting in an increased lumbar lordosis (negative wedging) and a slightly decreased thoracic kyphosis (reduced anterior wedging) relative to the earlier condition wherein there is a somewhat greater thoracic kyphosis linked to a minimal lordosis (Scoles et al. 1991; Latimer and Ward 1993). Failure of this vertebral growth pattern can result in a continuation of the adolescent spinal curvatures and vertebral wedging patterns resulting in the condition described as Scheuermann's or adolescent kyphosis (Scoles et al. 1991). It should be noted that the condition described as Scheuermann's kyphosis is largely confined to humans and that the excessive anterior wedging associated with the condition usually occurs at the depth of the developing thoracic kyphosis – T7, T8, and T9 – the vertebral elements under the greatest compressive stress as a consequence of their residing within the depth of the anteriorly directed thoracic concavity (Scoles et al. 1991; Christiansen and Bouxsein 2010; Cotter et al. 2011). Another uniquely human consequence of this positioning within the thoracic kyphosis is the observation that the seventh and eighth thoracic vertebrae are also the most commonly fractured vertebral elements (wedge fractures) usually owing to the bone mineral loss that accompanies human senescence (Christiansen and Bouxsein 2010; Cotter et al. 2011). These vertebral fractures can result in the initiation of the "vertebral fracture cascade" (Christiansen and Bouxsein 2010). Such fractures, while relatively common among humans, do not occur in ape vertebrae (Cotter et al. 2011). The inflexible and stiff thoracolumbar spine described in the African apes as well as their higher bone mineral density and accompanying greater compressive strength of the vertebral bodies (Cotter et al. 2011) provides a protective, stress shielding mechanism mitigating against vertebral fractures. It is interesting that several *Australopithecus* specimens for which there are relevant vertebral elements (A.L. 288-1; Sts 14; StW 431) also display a spinal condition similar in location and morphology to that seen in humans with Scheuermann's kyphosis (Cook et al. 1983; Scoles et al. 1991; Ward et al. 2012). This particular condition is not seen in the African apes and is related to the unique loading trajectories imposed upon the flexible and sinusoidally curved vertebral column of an habitual biped (Scoles et al. 1991; Ward et al. 2012).

Also included in the normal development of the human spine, are the changes in the secondary ossification centers for the transverse processes of the thoracic vertebrae (Latimer and Ward 1993; Jellema et al. 1993). These modifications alter the angular orientation of the transverse processes relative to the vertebral centra and in so doing, change the associated rib costal axes resulting in further declination and torsion of the ribs. These ontogenetic alterations are unique to *Homo* and *Australopithecus* (Latimer and Ward 1993; Jellema et al. 1993; Ward et al. 2012) and can have profound effects upon the overall shape of the thoracic cage (decreased anteroposterior dimension) and the mechanics of breathing (see below).

Associated with these alterations in rib orientation is the so-called "descent" of the manubrium, sternum, and clavicles (Todd 1912; Keith 1923; Ohman 1986; Basir et al. 2013). These ontogenetic alterations further reduce the relative anteroposterior dimensions of the thorax. This "descent of the shoulder" (Todd 1912) is a consequence of the enhanced lumbar lordosis and the associated increasing declination of the underlying rib cage. It is likely that the single vertebral facet on the first rib among hominins (Ohman 1986; Schmid et al. 2013) and the anterior attachment of *M. deltoideus* on the hominin clavicle (Ohman 1986; Ward et al. 2012) are also related to this process. Indeed, essentially all described *Australopithecus* clavicles (A.L. 333x-6/9, A.L. 438-1, A.L. 333-94, StW 431, StW 582, and KSD-VP-1/1) also show this hominin modification in muscular attachment suggesting a horizontally oriented clavicle, a descended shoulder and a *Homo*-like pectoral girdle (Ward et al. 2012).

It should be noted that in human females several associated features are exaggerated relative to males. For instance, among females, the manubrium and sternum are situated somewhat lower along the thoracic column (Keith 1923; Aiello and Dean 1990; Bastir et al. 2013). They also have greater declination and twisting to their ribs (Bellemare et al. 2003, 2006), as well as a greater lumbar lordosis (Masharawi et al. 2010). These differences between men and women occur during puberty and are driven by the increased lumbar lordosis in females and the subsequent changes in the thoracic spine and torso. It is, moreover, likely that these developmental processes are pathologically involved in the development of adolescent idiopathic scoliosis, a condition more common in females and one that is known only in humans and no other primate.

At this point, some discussion of the functional reasons for these described evolutionary modifications in the human vertebral column, shoulder, and thoracic cage is necessary. That is, why do humans alone descend their shoulders and decline and twist their ribs in a manner so unlike the African apes? A potential explanation resides in the early and dramatic modifications seen in the hominin pelvis (see Lovejoy et al. 2016). The "sagittalization" and the transverse narrowing of the hominin false pelvis that occurred in order to allow an abductor complex coupled with the long, flexible, lordosed lumbar column resulted in the tapering of the inferior thoracic cage. This "waisting" of the elongated hominin iliocostal space necessarily resulted in the reduction of the surface area of the hominin diaphragm relative to the condition in the apes, wherein the transversely broad inferior thorax provides an expansive area of attachment for this critical respiratory muscle (the diaphragm). This means that in order to produce similar inspiratory flow rates, hominins must spatially displace their relatively smaller diaphragms more as a consequence of the reduced radius of curvature and abbreviated cross sectional area of the muscle attachment area. To compensate for this somewhat reduced ability to diaphragmatically breathe, hominins declined (and twisted) their ribs allowing the addition of "thoracic" breathing, which is to say, they elevated the caudally angled ribs and in so doing increased the anteroposterior dimension of the thorax (the so-called "pump handle" effect). Thus, the addition of this thoracic/costal breathing mechanism to what would otherwise be primarily the diaphragmatic ventilator mechanism accompanied the structural changes in the pelvis and lumbar column required for habitual bipedality.

This hypothesis is supported by the observation that prior to the ontogenetic modifications of the thoracic cage described above, young children largely breathe utilizing the diaphragm and do not, and cannot, effectively use the thoracic mechanism until the ribs begin to decline. Similarly, the African apes with their horizontally oriented ribs (i.e., non-declined) are much less able to effectively utilize the anterosuperior rib elevating mechanism. Habitual, obligate bipeds have, as a consequence of their erect posture, also forfeited their ability to utilize the "visceral pump" used by quadrupeds (Bramble and Carrier 1983; Daley et al. 2013). This decoupling of the locomotor-respiratory systems in hominins has resulted in the much greater flexibility in breathing patterns in hominins (Daley et al. 2013). In addition, women with a greater decent and declination of the rib cage use a greater percentage of thoracic breathing than do men (Bellemare et al. 2003, 2006), offsetting their smaller diaphragms with enhanced volume displacements produced by thoracic rib elevation. In sum, the highly distinctive changes in the hominin rib cage can be viewed as a way of compensating for the modified pelvis and the long, flexible lumbar spine – both of which are necessary for habitual bipedality.

In light of these described ontogenetic alterations (and associated uniquely hominin pathological conditions), many of what would otherwise be viewed as isolated features in the adult human thorax are in fact associated morphologies that are functionally and structurally interrelated, so that even fragmentary skeletal elements such as the KSD-VP-1/1 ribs allow important inferences to be made about adjoining structures.

African Ape Thoracic Shape

As described above, the "barrel-shaped" human thorax contrasts markedly with the inverted "funnel-shaped" or "frustum-shaped" thorax of the African apes, wherein the cupola is transversely constricted and the inferior thorax is markedly broadened in order to conform to the transversely elongated (broadened) iliac blades (see Fig. 7.4). In dramatic contrast to humans who have narrowed the inferior portions of their rib cage (waisting) to conform to the narrow false pelvis, the African apes have markedly expanded their lower thoracic cage to conform to the "coronalization" of their ilia. The resultant broad and tall pelvis in combination with the reduced number of lumbar vertebral elements (average = 3.5 vertebral elements) results in the virtual immobility of the African ape vertebral column (Lovejoy 2005).

As discussed above, additional changes seen in the African ape thorax include the ligamentous and skeletal "entrapment" of the inferior most lumbar vertebrae between the elongated iliac blades and the rigid fixation of the lower ribs upon the lowermost thoracic vertebrae (Ward et al. 2012). All of these changes act to dramatically stiffen the African ape thoracolumbar spine, essentially reducing it to an immobile "poker spine." This condition is an adaptation for climbing in a large-bodied ape to reduce potentially damaging shear and bending stresses on the lumbar column during active and vigorous bouts of arboreality (Lovejoy 2005). Moreover, as a consequence of these adaptations and the lack of a long, flexible lumbar lordosis, African ape ribs neither decline inferiorly nor do they demonstrate torsion along the rib corpus (Jellema et al. 1993). Indeed, the ribs of African apes cannot decline inferiorly as a consequence of their markedly abbreviated iliocostal space. The reduced iliocostal region in the African apes also prohibits any lateral mobility of the vertebral column, further stiffening the spine. In addition, the orientation of the vertebral transverse processes and costal curvatures (Latimer and Ward 1993; Jellema et al. 1993; Ward et al. 2012) indicates less spinal invagination and reduced costal neck angles. It is noteworthy that chimpanzees have vertebral columns that are less anteriorly invaginated than are those of gorillas (Kagaya et al. 2008), raising the possibility that the smaller African apes have secondarily modified their thoracic skeletons. Inasmuch as the thoracic vertebral counts differ significantly in chimpanzees and bonobos (McCollum et al. 2009), further examination of their rib elements would be a valuable area for future research.

In view of the obvious shape differences between human and African ape thoraces, it seems prudent to pose the question as to why the African great apes have markedly narrowed their upper thoracic cages. Traditionally, this thoracic shape has been attributed to climbing (see Hunt 1991; Schmid 1991; Schmid et al. 2013) with the implication that *Australopithecus* (and perhaps other early hominins) maintained this so-called primitive "funnel-shaped" thorax and that, furthermore, this was evidence of adaptively significant amounts of arboreality in this genus. However, the KSD-VP-1/1 ribs and pelvis (see Lovejoy et al. 2016) fail to support this scenario and instead point to a complete lack of evidence suggesting that *Australopithecus* ever possessed a "funnel-shaped" thorax. Because features like rib declination and torsion are evidence of an elongated and flexible lordotic region in KSD-VP-1/1, and because these differ dramatically from the African ape inflexible poker spine, it seems highly unlikely that *Australopithecus* ever engaged in ape-like arboreality. In fact, these early hominins would have been physically incapable of African ape-like climbing (contra Stern and Susman 1983; Stern 2000), and quite obviously, show no adaptations to it.

In an earlier analysis of the KNM-WT 15000 *Homo erectus* thoracic cage by Jellema et al. (1993), the lack of relevant fossil material led to the erroneous conclusion that humans had expanded their cupular region in compensation for their narrowed caudal rib cage. With the discovery of KSD-VP-1/1 this now seems improbable and instead it now appears that early hominins *never* possessed a "funnel-shaped" thorax.

A re-examination of the mechanical reasons for the African ape morphology can now be addressed. Instead of being viewed as a climbing and suspensory adaptation, the narrowed operculum can instead be seen as an adaptation to the unique African ape locomotor form, knuckle-walking. While the transversely broad thorax in the lesser and greater apes realigns scapular orientation to permit enhanced shoulder mobility and suspensory climbing, it also places the glenohumeral joint in an exceptionally vulnerable position during terrestrial locomotion.

Owing to the extremely low coefficient of friction in any healthy synovial joint it is necessary that the major transarticular forces remain perpendicular to the adjoining chondral surfaces (Burstein and Wright 1994). This requirement has profound consequences for large-bodied animals seeking to combine arboreal and terrestrial locomotor modalities. In chimpanzees and gorillas, the scapulae are positioned on the dorsum of the thorax such that the glenohumeral joints are directed superiorly and laterally, an orientation suitable for heightened shoulder mobility and suspensory climbing. However, this same orientation becomes highly deleterious during knuckle-walking because, unless modified, it would subject the glenohumeral joint to considerable shear stresses caused by the superoposteriorly directed ground reaction forces during the knuckle-strike phase. This is problematic in an extremely mobile articulation like the glenohumeral joint wherein the surrounding soft tissue envelope that largely maintains the joint's integrity provides little resistance to shear forces. The stiff, extended elbow used during knuckle-walking and the abbreviated olecranon process in apes (Drapeau 2004; Simpson et al. nd) both greatly reduce the capacity of the *M. triceps brachii* to eccentrically and/or isometrically contract (dissipating reaction forces), further exacerbating this deficiency. The tapering of the upper thorax in the African apes is likely a parallelism (as is knuckle-walking) that ameliorates this fundamental problem by allowing the scapulae to rotate about the top of the "frustum-shaped" thorax (reduced radius of curvature). This permits the glenohumeral joint to align and face both ventrally and inferiorly thereby opposing ground collision forces and thus reducing potentially damaging shear stress across the shoulder. This altered position of the scapulae would also enhance the ability of M. *serratus anterior* to participate in impact attenuation through isometric and eccentric contraction.

The elevated shoulder orientation and oblique clavicular angle of the African apes (shrugged shoulders) are also related to reducing shear in the glenohumeral joint during knuckle-walking. The hands and wrists also demonstrate adaptations for this highly unusual mode of locomotion (Simpson et al. in review). As African apes are relatively large bodied and spend greater than 80% of their total locomotor repertoire (Doran 1997) engaging in bouts of knuckle-walking, it is not surprising that numerous skeletal and soft tissue adaptations to this activity are evident. Importantly, the fact that these highly specialized adaptations are not seen in KSD-VP-1/1 indicates that early hominins never knuckle-walked and, furthermore, that this unusual locomotor mode was independently derived in chimpanzees and gorillas (White et al. 2009).

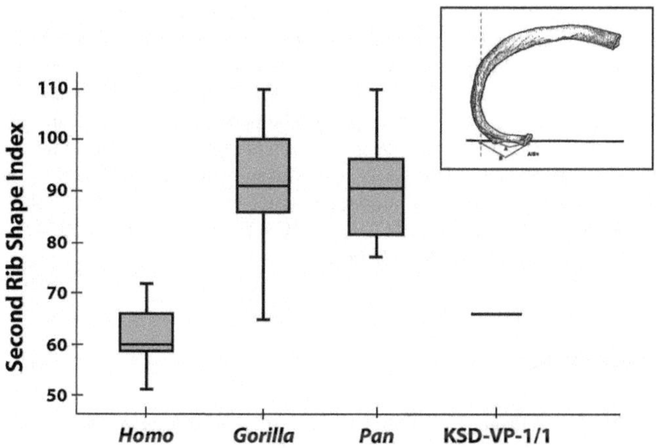

Fig. 7.5 Method of rib curvature index assessment used for the second rib (KSD-VP-1/1n) and comparisons with *Pan, Gorilla*, and *Homo* second rib indices. Although one *Gorilla* specimen (CMNH-B 1781; *N* = 33) fell within the human range, it is clear that KSD-VP-1/1n lies well within the human range. Modified from Haile-Selassie et al. (2010b)

The KSD-VP1/1 Thorax

The distinct and easily recognized anatomical contrasts between the thoracic skeletons of humans and the African apes provide a valuable context within which to examine the KSD-VP-1/1 rib remains (see Fig. 7.4). Moreover, this process should shed light on the thoracic skeleton of the CLCA (see Haile-Selassie et al. 2016) and also provide evidence regarding the evolutionary development of knuckle-walking and the "funnel-shaped" African ape thoracic cage.

The second rib has previously been demonstrated to be especially useful in discriminating the upper thoracic shapes between humans and African apes (Jellema et al. 1993; Haile-Selassie et al. 2010b). Thus, KSD-VP-1/1n, an essentially complete second rib, is particularly valuable in assessing the conformation of the cranial thorax in this specimen. As we noted earlier, the rib curvature in this specimen is more similar to humans than it is to chimpanzees (Fig. 7.5). The rib curvature index (Haile-Selassie et al. 2010b and see Fig. 7.5) clearly separates humans and KSD-VP-1/1n from the African apes indicating that the upper rib cage of this specimen was transversely expanded like humans and was not constricted as it is in *Pan* and *Gorilla* (see Fig. 7.4). Moreover, the neck/corpus angle indicates that the spine was invaginated deeply within the thorax, another distinguishing feature of hominins (Latimer and Ward 1993; Jellema et al. 1993; Lovejoy 2005; Ward et al. 2012). It is clear that the thoracic cupola of KSD-VP-1/1 was transversely broad and differed fundamentally from the condition in the African apes. As human costal declination and twisting of the rib corpus is especially apparent in the seventh and eighth ribs (Jellema et al. 1993), and these particular elements are also preserved in the KSD-VP-1/1 partial skeleton (KSD-VP-1q, 1 s, 1o, and 1p), additional information can also be gleaned about the caudal half of the rib cage of *Au. afarensis*.

Although the ribs are fragmentary and are not associated with thoracic vertebrae, it is nevertheless possible to orient the pleural surface of the rib heads into appropriate anatomical position and to estimate rib orientation from such an examination (see Jellema et al. 1993 for additional details). The preserved portions of KSD-VP-1/1o (seventh or eighth rib) and KSD-VP-1/p (eighth or ninth rib) indicate an oblique, anteroinferior orientation, and obvious declination of the rib corpus. Axial torsion along the costal corpus is apparent, and this is consistent with the rib declination and a long, flexible human-like lumbar vertebral column. This also suggests a lumbar lordosis and the progressive diminishment of the anteroposterior thoracic diameter relative to the transverse breadth of the thoracic cage. This latter observation also agrees with the pelvic morphology of the KSD-VP-1/1 partial skeleton (see Lovejoy et al. 2016). Because several of the characteristic features of the human thoracolumbar vertebral column are ontogenetically and structurally linked to the costal cage (see above), it is possible to reconstruct additional associated anatomies in the KSD-VP-1/1 partial skeleton.

It is evident that the KSD-VP-1/1 specimen did not have an ape-like, constricted upper thorax but rather had a transversely expanded operculum like that seen in modern humans. Moreover, the declination and torsion of the KSD-VP-1/1 costal elements indicate that this specimen had a long lower torso, a long and flexible lumbar column, and a high degree of spinal column invagination relative to the condition seen in the African apes. That the KSD-VP-1/1 thorax exhibits an anteroinferior declination of its costal elements is also clear, again paralleling the condition described in humans (Jellema et al. 1993; Bastir et al. 2013). All of the elements representing the thorax in KSD-VP-1/1 suggest a *Homo*-like rib cage and one that differs dramatically from the African ape condition.

These observations obviously indicate that the thoracic cage of KSD-VP-1/1 was not "funnel-shaped", but rather was transversely broad in its cranial half and relatively tapered (compared to the African apes) in its lower portion (see Fig. 7.4). It should be noted, however, that the relative mediolateral breadth of the pelvis in *Australopithecus* was greater than in modern humans, necessitating some amount of lateral flaring in the lower thorax (see Latimer and Ward 1993; Lovejoy et al. 2016). In light of this, the thoracic shape of early hominins (pre-*Homo sapiens*) should be more aptly described as "bell-shaped." This designation would eliminate the overly simplistic dichotomy of either barrel or funnel shaped and instead would recognize that *Australopithecus* would have flared slightly more at its thoracic bottom (less waisting than in modern humans, more waisting than in African apes) as a consequence of their broad bi-iliac breath induced by the platypelloidy of early (and even later) members of the species (Tague and Lovejoy 1986; Simpson et al. 2008; Lovejoy et al. 2016).

Following the same reasoning, Sawyer and Maley (2005) described the flaring of the caudal portion of the Neanderthal thoracic cage (owing to the broad false pelvis) as having created a "bell-shaped" thorax. With this in mind, perhaps the "barrel-shaped" descriptor should be reserved exclusively for modern *Homo sapiens*. It is now abundantly clear that these early hominins did not have "frustum-shaped" or "funnel-shaped" thoraces. Furthermore, it now is also obvious that they possessed *Homo*-like rib declination, and a long flexible, lordosed lumbar region. This latter feature, an elongated lumbar spine, clearly contravenes any selectively significant amount of African ape-like climbing in *Au. afarensis*. As importantly, the complete lack of a structural lordotic spine and the complete inability to even facultatively assume a lordosis in the African apes virtually precludes their assuming hominin-like bipedality, casting serious doubt upon the use of extant chimpanzees and/or gorillas as appropriate models for early hominin locomotion. Similarly, earlier suggestions that *Australopithecus* locomotion was confined to a BHBK posture during walking (Stern and Susman 1983; Stern 2000) are now contravened by all available evidence from a now dramatically improved fossil record for the genus. It is now clear that *Australopithecus* possessed long, flexible, lumbar columns that were fully capable of achieving structural lordotic curvatures; indeed, early hominins likely never went through an ape-like lumbar entrapment phase. Thus, earlier suggestions of ape-like arboreality in *Australopithecus* must be re-examined in this light.

Conclusion

The KSD-VP-1/1 thorax was "bell-shaped," clearly lacking an opercular constriction as in the African apes. *Australopithecus* did not have a frustum or "funnel-shaped" thorax. This conclusion is further supported by the somewhat less direct evidence provided by the *Ardipithecus ramidus* false pelvis (Lovejoy et al. 2009a, b). Following this, several points become clear. First, the African ape thoracic shape with its constricted cupola is independently derived in *Pan* and *Gorilla*. Furthermore, this thoracic form is likely an adaptation to terrestrial knuckle-walking in a large-bodied, suspensory, and vertically climbing hominoid. Second, there is now no reliable evidence that any hominin ever possessed a narrow, constricted upper thorax. Finally, a broad upper thorax is probably primitive and the CLCA did not exhibit a conically shaped thorax.

Acknowledgements We thank the Authority for Research and Conservation of Cultural Heritage (ARCCH) of the Ministry of Culture and Tourism of Ethiopia and administrative offices of the Afar Regional State of Ethiopia for field and laboratory research permits. We thank M. Decker, S. Simpson, and two anonymous reviewers for valuable comments. We also thank D. Su for help with figures. The Woranso-Mille project was financially supported by grants from The Leakey Foundation, The Wenner-Gren Foundation, The National Geographic Society, The Cleveland Museum of Natural History, and The National Science Foundation (BCS-0234320, BCS-0321893, BCS-0542037, BCS-1124705, BCS-1124713, BCS-1124716, BCS-1125157, and BCS-1125345).

References

Aiello, L., & Dean, C. (1990). *An introduction to human evolutionary anatamy*. London: Academic Press Harcourt Brace & Company.

Bastir, M., Martinez, D. G., Recheis, W., Barash, A., Coquerelle, M., Rios, L., et al. (2013). Differencial growth and development of the upper and lower human thorax. *PLoS ONE, 8*, e75128.

Bellemare, F., Jeanneret, A., & Couture, J. (2003). Sex differences in thoracic dimensions and configuration. *American Journal of Respiratory and Critical Care Medicine, 168*, 305–312.

Bellemare, F., Fuamba, T., & Bourgeault, A. (2006). Sexual dimorphism of human ribs. *Respiratory Physiology & Neurobiology, 150*, 233–239.

Bramble, D. M., & Carrier, D. R. (1983). Running and breathing in mammals. *Science, 219*, 251–256.

Burstein, A. H., & Wright, T. M. (1994). *Fundamentals of orthopaedic biomechanics*. New York: Williams & Wilkins.

Christiansen, B. A., & Bouxsein, M. L. (2010). Biomechanics of vertebral fractures and the vertebral fracture cascade. *Current Osteoporosis Reports, 8*, 198–204.

Cook, D. C., Buikstra, J. E., DeRousseau, C. J., & Johanson, D. C. (1983). Vertebral pathology in the afar Australopithecines. *American Journal of Physical Anthropology, 60*, 83–101.

Cotter, M. M., Loomis, D. A., Simpson, S. W., Latimer, B., & Hernandez, C. J. (2011). Human evolution and osteoporosis-related spinal fractures. *PLoS ONE, 6*, e26658.

Daley, M. A., Bramble, D. M., & Carrier, D. R. (2013). Impact loading and locomotor-respiratory coordination significantly influence breathing dynamics in running humans. *PLoS ONE, 8*, e70752.

Doran, D. M. (1997). Ontogeny of locomotion in mountain gorillas and chimpanzees. *Journal of Human Evolution, 32*, 323–344.

Drapeau, M. S. (2004). Functional anatomy of the olecranon process in hominoids and plio-pleistocene hominins. *American Journal of Physical Anthropology, 124*, 297–314.

Franciscus, R. G., & Churchill, S. E. (2002). The costal skeleton of Shanidar 3 and a reappraisal of Neandertal thoracic morphology. *Journal of Human Evolution, 42*, 303–356.

Gómez-Olivencia, A., Eaves-Johnson, K. L., Franciscus, R. G., Carretero, J. M., & Arsuaga, J. L. (2009). Kebara 2: New insights regarding the most complete Neandertal thorax. *Journal of Human Evolution, 57*, 75–90.

Haile-Selassie, Y., Saylor, B. Z., Deino, A., Alene, M., & Latimer, B. M. (2010a). New hominid fossils from Woranso-Mille (Central Afar, Ethiopia) and taxonomy of early *Australopithecus*. *American Journal of Physical Anthropology, 141*, 406–417.

Haile-Selassie, Y., Latimer, B. L., Alene, M., Deino, A., Gibert, L., Melillo, S. M., et al. (2010b). An early *Australopithecus afarensis* postcranium from Woranso-Mille, Ethiopia. *Proceedings of the National Academy of Sciences USA, 107*, 12121–12126.

Haile-Selassie, Y., Latimer, B. M., Lovejoy, C. O., Melillo, S. M., & Meyer, M. R. (2016). Conclusion: Implications of KSD-VP-1/1 for early hominin paleobiology and insights into the chimpanzee/human last common ancestor. In: Y. Haile-Selassie & D. F. Su (Eds.), *The postcranial anatomy of* Australopithecus afarensis: *New insights from KSD-VP-1/1* (pp. 179–188). Dordrecht: Springer.

Hunt, K. D. (1991). Mechanical implications of chimpanzee positional behavior. *American Journal of Physical Anthropology, 86*, 521–536.

Jellema, L. M., Latimer, B., & Walker, A. (1993). The rib cage. In A. Walker & R. E. Leakey (Eds.), *The Nariokotome Homo erectus Skeleton* (pp. 294–325). Cambridge: Harvard University Press.

Kagaya, M., Ogihara, N., & Nakatsukasa, M. (2008). Morphological study of the anthropoid thoracic cage: Scaling of thoracic width and an analysis of rib curvature. *Primates, 49*, 89–99.

Keith, A. (1923). Man's posture: Its evolution and disorders. *British Medical Journal, 1*, 451–454.

Latimer, B., & Ward, C. V. (1993). The thoracic and lumbar vertebrae. In A. Walker & R. E. Leakey (Eds.), *The Nariokotome Homo erectus skeleton* (pp. 266–293). Cambridge: Harvard University Press.

Lovejoy, C. O. (1974). The gait of australopithecines. *Yearbook of Physical Anthropology, 16*, 18–30.

Lovejoy, C. O. (2005). The natural history of human gait and posture II: hip and thigh. *Gait and Posture, 21*, 129–151.

Lovejoy, C. O., & McCollum, M. (2010). Spinopelvic pathways to bipedality: why no hominids ever relied on a bent-hip–bent-knee gait. *Philosophical Transactions of the Royal Society of London, 365*, 3289–3299.

Lovejoy, C.O., Latimer, B. M., Spurlock, L., & Haile-Selassie, Y. (2016). The pelvic girdle and limb bones of KSD-VP-1/1. In Y. Haile-Selassie & D. F. Su (Eds.), *The postcranial anatomy of* Australopithecus afarensis*: new insights from KSD-VP-1/1* (pp. 155–178). Dordrecht: Springer.

Lovejoy, C. O., Simpson, S. W., White, T. D., Asfaw, B., & Suwa, G. (2009a). Careful climbing in the Miocene: the forelimbs of *Ardipithecus ramidus* and humans are primitive. *Science, 326*, 70e71–70e78.

Lovejoy, C. O., Suwa, G., Simpson, S. W., Matternes, J. H., & White, T. D. (2009b). The great divides: *Ardipithecus ramidus* reveals the postcrania of our last common ancestors with African apes. *Science, 326*, 100–106.

Masharawi, Y., Dar, G., Peleg, S., Steinberg, N., Medlej, B., Kay, H., et al. (2010). A morphological adaptation of the thoracic and lumbar vertebrae to lumbar hyperlordosis in young and adult females. *European Spine Journal, 19*, 768–773.

McCollum, M. A., Rosenman, B. A., Suwa, G., Meindl, R. S., & Lovejoy, C. O. (2009). The vertebral formula of the last common ancestor of African apes and humans. *Journal of Experimental Zoology Part B Molecular and Developmental Evolution, 314B*, 123–134.

Ohman, J. C. (1986). The first rib of hominoids. *American Journal of Physical Anthropology, 70*, 209–229.

Openshaw, P., Edwards, S., & Helms, P. (1984). Changes in rib cage geometry during childhood. *Thorax, 39*, 624–627.

Ryan, T. M., & Sukhdeo, S. (2016). KSD-VP 1/1: Analysis of the postcranial skeleton using high-resolution Computed Tomography. In Y. Haile-Selassie & D. F. Su (Eds.), *The Postcranial Anatomy of* Australopithecus afarensis*: New Insights from KSD-VP-1/1* (pp. 39–62). Dordrecht: Springer.

Sawyer, G. J., & Maley, B. (2005). Neanderthal reconstructed. *The Anatomical Record (part B: New Anatomy) 238B*, 23–31.

Schmid, P., Churchill, S. E., Nalla, S., Weissen, E., Carlson, K. J., et al. (2013). Mosaic morphology in the thorax of *Australopithecus sediba. Science, 340*, 1234598.

Schmid, P. (1991). The trunk of the *Australopithecus*. In Y. Coppens & B. Senut (Eds.), *Origine(s) de la bipedie chez les hominides* (pp. 225–234). Cahiers de Paleoanthropologie, Editions du CNRS. Paris.

Schultz, A. H. (1961). Vertebral column and thorax. In H. Hofer, A. H. Schultz, & D. Starck (Eds.), *Primatologia-Handbook of Primatology* (Vol. 4, pp. 1–66). Basel/New York, NY: Karger.

Scoles, P. V. (1991). Vertebral alterations in Scheuermann's kyphosis. *Spine, 16*, 509–515.

Simpson, S. W., Quade, J., Levin, N. E., Butler, R., Dupont-Nivet, G., Everett, M., et al. (2008). A female *Homo erectus* pelvis from Gona, Ethiopia. *Science, 322*, 1089–1092.

Simpson, S. W., Lovejoy, C. O., & Latimer, B. M. (in review). Why do knuckle-walking African apes knuckle-walk? *The Anatomical Record*.

Stern, J. T. (2000). Climbing to the top: A personal memoir of *Australopithecus afarensis*. *Evolutionary Anthropology, 9*, 113–133.

Stern, J. T., & Susman, R. L. (1983). The locomotor anatomy of *Australopithecus afarensis*. *American Journal of Physical Anthropology, 60*, 279–317.

Tague, R. G., & Lovejoy, C. O. (1986). The obstetric pelvis of AL 288-1 (Lucy). *Journal of Human Evolution, 15*, 237–255.

Todd, T. W. (1912). The decent of the shoulder after birth. *Anatomischer Anzeiger, 41*, 385–397.

Ward, C. V., Kimbel, W. H., Harmon, E. H., & Johanson, D. C. (2012). New postcranial fossils of *Australopithecus afarensis* from Hadar, Ethiopia (1990-2007). *Journal of Human Evolution, 63*, 1–51.

White, T. D., Asfaw, B., Beyene, Y., Haile-Selassie, Y., Lovejoy, C. O., & Suwa, G. (2009). *Ardipithecus ramidus* and the paleobiology of early hominids. *Science, 326*, 75–86.

Chapter 8
The Pelvic Girdle and Limb Bones of KSD-VP-1/1

C. Owen Lovejoy, Bruce M. Latimer, Linda Spurlock, and Yohannes Haile-Selassie

Abstract The pelvis and limb bones of KSD-VP-1/1 are analyzed with respect to their implications for the history of morphology, gait pattern, and *bauplan* of *Australopithecus afarensis*. The pelvis is essentially of modern aspect, and differs little from that of modern humans save in a few minor respects, such as, for example, size of the auricular surface and acetabulum. Its ischial conformation is the direct evidence of a strong history of dynamic running in its immediate ancestors, and its general shape confirms the overall *Australopithecus* pattern of platypelloidy. The implications with respect to the history of the hominin *bauplan* are discussed, as are those impacted by recent developments in our understanding of developmental biology. These are reviewed in light of other recent analyses of early hominin morphology and locomotion.

Keywords *Australopithecus* • *Ardipithecus* • Pelvis • Bipedality • Hominin *bauplan*

Introduction

The partial *Australopithecus afarensis* skeleton (KSD-VP-1/1) from Woranso-Mille, Ethiopia (Haile-Selassie et al. 2010a, b), has proved to be one of the most important discoveries in human paleontology over the past 50 years. Articulated fossil specimens are now known to be *the* key elements that preserve sufficient information with which to solve major human evolutionary issues that cannot be resolved using only the typical single, isolated anatomical site (e.g., femur or humerus).

KSD-VP-1/1 has been described in considerable detail (Haile-Selassie et al. 2010a) and the reader will be referred to those descriptions in the discussions that follow. Here, additional studies and data relating to the limb skeleton and pelvis are reviewed. Review and discussion of the thorax are provided separately (see Latimer et al. 2016).

Humerus

KSD-VP-1/1b is the distal two-thirds of a right humerus (Figs. 8.1 and 8.2). The specimen has suffered extensive exfoliation, with little preservation of morphology save that of the distal end and the clear presence of a highly robust deltopectoral crest, a typical hominin character, and one that distinguishes human and other hominin humeri from those of both the African apes and orangutans (Lovejoy et al. 2009a). The trochlea and capitulum have also suffered some exfoliation but are sufficiently preserved to allow reasonably accurate estimates of the breadth of the distal articular surface using Computed Tomography (CT) data.

These data include a distal articular breadth of 48 mm, a trochlear breadth of 27 mm, and a total distal joint breadth of 58.8 mm. Ryan and Sukhdeo (2016) provide a number of comparative metrics for *Homo*, the African apes, and *Australopithecus*. The KSD-VP-1/1b distal articular breadth well exceeds those of MH2 and StW 431, both relatively large specimens, and obviously the value for A.L. 288-1 ("Lucy"), as well as the mean for early *Homo* (see Ryan and Sukhdeo (2016) for a list of included specimens), although KNM-WT 15000 was included in that mean and is subadult.

C.O. Lovejoy (✉) · L. Spurlock
Department of Anthropology and Division of Biomedical Sciences, Kent State University, Kent, OH 44242, USA
e-mail: olovejoy@aol.com

L. Spurlock
e-mail: lspurloc@kent.edu

C.O. Lovejoy · B.M. Latimer · L. Spurlock · Y. Haile-Selassie
Department of Physical Anthropology, Cleveland Museum of Natural History, 1 Wade Oval Drive, Cleveland, OH 44106, USA
e-mail: yhailese@cmnh.org

B.M. Latimer
Department of Orthodontics, School of Dental Medicine, Case Western Reserve University, Cleveland, OH 44106, USA
e-mail: bxlatimer@aol.com

Yohannes Haile-Selassie and Denise F. Su (eds.), *The Postcranial Anatomy of Australopithecus afarensis: New Insights from KSD-VP-1/1*, Vertebrate Paleobiology and Paleoanthropology, DOI 10.1007/978-94-017-7429-1_8

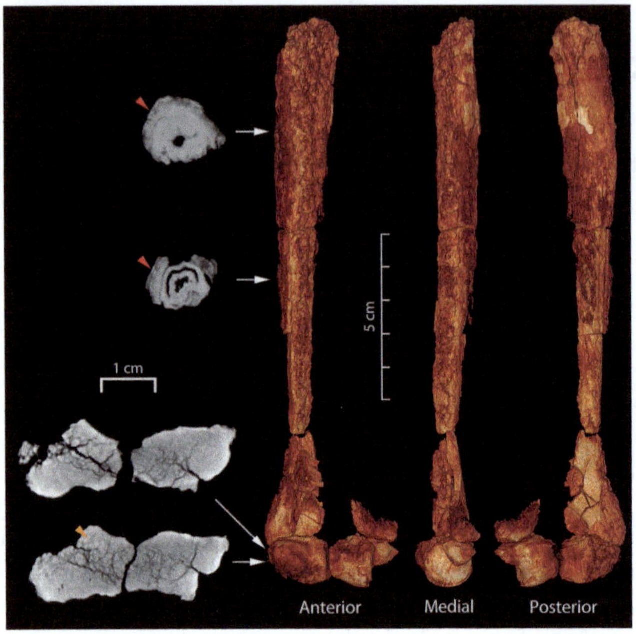

Fig. 8.1 Three CT-based views of the KSD-VP-1/1 humerus with some cross-sectional data provided (from Ryan and Sukhdeo 2016). The substantial damage to the shaft is evident. For further information, see Ryan and Sukhdeo (2016) and Haile-Selassie et al. (2010a)

Although much of the shaft of the humerus is damaged and exfoliated, it seems likely that its closest morphological analog, in terms of robusticity, is the MAK-VP-1/2 humerus (White et al. 1993), although it is also similar to those from the A.L. 333 locality at Hadar (Lovejoy et al. 1982). All of these specimens exhibit demonstrable robusticity of the deltoid tuberosity that exceeds on average those of the living African apes, when adjusted for body mass. For further discussion of the importance of what is likely a primitive trait in hominins but derived in other more suspensory hominoid species, see Lovejoy et al. (2009a). Although the distal end of the hominoid humerus has received much attention because of its frequent preservation in the fossil record, its morphology is generally not, in our view, likely to be informative about locomotion (or phylogeny) at least within hominoid postcrania. The hominoid elbow joint has been argued to be more robust in hominoids than in other anthropoids because of an enlarged trochlea, relative to the size of the capitulum (see for example, McHenry and Corruccini 1975; Rose 1988, 1993; Begun 1992; Benefit and McCrossin 1995; Harrison and Rook 1997). However, in a recent paper, Selby and Lovejoy (2014) demonstrate that these relationships are largely accountable by body size alone. Lague (2014) has presented an extensive analysis of hominin distal humeri that incorporates a digital profile of each and considers allometric and other potential variables, especially progressive change from a presumed primitive pattern as seen in KNM-KP 271 (attributed to *Au. anamensis*) to a more novel one as seen in modern humans. We agree with one of his potential conclusions that these changes are likely not to have been of sufficient functional significance to have been under active selection, but must add that such changes may also reflect loading patterns in the sense of what we have previously defined as Type IV changes (those due to cartilage modeling) (see Table 8.1). The brachioradialis crest in KSD-VP-1/1b is of relatively moderate size and quite typical of *Australopithecus*.

Ulna

KSD-VP-1/1a is a well-preserved right ulna with approximately 60% of the shaft present (Figs. 8.3, 8.4 and 8.5). The bone's proximal portion is much better preserved than its more distal shaft, which has suffered extensive exfoliation. These conditions have been confirmed using CT imaging which has in turn made it possible to provide a number of metric characters (see Ryan and Sukhdeo 2016; Fig. 8.3; Table 8.2). The cortex and trabeculae of the shaft are also poorly preserved and therefore do not provide reliable anatomical data.

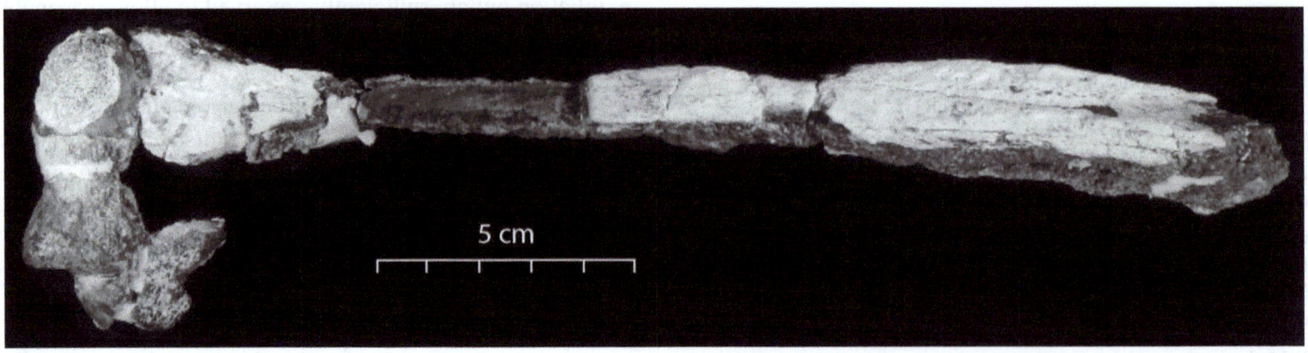

Fig. 8.2 Anterior view of the KSD-VP-1/1 humerus. The surface of the capitulum has been partially sheared, although its overall breadth can be estimated with high reliability. Exfoliation to the shaft has removed much of the anatomical detail – however, the marked expansion proximally in the region of the deltoid tuberosity suggests that it was highly rugose as it is in other hominin specimens

Table 8.1 Analytical trait types (from Lovejoy et al. 2002)

Type I. A trait that differs in two taxa because its presence and/or expression are downstream consequences of differences in positional information and their resultant effects on local pattern formation. Type 1 traits are fixed by directional and/or stabilizing selection because their primary functional features have a real effect on fitness, and result largely from a direct interaction between genes expressed during field deployment and the functional biology of their adult product. Example: The superoinferior shortening of the ilium in hominins
Type II. A trait which is a collateral byproduct of other field changes. Type 2 traits differ in two taxa because of differences in pattern formation (as in Type 1's), but themselves have no direct functional consequences. Unlike Types 4 and 5, they represent true field-derived pleiotropy
Type IIA are traits whose primary antecedent shift occurred under the action of natural selection. Example: Morphology and position of the greater trochanter and femoral tubercle
Type IIB are traits whose "parent" changes were probably not under positive or negative selection. Example: Morphology and structure of the trochanteric fossa in hominoid femora
Type III. A trait which differs in two taxa because of modification of a systemic growth factor which affects multiple elements, such as an anabolic steroid. Example: Body size and its allometric effects. Allometric shifts probably usually reflect slight changes of systemic control factors during development, e.g., small modulations of GH, its receptors, and/or its related factors that can generate fully coordinated morphological change
Type IV. A trait which differs between taxa because its presence/absence and/or "grade" are attributable exclusively to phenotypic effects of the interaction of "systematic assembly mechanisms" (e.g., the PTHrP-IHH loop) (Lovejoy et al. 1999) and environmental stimuli. Such traits have no antecedent differences in pattern formation, and therefore have no value in phyletic analysis. They are epigenetic and are not pleiotropic. However, they do provide significant *behavioral* information, and are therefore of expository or evidentiary value in interpreting fossils. They result from habitual behaviors during ontogeny. Example: The bicondylar angle of the femur
Type V. Traits arising by the same process as Type 4's but which have no reliable diagnostic value with respect to significant behavior. Such traits are not consistently expressed within species and often show marked variation of expression within individuals and local populations. Example: Femoral anteversion

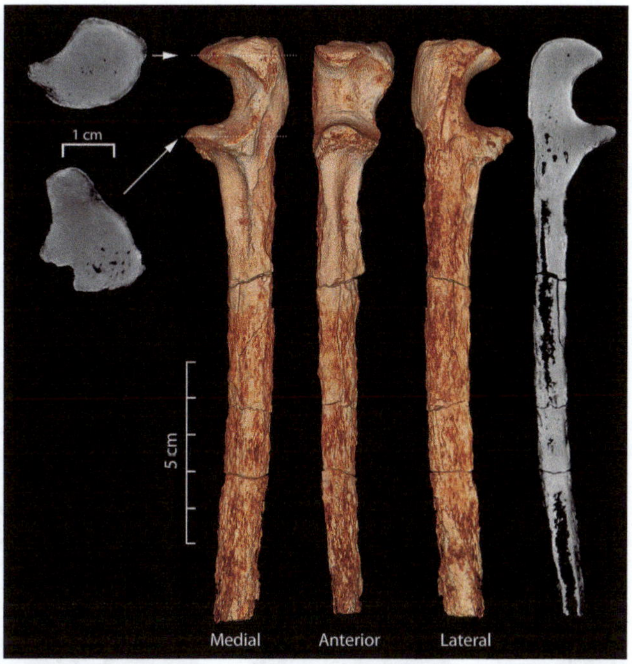

Fig. 8.3 CT scan views of the ulna of KSD-VP-1/1 (from Ryan and Sukhdeo 2016). Note that preservation of the proximal portion is generally good, but that there is surface degradation more distally. Two CT cross sections and one long section are also provided, showing general absence of trabecular data

Table 8.2 Crural index in hominoids

Taxon	Mean
H. sapiens	0.816 (0.038)
P. troglodytes	0.847 (0.019)
G. gorilla	0.829 (0.023)
Mean: All three taxa	0.831

Crural index is calculated by tibia maximum length/femur maximum length
N 50 for each taxon

The ulna can be articulated with KSD-VP-1/1b (see earlier) and when this is done it shows virtually no significant "carrying angle." Shaft curvature is minimal (Fig. 8.4). The specimen does not exhibit a flexion tubercle (Lovejoy et al. 2009b). Its ulnar tuberosity is well preserved and similar to those of modern humans.

Although, as just noted, proximal elbow morphology is very conservative in extant hominoids, an exception is the orientation of the trochlear notch, which faces distinctly superiorly in apes and to some degree in modern humans as well, but surprisingly more directly *anteriorly* in most early hominins (Figs. 8.4 and 8.5) (Lovejoy et al. 2009a; Drapeau 2004, 2008). This may well be associated with detailed changes in the distal humerus discussed earlier by Lague (2014) (Fig. 8.5). Although the etiology of this trait's variation (i.e., whether the notch faces superiorly or more anteriorly) is unknown, it would seem reasonable to opine that it might be a product either of habitual transarticular loading during ontogeny (if the trait is a consequence of cartilage modeling) (Hamrick 1999; Lovejoy et al. 1999), or reflects an underlying genomic shift in joint specification that is consistent with the most favorable habitual loading regimen experienced during adult life.

In either case it seems equally reasonable to conclude that an anterior-facing notch most likely reflects a predominance of transarticular loading during variable degrees of elbow flexion, rather than reflecting any single joint position. It also seems likely that a distinctly superior-facing ulnar notch is the product of a high degree of loading imposed by frequent suspension either during development, if it is the consequence of cartilage modeling (Type IV trait), or adult life, if the trait is a genuine adaptation (Type I) (Table 8.1). The trochlear notch

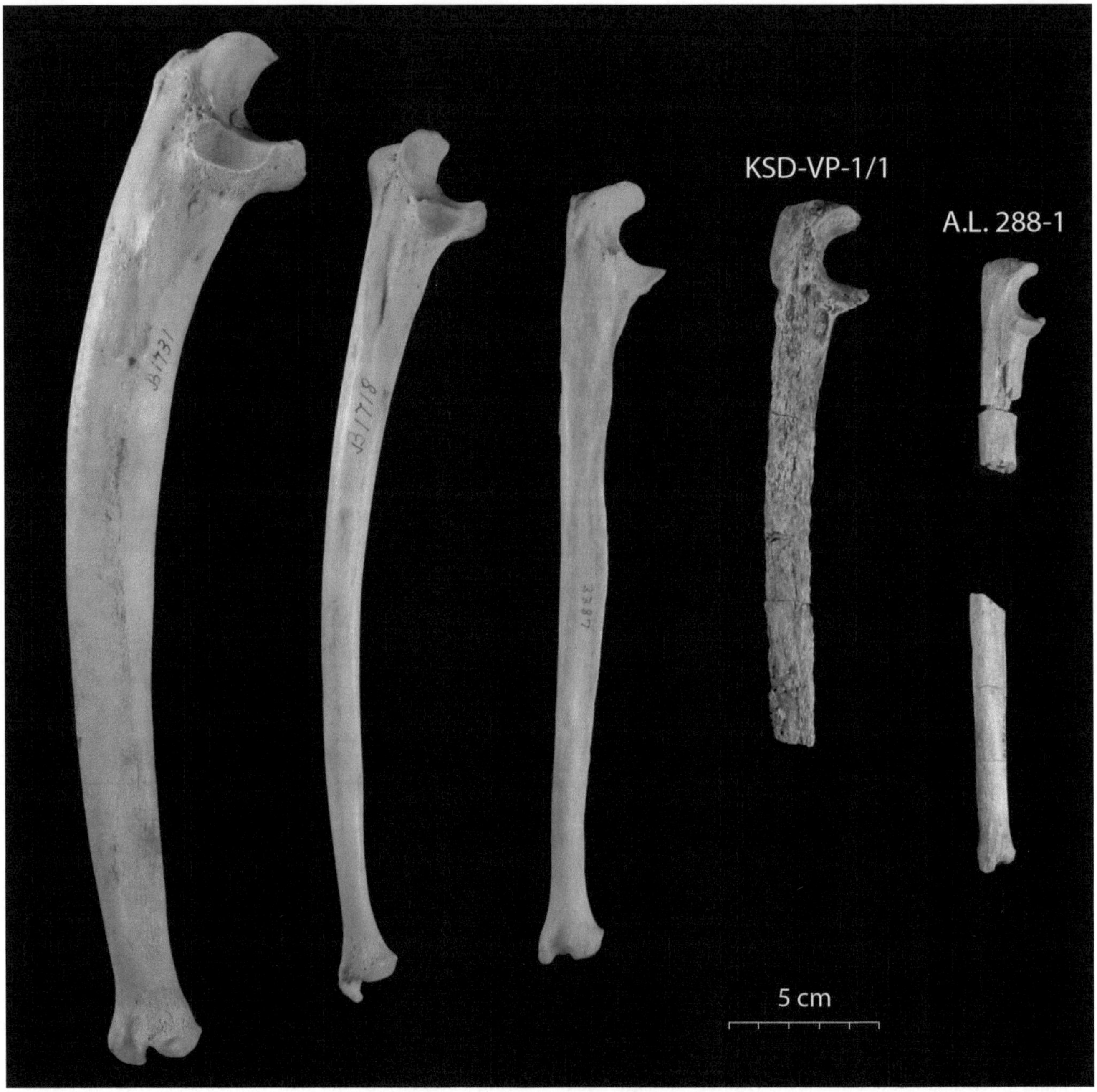

Fig. 8.4 Lateral view of the ulnas of (left to right) *Gorilla*, *Pan*, *Homo* (all males) shown with those from KSD-VP-1/1 and A.L. 288-1 (casts). The straight profile of the shafts of the three hominins contrasts strongly with those from the two apes. However, the more anterior-facing trochlear notch in the two early hominins also contrasts strongly with the more superior-facing notches in the two apes. The human specimen appears somewhat intermediate. These views are consistent with metric data that also reveal a strong difference in trochlear notch orientation – a feature is that it is also similar to these hominin specimens in *Ar. ramidus*. See also Fig. 8.5

of KSD-VP-1/1a (10°; see Table 8.1) is similar to those of A.L. 288-1 and ARA-VP-6/500, and its distinctly anterior-facing orientation strongly suggests the absence of any habitual suspension. It is not clear why humans and African apes are convergent with respect to this trait.

Ryan and Sukhdeo (2016) provide an extensive collection of metrics from the well-preserved proximal end of the specimen, including those related to the trochlear and radial notches, and several from the well-preserved proximal portion of its shaft nearer to its joint surface than the exfoliated

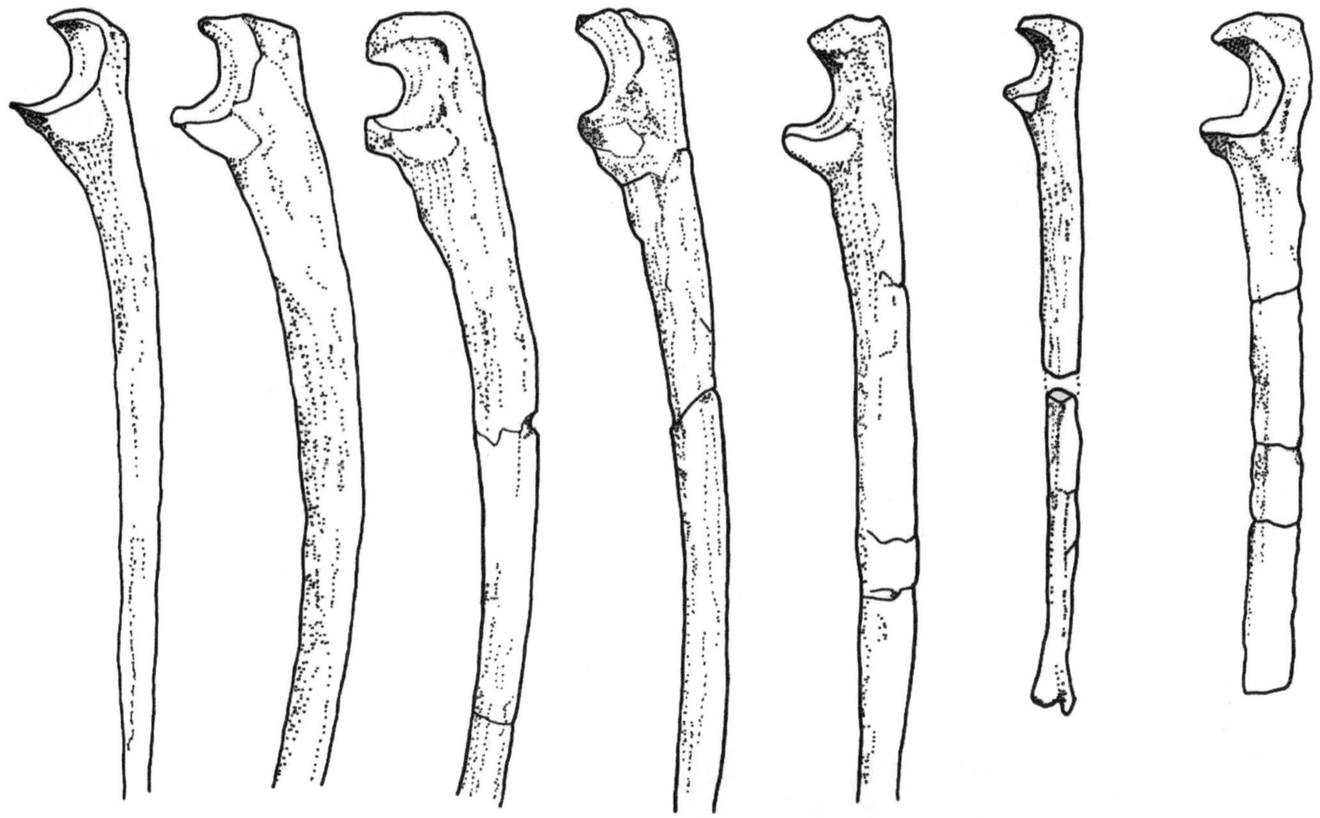

Fig. 8.5 Drawings of several ape and hominin proximal ulnas to show differences in trochlear notch orientation. All specimens except KSD-VP-1/1 redrawn after Drapeau et al. 2004. From left to right Modern human, modern chimpanzee, O.H. 36, L 40-19, A.L. 438-1, A.L. 288-1, KSD-VP-1/1. Note that both the chimpanzee and human specimens have ulnae whose trochlear notch faces much more superiorly than do those of any of the early hominins shown

region that begins more distally. They also provide comparative metrics from a variety of other hominin and extant African ape specimens. They used Churchill et al.'s (see Ryan and Sukhdeo 2016) method to calculate trochlear notch orientation (100 × olecranon AP diameter/coronoid height), which in KSD-VP 1/1a is 93.5 and is substantially higher (about 2.5 standard deviations) than Churchill's *Australopithecus* sample mean (82.6). Their method differs from ours, which is a simple assessment of the angle between the posterior shaft axis and one formed by a line connecting the most anterior points on the superior and inferior projections of the trochlear notch (see Lovejoy et al. 2009a for illustration). Indeed, comparison of the KSD-VP-1/1 ulna to others from *Australopithecus* (Figs. 8.4 and 8.5) reveals considerable diversity in overall detail but consistent with the more proximal orientation of the trochlear notch in modern humans and *Pan*. This extensive variation, again, might reflect other complex changes in the distal humerus outlined by Lague that we discussed previously; although we have no functional explanation save the possibility that if these various dimensions are largely Type IV dependent, they might be expected to substantially vary depending on individual life history.

On the other hand, the values obtained by Ryan and Sukhdeo (2016: 39–62) for the other metric dimensions of the trochlear notch and various other aspects of the proximal articular surface of KSD-VP 1/1a "are generally close to the values found in early *Homo*, and within the ranges of *Homo sapiens* and *Pan troglodytes*."

Judging from descriptions provided by Drapeau et al. (2005), shaft curvature is quite variable in early hominins, and is especially pronounced in O.H. 36 and moderate in L40-19, whereas the three sufficiently complete specimens of *Au. afarensis* now available (A.L. 288-1, A.L. 438-1 and KSD-VP-1/1) show minimal to moderate curvature and are, in this regard, quite unlike those of the African apes and more similar to the ulnas of modern humans. Drapeau et al. (2005:617) concluded that "If ulnar curvature of A.L. 438-1 and A.L. 288-1 is representative of normal variation in *Au. afarensis*, then the 'middle range' value for this species suggests a less curved ulna than typical for great apes and more in line with what is seen in modern humans." We agree with these authors that the functional significance of such curvature is poorly understood and direct the reader to their discussion of this issue.

Femur

KSD-VP-1/1c is a left distal femur with considerable damage to its distal joint surface (Figs. 8.6, 8.7 and 8.8). Although CT scanning reveals some internal gaps imposed during fossilization (Fig. 8.6), preservation is sufficiently good to assess general condylar shape, and as previously reported, the lateral condyle is definitively elliptical, although its shape index is in the upper range of modern humans and is near the lower range of *Pan*, and may have been significantly (albeit slightly) altered by the geological post-mortem crushing of the specimen (Table 8.3). In general, however, the femur does not differ significantly from many modern human lateral condyles of similar size. The medial condyle has lost a significant portion and is much less well preserved.

The lateral condyle does exhibit the typical elevated lateral wall for patellar retention, implying selection imposed by habitual knee valgus. There is a deep patellar groove and a deep, well-defined popliteus groove as well. We previously estimated the length of the specimen to be between 418 and 438 mm, using the crural index and known tibial length (see below) (Haile-Selassie et al. 2010a). A more detailed account of various aspects of preservation of the specimen is provided by Ryan and Sukhdeo (2016). These authors have also reconstructed the preserved portion of the femoral shaft virtually (Fig. 8.7). Although the cortex and matrix are not completely separable, they have also provided various geometric properties based only on the external form of the preserved shaft. These must be treated with great caution, however, since total cross-sectional area requires normalization using both estimated femoral length and estimated body mass, in turn ultimately based on an estimate of acetabular diameter by Haile-Selassie et al. (2010a).

Table 8.3 Ratio of vertical and horizontal tangent lengths of the lateral condyle (data from Lovejoy et al. 2009b)

Taxon/Specimen	*N*	Mean ratio	Range
H. sapiens	15	47.5 (3.9)	41–53
P. troglodytes	15	65.9 (8.2)	55–82
A.L. 333-1		46	
A.L. 129-1		48	
KSD-VP-1/1		*56*[a]	

Estimated values are in *italics*
[a]The lateral condyle has suffered significant damage, though its overall dimensions appear intact. However, the exact location of the meniscal groove must be estimated because of surface damage to the condyle in its likely location

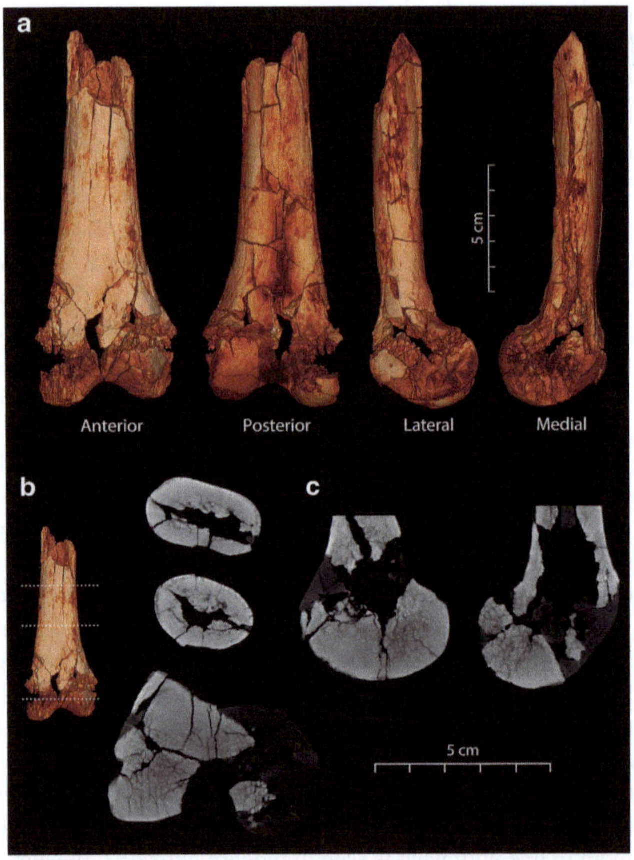

Fig. 8.6 CT scan images of the femur of KSD-VP-1/1 (from Ryan and Sukhdeo 2016). Views of entire specimen as labeled. Cross-sectional views shown in **b** and **c**. Note that lateral condyle is well preserved with only slight distortion, whereas the medial condyle is much less well preserved

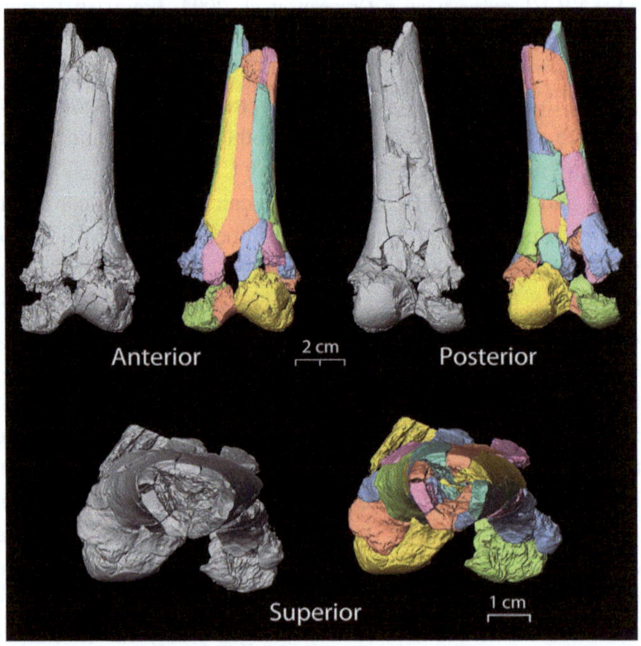

Fig. 8.7 CT scan reconstruction of the distal femur of KSD-VP-1/1 (from Ryan and Sukhdeo 2016). Note that alignment is good and that a small bicondylar angle is present as expected. Superior view shows a shaft that is generally less elliptical than in humans, likely reflecting the relatively short anteroposterior dimension of the distal femoral surface, in turn suggesting that the quadriceps moment arm had not yet reached modern dimensions. For discussion, see text

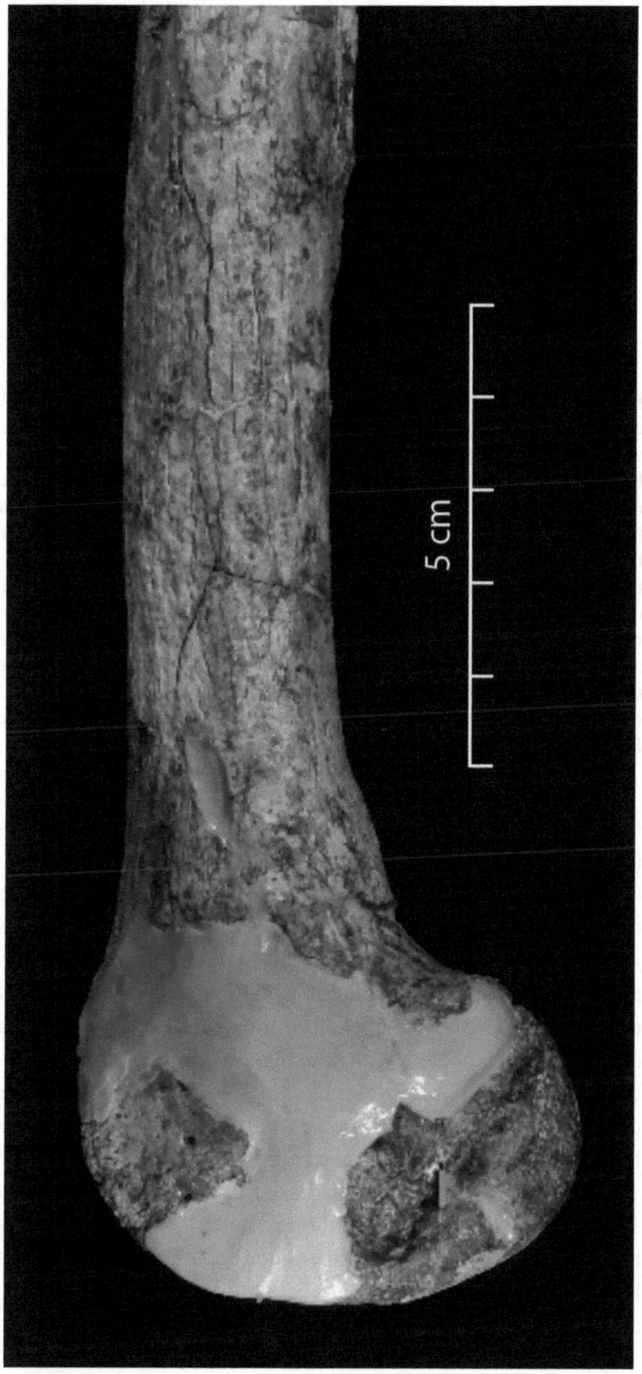

Fig. 8.8 View of lateral condyle of the femur of KSD-VP-1/1 (original specimen) showing the method of preservation required to maintain the intactness of the specimen. While this treatment has made physical assessment of detail somewhat more difficult, it has preserved all portions of the specimen, which has allowed CT reconstruction and restoration of the original to near perfect condition with respect to metric dimensions. See also Fig. 8.7

Preservation is not sufficient to obtain a reliable index of the specimen's patellar moment arm, which in other *Australopithecus* specimens is generally lower than those of modern humans (Lovejoy 2007b). Ryan and Sukhdeo (2016) used CT data to reconstruct several shape parameters of the femoral shaft, including Imax/Imin and Polar Moment (J) (see Fig. 4.12a–c in Ryan and Sukhdeo 2016). These fall within the *Pan* range or are intermediate between *Pan* and *Homo*.

In general, the shape of the adult diaphysis at any point is more likely to reflect shape of the growth plate (epiphysis) at earlier points during the bone's development, than a response to geometrical impact of external forces – especially, in regions that are not subject to either important torsional or bending loads that might pose significant fracture risk (i.e., those areas increasingly distant from mid-shaft). This, in concert with the general *Australopithecus* pattern of having a quadriceps moment arm, significantly shorter than those of modern humans is therefore the most likely explanation for the KSD-VP-1/1's tendency to fall with *Pan* rather than *Homo*, or to fall between the two taxa, since *Pan* also has a much lower quadriceps moment arm than do modern humans (Lovejoy 2007b), and its distal epiphysis will likely reflect this difference from *Homo* throughout development of the bone in the two species.

Posteriorly, the supracondylar lines, though badly damaged, can be traced proximally to form the beginning of what appears to be a moderately elevated linea (see superior view of the shaft provided by Ryan and Sukhedo (2016)). At the point where the shaft is broken away, the linea has an approximate breadth of 12 mm (estimated from the cast). This appears consistent with the same measurement estimated from the CT scan. In other assessable morphological characters, the specimen, as with the remainder of the skeleton, conforms to the *Au. afarensis* pattern.

Tibia

KSD-VP-1/1e is a left tibial shaft, which has been badly damaged for almost its entire length (Figs. 8.9 and 8.10; Table 8.4). It lacks most of its proximal end, having been fractured just proximal to the tibial tuberosity, but it does preserve the portion of tibial plateau that is usually occupied by a fat pad and bursa underlying the patellar tendon.

The distal end preserves most of the plafond, which, although displaying a significant traumatic pathology (see below), nevertheless allows an accurate estimate of overall tibial length accurate to within ± only a few mm at most. When compared to human tibias of equivalent distance between these two landmarks (bursal impression proximally and plafond surface distally), the specimen is approximately 355 mm in equivalent maximum length (Haile-Selassie et al. 2010a). This dimension is especially important because it allows comparison to the forelimb and allometric assessment compared to the skeleton of A.L. 288-1 (Haile-Selassie et al. 2010a). The shaft is dramatically platycnemic, but otherwise

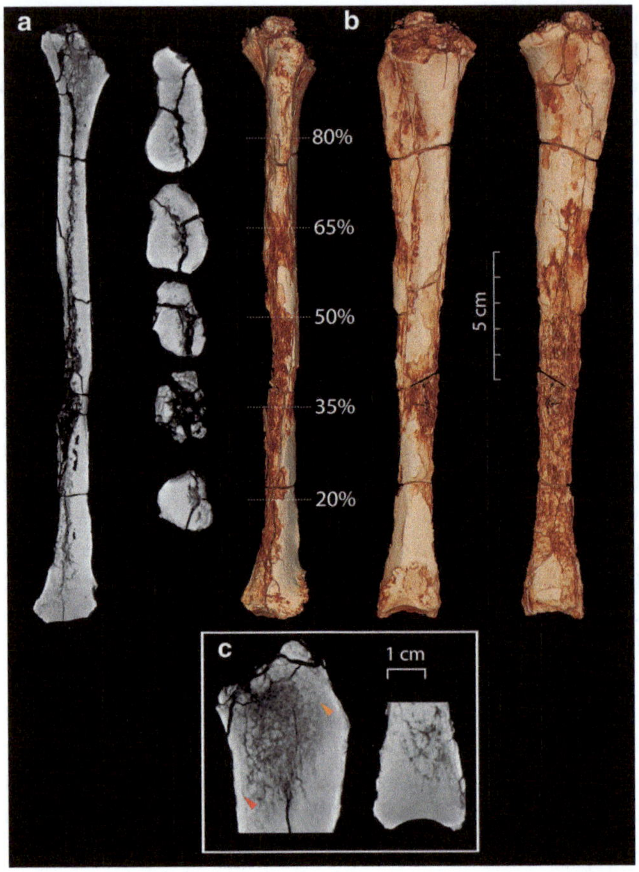

Fig. 8.9 CT scan images of the tibia of KSD-VP-1/1 (from Ryan and Sukhdeo 2016). As described in the text, although the specimen is not entirely complete, the anatomical landmarks required for an accurate reconstruction of maximum tibial length are preserved. The pathology affecting the distal end is evident on the anterior and lateral views of the distal end. See also Fig. 8.10

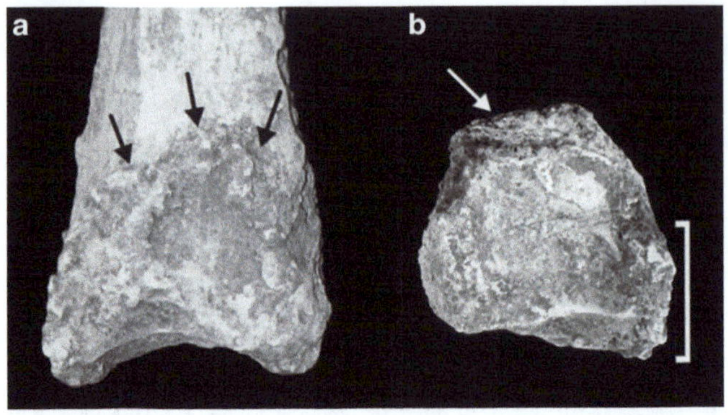

Fig. 8.10 Lateral **a** and Inferior **b** views of KSD/VP-1/1e. **a** A continuous perimeter of reactive bone surrounds the fibular facet (black arrows in **a**). The reactive bony outgrowth has elongated tibial length by about 1 cm. As discussed in Haile-Selassie et al. (2010a), these are most likely indications of a remote non-united, fibular fracture (see text for further discussion). **b** Inferior view shows a distal projection of the bony outgrowth (white bracket) and the severed medial malleolus (post-mortem). The form of this structure, however, can be deduced because of its near-surgical transection. Despite the effects of the fracture, the markedly concave tibial plafond is still evident in **a**, though metric assessment is made dubious by the florescence of reactive bone, since the fracture appears significantly remote and is likely to have affected distal tibial growth (see text for further discussions). Modified from Haile-Selassie et al. (2010a)

has experienced too much exfoliation to allow metric assessment or to assess the details of muscle insertion. For further discussion of the condition of the shaft, see Ryan and Sukhdeo (2016).

Although the medial malleolus has been sheared away from its base, the distal end is otherwise well preserved. The tibiofibular syndesmosis exhibits extensive pathology (Fig. 8.10), and details are provided in Haile-Selassie et al. (2010a). Specifically, the fibular facet's perimeter presents as a "ring" of elevated, reactive bone that extends medially to involve the entire lateral half of the anterior tibial surface. This has significantly enlarged non-articular portions of the lateral shaft and extended the bone's length inferiorly by about half a centimeter (this is not included in our estimated length of the bone). The plafond's subchondral surface, however, is unaffected. The most likely explanation is a remote, non-united fibular fracture. Tibial involvement suggests a pre-adult event (see below), a typical outcome in modern human cases (Wiltse 1972; Gibson and Prieskorn 2007).

DeSilva (2009) has noted that the radius of the curvature of the talocrural joint substantially discriminates various genera within hominoids. Generally, ape plafonds are flatter, probably reflecting substantial dorsiflexion during vertical climbing, whereas those of hominins have shorter radii of curvature as expected with much more limited dorsiflexion during the overall gait cycle, i.e., they do not reflect evidence of significant climbing. Relative plafond depth (RPD) varies between 12 and 14 in apes, whereas in humans the median value is 16. We previously assessed its value in KSD-VP-1/1e as lying well above 20, although an exact measurement is not possible because of the specimen's pathology. This is of little consequence since the specimen's RPD is clearly well above the range for African apes, a character consistent with a virtual absence of the vertical climbing seen in African apes. Whether these values in African apes and hominins reflect cartilage modeling (Type IV) or instead reflect the genomic constraints associated with joint formation (Type I or II) remains unknown. However, in any case, it is clear that the KSD individual did not engage in frequent climbing, a conclusion that is also suggested by at least 11 other early hominin tibias (DeSilva 2009).

The shape of African ape distal tibial metaphyses is also distinctive. They tend to be broad and rectangular, with

Table 8.4 Tibia length relative to acetabular diameter and other metrics in KSD-VP-1/1 and in other hominoids (n = 25) used in this analysis

Character	KSD-VP-1/1	A.L. 288-1	*H. sapiens*	*P. troglodytes*	*G. gorilla*
Tibia maximum length	355	227–235	376.2 (27.3)	254.3 (12.4)	288.6 (32.1)
Tibia physiological length	343		363.1 (27.0)	238.0 (12.0)	271.5 (31.2)
Acetabulum diameter	48	38[c]	54.4 (4.4)	40.8 (2.9)	54.3 (6.6)
Femoral head diameter	*34.7*[a]	28.6	45.7 (4.1)	33.5 (2.3)	45.3 (5.8)
Femoral length	*423–438*[b]	280	460.4 (26.2)	299.0 (14.6)	347.2 (37.1)
Tibia/acetabulum ratio	0.135	*0.162–0.167*	0.145 (0.011)	0.161 (0.011)	0.188 (0.012)

Estimated values are in *italics*

[a]Estimated by ratio using A.L. 288-1

[b]Estimated by crural index values of 81 and 84

[c]The value listed in Johanson et al. (1982) is 34.7 but represents the inside diameter; the method used here is that described by Ward (1991)

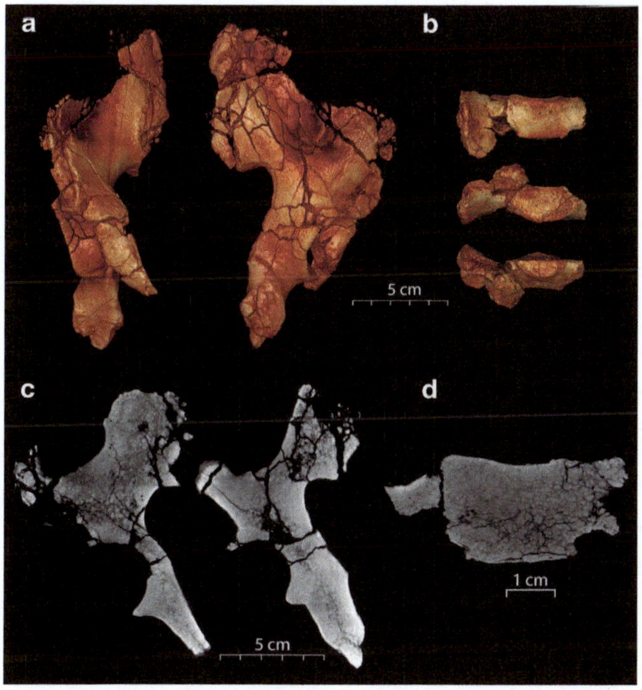

Fig. 8.11 CT scans of innominate and sacrum of KSD-VP-1/1 (from Ryan and Sukhdeo 2016). Note that there has been breakage and slight shifting of a number of pieces of the innominate, although most have been maintained in relatively good alignment within the enclosing matrix. However, the most superior portion of the ilium presented fragments that required significant realignment, although this could be achieved because the mating surfaces were relatively well preserved

relatively more massive medial malleoli than human tibias. In the case of the KSD specimen, we previously estimated its malleolar index at 47 and a simple AP/ML ratio of 100, with little or no impact from the specimen's pathology. Both values are well within the human ranges.

Haile-Selassie et al. (2010a) also previously reported that KSD-VP-1/1e exhibits an unusual lateral declination of its distal articular surface, i.e., whereas the plafond/shaft angle is 90° in humans, it is 100° in KSD-VP-1/1e. However, reference to the clinical literature suggests that this outlying value probably reflects the specimen's juvenile fibular fracture [see Haile-Selassie et al. (2010a) for further discussion].

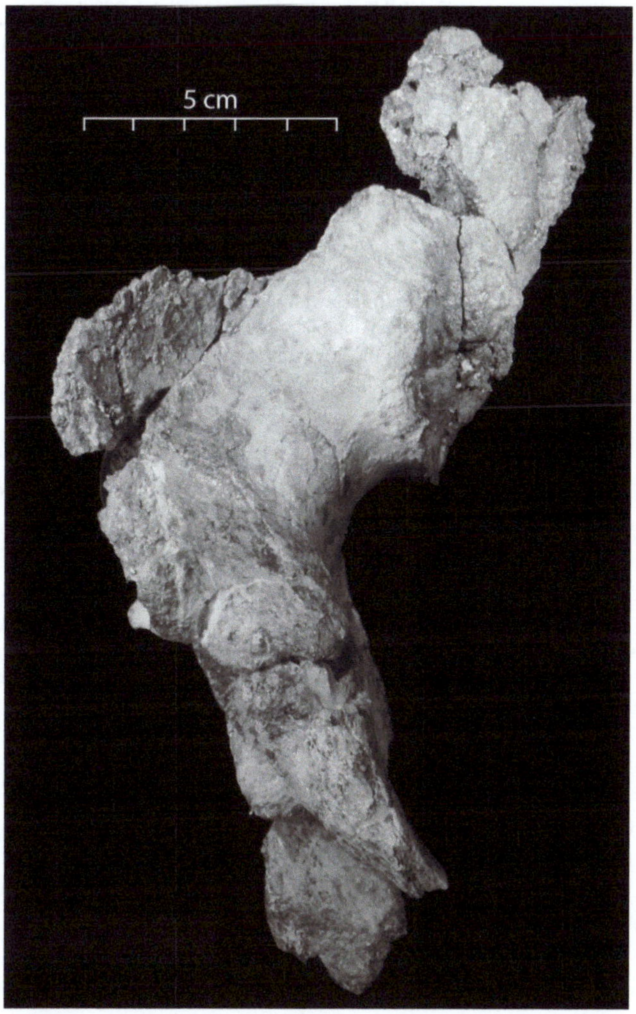

Fig. 8.12 Frontal view of the original innominate (os coxa) of KSD-VP-1/1. Displacement of the superior pubic ramus is evident. Compare to Figs. 8.13 and 8.14

Ryan and Sukhedo provide a measure of the tibial arch angle as defined by DeSilva (2010) and KSD-VP-1/1e appears to be well within the human range (for further discussion, see Ryan and Sukhdeo 2016).

Pelvis (Os Coxa and First Sacral Segment)

Perhaps the most important bone with respect to the interpretation of gait pattern in *Australopithecus*, an os coxa, was recovered from the KSD-VP-1/1 (1d). Although the specimen lacks most of its anterior portion and the inferior pubic ramus, the preserved portions are substantial and include a portion of its superior pubic ramus, the dorsal half of the ischial tuberosity, which is unusually informative (see below), most of the body of the ilium (although its upper limits have been fragmented with some significant shifting and actual translation of elements with re-adherence via matrix), and an only slightly damaged acetabulum (Figs. 8.11 and 8.12). Therefore, the innominate is sufficiently complete to be instructive with respect to the locomotion of hominins of this age (i.e., circa 3.6 Ma) (Table 8.5). An articulating

Table 8.5 Partial list of cardinal characters of preserved portion of the KSD-VP-1/1 os coxa compared to those of other hominoids

Character	*Australopithecus*		*Pan*	*Homo*
	KSD-VP-1/1	A.L.288-1		
Size and form of Ilium	Dramatic anteroposterior expansion with extensive origin for lesser gluteals	As in KSD-VP-1/1	Narrow superoinferiorly elongated, with narrow g. Minimus origin along anterior edge	As in KSD-VP-1/1 but somewhat anteroposteriorly shorter because of change in birth canal architecture
Size and form of Ischial Tuberosity	Greatly reduced with marked angulation of hamstring/adductor surface	As in KSD-VP-1/1	Massive, with arc-like undifferentiated hamstring/adductor surface	As in KSD-VP-1/1 but usually with slightly narrower gap between acetabulum and ischial tuberosity
Lesser sciatic notch	Greatly abbreviated distance between ischial spine and auricular surface with narrow acute angulation	As in KSD-VP-1/1 but with less acute angle of notch	Elongate and only marginally defined	Varies with sex. Most *Homo* males as with KSD-VP-1/1; most females as with A.L. 288-1
Iliac fossa	Extremely broad with sigmoid curvature when viewed from dorsal aspect	As in KSD-VP-1/1	Narrow, with almost no curvature when viewed from dorsal aspect	Less broad than in *Australopithecus* and with more pronounced sigmoid curvature
Sacrum	Very broad, with high alae/sacral centrum area ratio	As in KSD-VP-1/1	Narrow overall with narrow alae	Broad as in *Australopithecus* but with reduced alae/centrum ratio
Retroauricular region	Very broad	Moderate to narrow	Extremely narrow	Very broad
Obturator internus groove	Deep and well defined	As in KSD-VP-1/1	Not present	As in KSD-VP-1/1
Iliopsoas groove	Deep, well defined, with steep sides	As in KSD-VP-1/1	Not present	As in KSD-VP-1/1
AIIS by separate ossification center	Adult, but almost certainly	Adult, but almost certainly	No	Yes
Iliac pillar	Anteriorly located and distinct	As in KSD-VP-1/1	No	Pillar well defined but located more posteriorly
Iliopubic eminence	Damaged, but highly probable	No	No	Typical and very frequent
Lunate surface of acetabulum	Covers all of acetabulum except acetabular notch	As in KSD-VP-1/1	Covers most of acetabulum except acetabular notch	As in KSD-VP-1/1
Anterior inferior iliac spine (AIIS)	Large, sigmoid shaped, projecting	As in KSD-VP-1/1	Limited to rugose enthesis with only moderate elevation above iliac surface	As in KSD-VP-1/1
Auricular surface	Narrow and moderately elongate	Broader and less elongate; more *Homo*-like	Highly elongate and very narrow	Broad and reniform with distinct superior and inferior demifaces
Acetabulum size	Moderate	As in KSD-VP-1/1	Small to moderate	Very large
Superoinferior height of os coxa	Short and broad	As in KSD-VP-1/1	Very tall with greatly elongated supra-acetabular region of ilium	As in KSD-VP-1/1
Posterior superior iliac spine	Massive	Much more gracile than KSD-VP-1/1, but likely a consequence of diminutive body size	Little of no expansion of iliac crest at posterior terminus	As in KSD-VP-1/1

second (but see below) sacral segment was also recovered, which is again informative because of its geologic age. Most of these features have been thoroughly presented in Ryan and Sukhdeo (2016) and can be visualized in the figures they provide, but see below.

The retroauricular area of the innominate has undergone sufficiently substantial translation and repositioning of elements to alter its general anatomical appearance. As just noted, the relevant cracks are easily visible on conventional X-ray and CT scan pilot images (Figs. 8.11 and 8.13), and the reader should always remain cognizant of this damage in the descriptions that follow below. The affected region includes the posterior superior iliac spine (PSIS). Some successful repositioning of fragments has been possible since its original publication by Haile-Selassie et al. (2010a). This has made the specimen more readily interpretable.

In addition to these alterations, a fragment of the superior pubic ramus has been broken and bent anteroinferiorly, causing it to "point" inferiorly away from the acetabulum whose diameter has thereby also been slightly reduced by this damage (see Figs. 8.13 and 8.14). The deflection of the superior pubic ramus can be readily seen in the previously mentioned CT scans.

As described earlier, the os coxa conforms to what may now be termed the "classic" *Australopithecus* pattern (Table 8.5). The most salient features that are relevant to gait pattern and locomotion include a voluminous obturator internus groove that separates the inferoexterior face of the acetabulum from the ischium, a massive anterior inferior iliac spine (AIIS) which lies above a very deep iliopsoas groove, and a robust iliac pillar which is located somewhat more anteriorly than its typical human homolog. All of these features are now known to be quite typical of the *Australopithecus* pattern (Lovejoy 1974, 1975, 1979, 1988; Johanson et al. 1982; Lovejoy et al. 2009a, b).

A distinct greater sciatic notch is preserved with only minor damage in its region. The notch is relatively narrow, suggesting along with the specimen's obvious great mass, that it was male. Indeed, Simpson et al. (2008) devised a method of discriminating sex in os coxae based on the shape of the sciatic notch, with "J"-shaped notches being more frequently male than more symmetric "V"-shaped (largely female) notches. They also devised a numerical method of evaluating this shape by inscribing a line connecting the ischial spine to the tubercle for the inferior sacroiliac ligament. The distance along this line

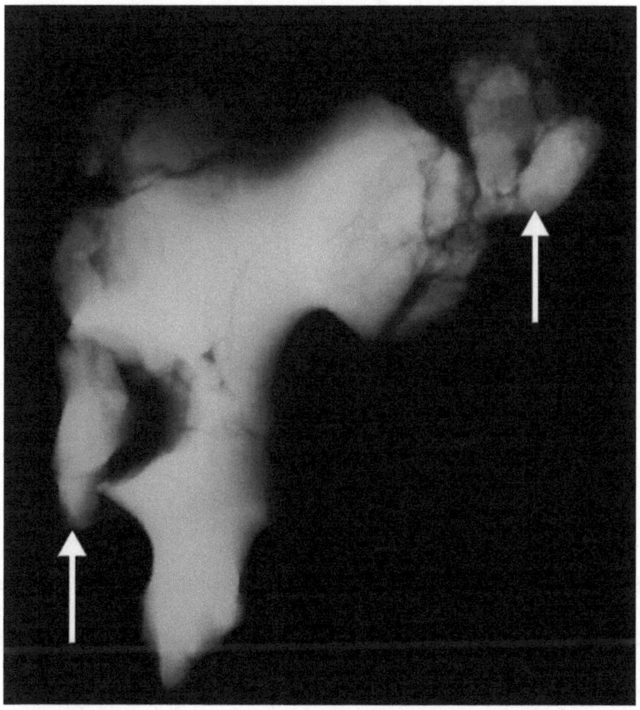

Fig. 8.13 X-ray of the innominate of KSD-VP-1/1. Displacement of the superior pubic ramus and extensive damage to the retroauricular region are evident (white arrows). Modified from Haile-Selassie et al. (2010a)

Fig. 8.14 Acetabulum of KSD-VP-1/1 (original specimen). Displacement of a superomedial portion, leaving a significant vertical gap. The remainder of the specimen appears mediolaterally crushed, requiring an estimate of its original dimensions (see text). The same kind of preservation is apparent in the distal femur (see Fig. 8.8)

at which the deepest inflection of the sciatic notch occurs is used to define a ratio of the two distances which then serves as an index. Minor damage to the region in question in KSD-VP-1/1 makes our calculation only approximate but at the same time very likely to be correct within a percentage point or two with a value of 0.20–0.23, which is outside their female range in *Homo sapiens* (Fig. 8.15). These values are dramatically lower than that for the BSN49/P27 pelvis, which is assigned as a female *Homo erectus*, but still preserves the fundamental *Australopithecus* pattern.

Although an extended effort to fully reconstruct the complete pelvis of KSD-VP-1/1 is underway, completion will take significantly more time and for the present contribution we have provided figures showing the current state of the reconstruction (Fig. 8.16). These show the fundamental morphology of the pelvis and its remarkable similarity to even modern human pelves, considering its great age. As seen in photos of the original specimen, the superior pubic ramus was broken, slightly dislodged, and bent inferiorward during fossilization. On the current figure, we have indicated the ramus as reconstructed but the figure is based on the lateral portion, which is well preserved and merely dislodged (although there is no symphysis pubis). The figure also shows the preserved portion of the ischial ramus which plays an important role in interpreting the level of adaptation of this pelvis to upright locomotion (see below), and indicates the relationship between the sacrum with its (presumed) sacralized lumbar (see below) and the well-preserved auricular surface of the os coxa. A top view of the current reconstruction is also provided (Fig. 8.16b).

Among the most impressive aspects of the specimen, and the *Australopithecus* pattern in general, is that it so closely mirrors the human pattern, albeit with the exception of those largely minor ways in which *Australopithecus* specimens differ regularly from *Homo*. In general, it is worth noting that KSD-VP-1/1d has a more "modern human-looking" overall form than does A.L. 288-1, perhaps because of the fact that its general body mass is likely well within the modern range, and the specimen appears substantially more robust than the specimen from Hadar, being that of a large male rather than a female whose body mass is likely close to the lower limit of the species. That aspect may account for the latter's rather extreme platypelloidy.

Also, of special interest in respect to additional comparison with A.L. 288-1 is that the lunate surface of the acetabulum of KSD-VP-1/1d is well formed and extends more anteriorly than it does in A.L. 288-1, and most importantly, is indistinguishable from the typical modern human conformation. Along with correction of the erroneous description of the lunate surface of the latter specimen as not having an anterior extension as in *Homo* (Stern 1983, 2000; for explanation see

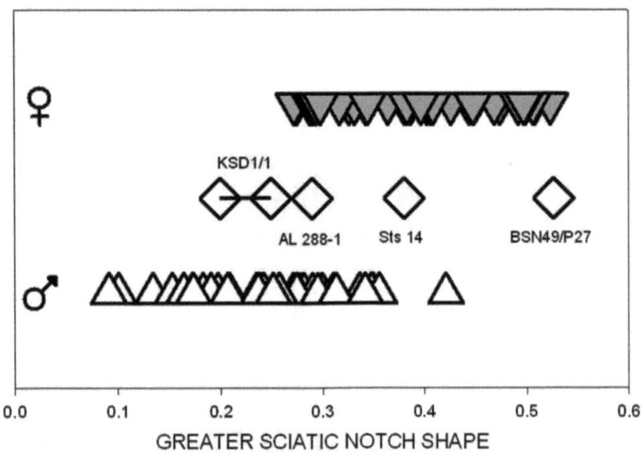

Fig. 8.15 The distributions of samples of human male and females are shown as individual data points, as are A.L. 288-1, Sts 14, and BSN49/P27 (the latter is *Homo erectus*). A range is given for KSD-VP-1/1 because of slight uncertainty with respect to the position of the tubercle for the inferior sacroiliac ligament. See text and Simpson et al. (2008) for methodology and sample composition. Figure originally published in Haile-Selassie et al. (2010a)

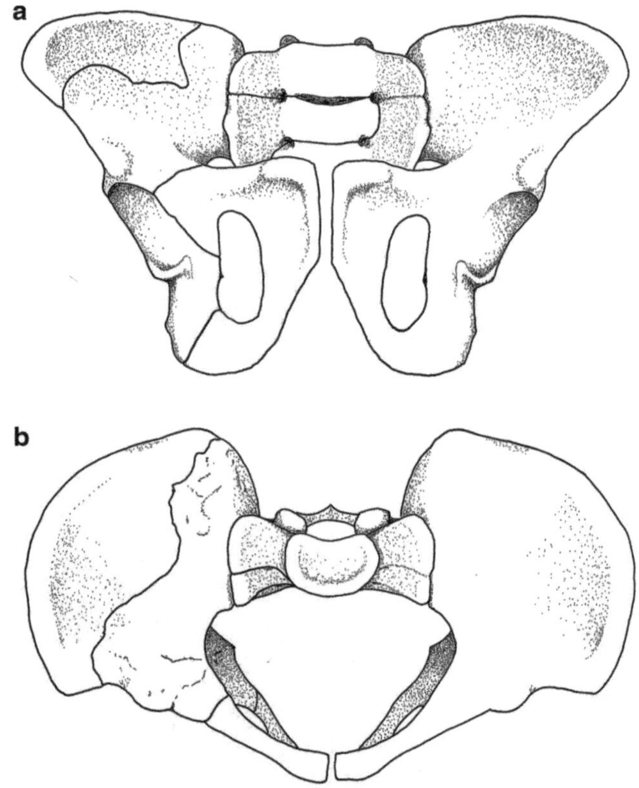

Fig. 8.16 Anterior (**a**) and superior (**b**) views of the current plastocene reconstruction of the pelvis of KSD-VP-1/1. Missing portions as described in the text are restored as indicated, and the left os coxa is mirror imaged. See Fig. 8.18 for details of the sacrum reconstruction. Although the os coxa's superior pubic ramus is shown as reconstructed, most of its body was preserved but fractured and dislodged downward (see Figs. 8.11, 8.12, 8.13 and 8.14). Most striking is the very small pelvic inlet relative to the size of the ilia and sacrum, and the otherwise very human aspect of the overall pelvis. The pelvis is almost certainly male (See Fig. 8.15 and the text)

Lovejoy 2005), KSD-VP-1/1d should put to rest the argument that the acetabulum of *Australopithecus* was somehow less developed than that of *Homo*. Consistent with the overall total morphological pattern of *Australopithecus*, the acetabulum appears somewhat small. Haile-Selassie et al. (2010a) previously estimated a vertical acetabular diameter of approximately 48 mm relative to other dimensions of the skeleton (Fig. 8.14). For example, a simple ratio with tibial length is 0.135, about one SD below the human mean. This is, again, quite typical of *Australopithecus*.

While the dimensions of the sacrum fragment (Figs. 8.17 and 8.18) also conform to the *Australopithecus* pattern (see below), its relationship to the well-preserved auricular surface of the os coxa is unusual for *Australopithecus* in that it presents, to our knowledge, the first evidence of sacralization of the most caudal lumbar element (or conversely, lumbarization of the most cranial sacral element), that is, the likely presence of a "lumbosacral transitional vertebra" (LSTV) at the L5-S1 junction.

Traditionally, assignment of an instance of LSTV as lumbar sacralization or sacral lumbarization requires both a complete sacrum and lumbar column, as the ultimate candidate vertebra is usually considered to be sacralized if there are four independent lumbars or is considered lumbarized if there are six (Konin and Walz 2010; Ucar et al. 2013). In the case of KSD-VP-1/1, assignment to category remains uncertain as no lumbars were preserved and the sacrum is not complete. Nevertheless, articulation of the preserved portion of the first (and only) sacral segment with the os coxa's auricular surface shows that a considerable amount of the dorsal aspect of the os coxa's auricular surface remains vacant (see Fig. 8.18). This in turn implies that the missing portion of its ala was substantially elevated above the surface of its preserved vertebral body – so much so that it would have clearly resembled a typical LSTV in *Homo sapiens* (Fig. 8.18) if fully preserved, or, conversely, that the immediately more cranial lumbar exhibited at least one sacralized transverse process.

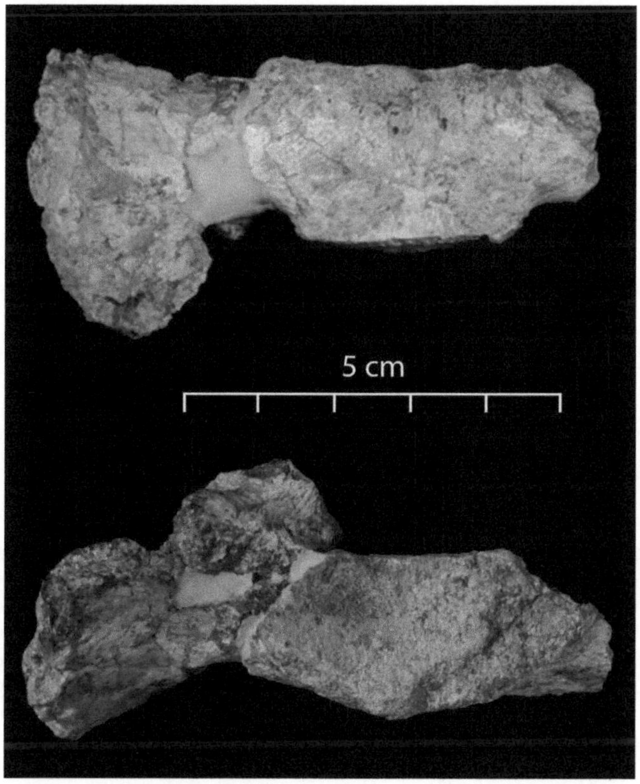

Fig. 8.17 Top and anterior views of the second sacral segment of KSD-VP-1/1. Damage to the surface of the right ala clearly suggests loss of a previously fused element from a missing alar-like transverse process of the next most craniad vertebra (sacralization of the last lumbar). For discussion, see text and Fig. 8.18

Sacralization is far more frequent than lumbarization in modern humans (17.2 vs. 1.7%; Konin and Walz 2010), and based on these frequencies alone, KSD-VP-1/1 is more likely to be a case of the former, i.e., an individual with six lumbars, the most caudal of which was sacralized, reducing the count to the typical human one of five – a process that we have argued is likely responsible for the dominance of five lumbars as the modal condition in modern humans from an ancestral condition of six (McCollum et al. 2009; Lovejoy and McCollum 2010). This is, however, a tenuous assumption, as the specimen could have had five lumbars with the most caudal one being sacralized and reducing its ultimate count to four. However, the latter is a very rare occurrence in modern humans. Vertebral developmental expression *is* known to be controlled by a complex *Hox* code (Wolpert 1998), and what this new finding *does* establish is that the lower vertebral column was probably under active selection for modification.

Like the os coxa (Figs. 8.16), the sacrum's second segment (or its first, depending on definition – see above) also conforms to the *Australopithecus* pattern. This includes the observation that, as previously noted, its auricular surface appears slightly narrower than its homolog in *Homo*, and that its sacral alae are unusually mediolaterally broad relative to centrum size as they are in other *Australopithecus* specimens including A.L. 288-1, Sts-14, and BSN49/P27 (Simpson et al. 2008; Haile-Selassie et al. 2010a) (Fig. 8.19).

Despite its post-mortem damage, the os coxa's postauricular region was clearly rugose and extensive, suggesting powerful and extensive posterior reinforcement of the sacroiliac joint. We previously suggested that this combination of a narrow auricular surface and an exceptionally rugose retroauricular area might imply that sacral nutation and counternutation were significantly greater in *Australopithecus* than in *Homo* and may have served as an important energy dissipating mechanism during running in *Australopithecus* (Lovejoy 2007a).

None of the minor morphological differences between the os coxae of *Australopithecus* and *Homo* suggest any

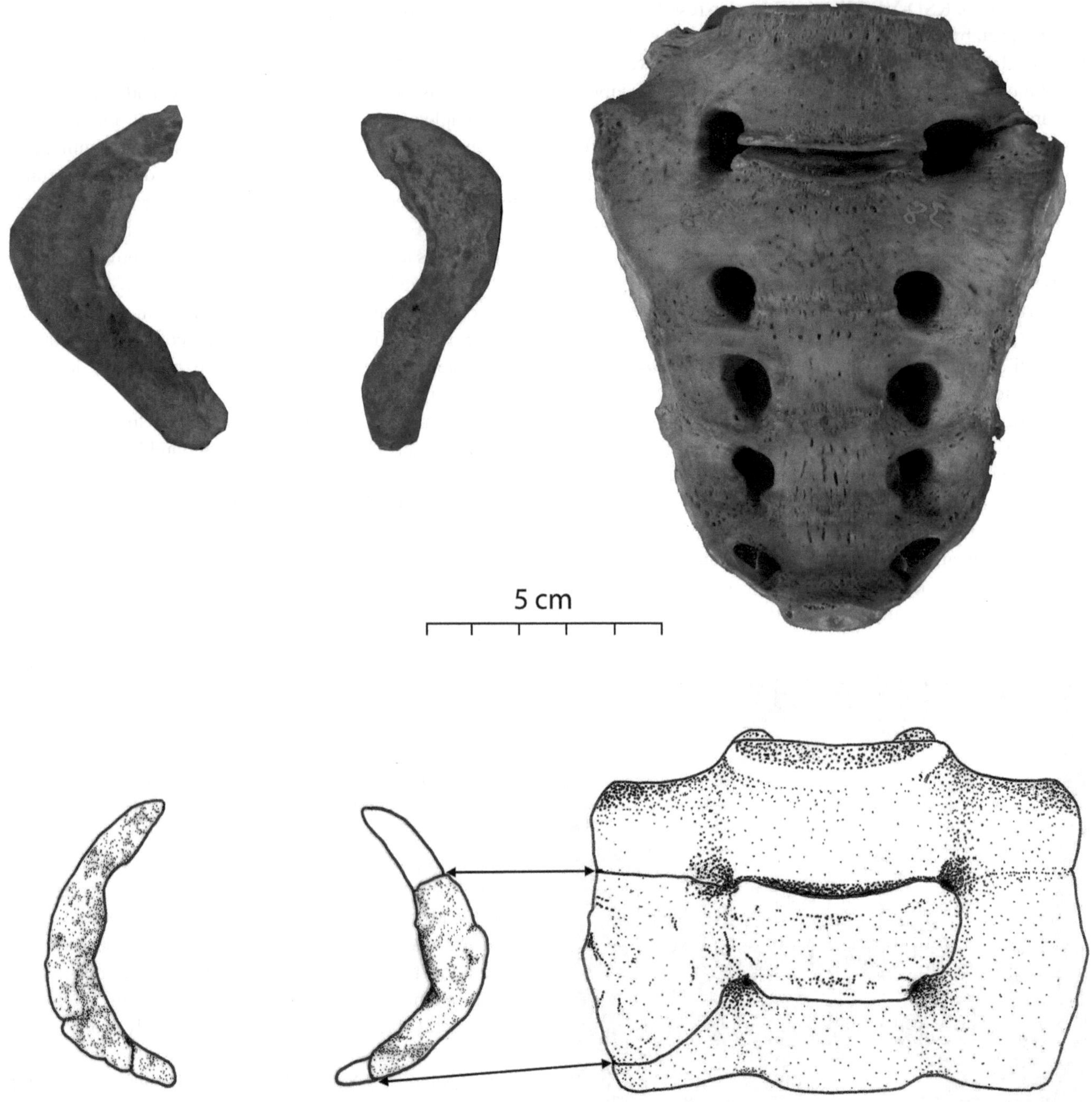

Fig. 8.18 *Top* Modern human sacrum (reversed) (Hamann-Todd CMNH specimen 738) with typical unilateral fusion of the first sacral segment/last lumbar (lumbar sacralization). *Bottom* Reconstruction of the first three elements of the KSD-VP-1/1 sacrum. To the far left of both specimens is the auricular surface of its associated ilium (CMNH-738 and KSD-VP-1/1d). To the near left of each is the auricular surface of its sacrum, which is reconstructed for KSD-VP-1/1. The missing portions of the KSD sacrum are based on the preserved auricular surface of the ilium as shown to the far left. The extents of sacral segments one and three are based on the fully preserved auricular surface of its ilium (top far left). Multiple specimens of modern humans were selected that closely matched the sacral and iliac morphology of KSD-VP-1/1 and these were used as general models for the missing sacralized lumbar portion of the sacrum. The arrows showing the preserved portion of the KSD-VP-1/1 sacrum and its corresponding auricular surface are not parallel because the sacrum is shown in slight rotation, whereas its auricular surface is not. Compare to Fig. 8.17

significant differences in the actual locomotor pattern in the two genera. Rather, virtually all are consistent with relatively subtle changes in overall morphological pattern related more to a combination of body size and the effects of pelvic structure related to adaptations to more facile parturition (Rosenberg and Trevathan 2002).

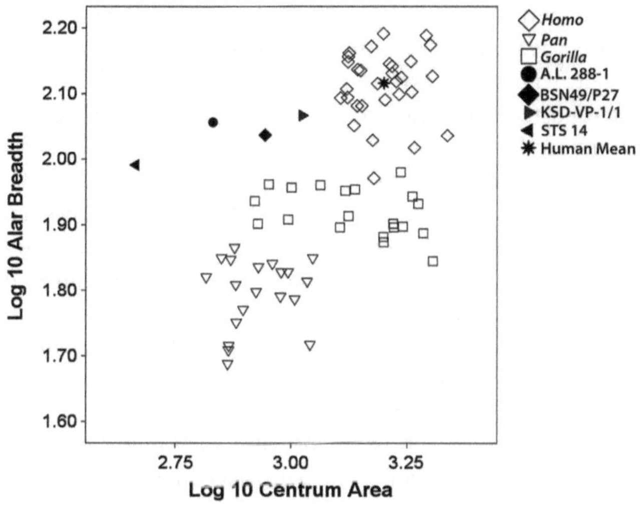

Fig. 8.19 Log–log plot of relative alar breadth of the sacrum in *Australopithecus* and extant hominoids. Alar breadth was calculated as 1/2 the maximum breadth of the sacrum less the transverse diameter of the S1 centrum. S1 centrum area is the product of its AP and ML dimensions. The human mean has also been plotted. Other symbols are provided in the legend. The S2 of KSD-VP-1/1 (plotted here as an S1) lies nearer the human cloud than do A.L. 288-1 and Sts 14, which might be an allometric effect of the smaller examples of *Australopithecus*. However, the more likely explanation is that the KSD-VP-1/1 sacrum's actual first segment was a sacralized lumbar, and lumbar centrum mass does not appear to exhibit the inferiorly enlarging condition seen in later hominins. For data sources, see Haile-Selassie et al. (2010a)

It should be also noted here that recent claims that the *Australopithecus* pelvis became modified in favor of some aspects of gait pattern rather than issues of parturition have been made using the somewhat fragmentary remains of "*Australopithecus sediba*" from the site of Malapa (Kibii et al. 2011). Not only are these claims based on a single specimen and a tenuous estimate of cranial capacity at birth, but they are also based on a pelvic reconstruction that relies on numerous, likely incorrect, assumptions. Review of that reconstruction shows that the ilia are grossly internally rotated (no auricular surface of even the lateral part of the sacrum was preserved and had to be created), the two portions of the superior pubic ramus are misaligned in superior view (i.e., they lack continuity of the arcuate line curvature), the pubic symphysis is far too elevated (its position depends on the length of the lower ischial ramus which for the most part is not preserved and its length certainly unknown) and the symphyseal joint, moreover, is externally rotated, and the reconstruction cannot be judged from acetabular position because much of the latter is also not preserved. Moreover, as well established by Rosenberg and Trevathan (2002), hominin parturition also involves passage of the broad shoulders of the fetus so as to prevent shoulder dystocia, and modifications of the hominin pelvis for bipedality greatly alter such relationships, regardless of cranial capacity in the infant. It is the pelvis and not the cranium that was so greatly modified for bipedality and was thereby the principal venue of natural selection.

The slightly more anterior position of the iliac pillar in KSD-VP-1/1 largely reflects the more elliptical shape of the overall pelvis in *Australopithecus*, a character made quite obvious by more complete specimens including A.L. 288-1, Sts 14, and BSN49/P27. Such "ellipticity" places the juncture of the internal and external oblique muscles relatively more anteriorly, and it is from their region of juncture that the iliac pillar is formed to descend externally on the surface of the iliac blade during development and growth.

Regularly, variant features such as, for example, the rugosity of this pillar and its trabecular contents are frequently assumed to reflect dynamic loading patterns during life, and while bone quantity and quality are certainly dependent upon its sensitivity to loading, the evidentiary value of variations in common anatomical landmarks (such as the iliac pillar) has long been overemphasized since general form is much more dependent upon positional information directing the embryogenesis of bone formation than to simple, direct reaction of external loading during life (Cunningham and Black 2009; Abel and Macho 2011).

However, the size and structure of the pillar can easily be seen to instead be a feature of the ilium's development as "programmed" by the positional information of the iliac anlagen (Cunningham and Black 2009). That is, even in fetal specimens the special enlargement of the iliac crest from which the pillar descends is already apparent, long before any significant stress occurs within the structure, and it is this "swelling" of the crest, typically called the *tuberosity of the crest*, which is present throughout the development of the iliac crest apophysis. The pillar is therefore largely merely the "track" or "wake" of the development of the iliac crest and its emerging tuberosity. Its more anterior position in KSD-VP-1/1 and in *Australopithecus*, in general, therefore largely reflects (compared to *Homo*) the more spherical conformation of the birth canal in pelves of the later Pleistocene, although this pattern now seems to have emerged much more recently than previously thought (Simpson et al. 2008).

The Pelvis of KSD-VP-1/1 and that of *Ardipithecus ramidus*

One of the most instructive comparisons that can be made during an effort to interpret the gait pattern of *Australopithecus* is with the pelvis of *Ardipithecus ramidus*, which is notably less advanced in its adaptation to upright walking and running than is that of *Australopithecus*. Indeed, comparisons made between these two pelves (and postcranial skeletons in general) and between these two taxa in general can be quite informative in demonstrating the advanced nature of upright walking in the more recent of the two.

Perhaps the most important of the revelations provided by the *Ar. ramidus* specimen ARA-VP-6/500 is the striking contrast between the two portions of the pelvis that are separated by the acetabular region, i.e., the superior ischial ramus below (separated from the acetabulum by the ischioacetabular sulcus) and the lower part of the ilium (i.e., the iliac isthmus) above. In *Ar. ramidus*, the upper portion of the pelvis, i.e., the ilium, shows a strong relative increase in anteroposterior/mediolateral breadth that is not present in other hominoid pelves (save those of later hominins) whether they are from the earlier Miocene or are only known from extant species (i.e., gorillas, chimpanzees and bonobos, orangutans). This is due to two factors. The first is a substantial increase in the anteroposterior expansion of the breadth of iliac isthmus (i.e., the region immediately superior to the acetabulum which bears the uniquely structured AIIS). The second is a dramatic reduction in the height of the isthmus, which is long in all quadrupedal higher primates both living and extinct, including Middle Miocene and living apes and as well as all cercopithecoids (White et al. 2014).

In addition, while the uppermost portion of the original specimen (ARA-VP-6/500) is partially fractured and thereby relatively "jagged" in its preserved profile, the presence and location of a distinct AIIS reveals the limits of the upper pelvis quite clearly, along with the location of the auricular surface which is also well marked in the original (White et al. 2014). These observations can be confirmed simply by comparing the distance between the top of the superior pubic ramus and the auricular surface, both of which have locations that are fully known from the original specimen. In general, this relatively broad and shortened ilium is fully commensurate with an earlier form of bipedality, even though the foot reveals the first ray to still have been fully abducent (White et al. 2009, 2014).

These features of the ilium of ARA-VP-6/500 bear a striking contrast to the portion of the pelvis *below* the acetabulum, which exhibits a long and robust ischial body, the minimum length of which can be readily determined from the original specimen, even though it bears significant damage (Lovejoy et al. 2009b, c). That is, the transect from the edge of the lower acetabular rim to the site of the origination of the ischial tuberosity surface can be readily and confidently determined and is essentially as long as it is in living pongids. This is fully commensurate with what must have been retention of the primary muscles of the hip involved in climbing and arboreal scrambling, especially the adductors, and what must have been exceptionally powerful hamstrings (the proportion of hamstring/adductor mass relative to quadriceps mass must have far exceeded that of later hominins and was presumably largely similar to these ratios in living pongids) (Lovejoy et al. 2009c).

The pelvis of KSD-VP-1/1 stands in remarkable contrast to these features of the lower pelvis of *Ardipithecus*. Paramount within its more advanced morphology is a fully anteroposterorily expanded ilium that is now fully shortened as in all later hominin specimens such as A.L. 288-1 ("Lucy"). The auricular surface and much of the body of the os coxa are fully preserved in KSD-VP-1/1 (see earlier), and these reveal a fundamentally modern pelvic construction, and one that confirms its virtually complete adaptation to habitual and obligate bipedality.

Arguments that the relative robusticity of minor elements of pelvic construction, as have been made with respect to specimens from Malapa (see below), for example, are largely irrelevant to such issues because such minor perturbations in overall structure are well beyond the capacity of natural selection to so exactly modify lateral plate (and somitic) positional information that "codifies" condensation, anlagen formation, and growth, as to not be responsive to variation correlated with actual fitness (Lovejoy 2014).

Take, for example, the following supposition about structural details reported for the Malapa specimen: Kibii et al. (2011:1409) stated that "As in *Homo*, the weight transfer region of the iliac body of the Malapa hominins is more robust than that of other australopiths, manifested by greater thickness of the acetabulosacral buttress and shortening of the acetabulosacral load arm." Two elements of this argument that have little or no basis in the modern understanding of morphogenesis are the claims relating to "acetabulosacral buttressing" and the length of the "acetabulosacral load arm," which are both cited as though they were structural entities that are both (1) modifiable by *localized* particular shifts in the positional information giving rise to the pelvic condensations and anlagen, and (2) subject to individual elevation in population frequency by the specific action of selection – which in each case could have only become a factor of morphological determination by regular *failure* of such elements during locomotion. Regular examination of pelves of virtually all higher primates has yet, to our knowledge, ever shown structural failure in any way correlated with such localized morphological variation. Indeed, this observation applies to many of the various individual claims that have been made with respect to differences in the pelves of early hominins.

Take, as a further example, the following assertion by Kibii et al. (2011:1407) that some early Pleistocene *Homo* pelves (KNM-ER 3228, KNM-ER 5881, O.H. 28, and KNM-WT 15000) "share with modern humans a suite of features that include a thick iliac body, reduced (relative to body size) distance from the auricular to acetabular joint surfaces, expansion of the retroauricular area, a large iliac tuberosity, vertically set iliac blades, a sigmoid-shaped iliac crest with moderate- to well-developed fossae for iliacus and gluteus medius, a distinct iliac pillar, a sinusoidal anterior iliac border, and a narrow tuberoacetabular sulcus." Such a list is merely a catalog of traits, which in many cases appear to regularly differ in individual specimens of both early *Homo* and modern humans. None is likely to be of any mechanical or biological significance other than as simple pleiotropic correlates of overall pelvic structure and not each in themselves the product of selection (see previous discussion of the morphogenesis of the iliac pillar, for example). Indeed, all are likely to be merely Type 2 corollaries of the simple general trend toward enlargement of the birth canal.

Moreover, none of these can be taken as a meaningful adaptation, in and of itself, unless the two criteria listed previously can be met: (1) some indication that there is sufficient variation in underlying positional information to accomplish such singular changes in structure such that selection could lead to their fixation, and (2) demonstration that such changes are *specifically* responsive to actual *structural failure*. Whenever a substantial alteration in musculoskeletal anatomy takes place, there are almost certainly two highly distinguishable and potentially identifiable shifts in morphological structure – those which are true targets of selection (i.e., whose variation can be shown to directly affect *fitness*) and those which are simply byproducts of such shifts, but which are not themselves subject to selective influence and are therefore merely non-directionally associated with the first – essentially cases of morphological "hitchhiking" (Lovejoy et al. 2002; Lovejoy 2014). In fact there is no reason, a priori, to presume that the characters listed in the above citation in any way represent a morphological advance simply because they appear in modern humans and not specimens from the earlier Pleistocene. That is, because a feature is found in modern humans and not in an earlier hominin species, is not of itself evidence that in any way it can be considered a structural "improvement" or "adaptation."

In fact, in most cases such changes are more likely to be due to what may be termed "morphological drift" since they are most likely to be simply the product of one or two major structural changes (such as an enlargement of the birth canal or the prevention of an injury of known or reasonably functional etiology, such as prevention of the avulsion of the ischial apophysis – see below).

In summary, unless a structure consistently *fails* with some regularity, there is virtually no chance that selection will lead to modifications of that structure with respect to such minor details of biological construction, and much of pelvic morphology is virtually never subject to simple structural failure (i.e., independent of unanticipated and unpredictable external trauma); *ergo*, selection was not responsible for minor variants in its composition – these are instead almost certainly correlates of other, more significant changes in the rest of the bone. To argue the contrary case is to invoke unwarranted adaptationism (Gould and Lewontin 1979).

The KSD-VP-1/1 Ischium and Its Relevance to a Fully Evolved Running Gait

Of great interest in regard to the issue of the *degree* to which *Au. afarensis* might have engaged in habitual and effective bipedality is the conformation of the ischial body and especially its ischial tuberosity. The conformation and positioning of the latter differ dramatically in hominins and apes. In the former, the surface for attachment of the three hamstrings has been shifted to face more superiorly such that hamstring attachment will produce less direct tension at its surface during the latter portion of swing phase (just prior to heel strike). This has resulted in a substantial reduction in the actual length of the ischium, relative to the acetabulum (Fig. 8.20). Such repositioning of this portion of the tuberosity surface has introduced a distinct angulation of the ischial tuber, dividing it into two distinct portions. This duality is entirely absent in the homologous structure in pongids. These changes to the hamstrings have generally been referred to as "shortening of the hamstring moment

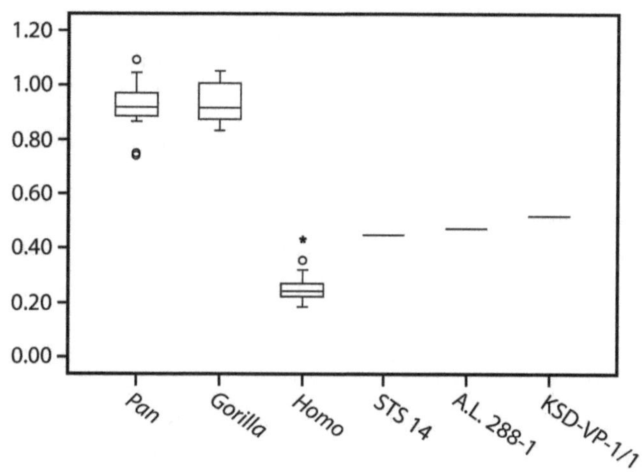

Fig. 8.20 Relative ischial length in KSD-VP-1/1d and extant hominoids. Angulation of the ischial apophysis in order to reduce the probability of avulsion has resulted in a substantial reduction in the length of the ischium and its moment arm (for discussion see text). Here, the minimum distance between the edge of the lower rim of the acetabulum and the superior-most point on the ischial tuberosity surface has been divided by the acetabular diameter for samples of extant hominoids (N = 25 each species). The positions of the three *Australopithecus* specimens are close to those of *Homo*, but still lie slightly outside the range of the sample shown here, although this ratio in other human samples sometimes overlaps those of *Australopithecus*. This is likely also related to some enlargement of the acetabulum in later *Homo*. Each box shows the median, interquartile range (the box length), outliers (open circles), and extreme values (asterisks) within a category. Outliers are cases with values between 1.5 and 3 box lengths from the upper or lower edge of the box. Extreme values are those with values more than 3 box lengths from the upper or lower edge of the box. Figure originally published in Haile-Selassie et al. (2010a)

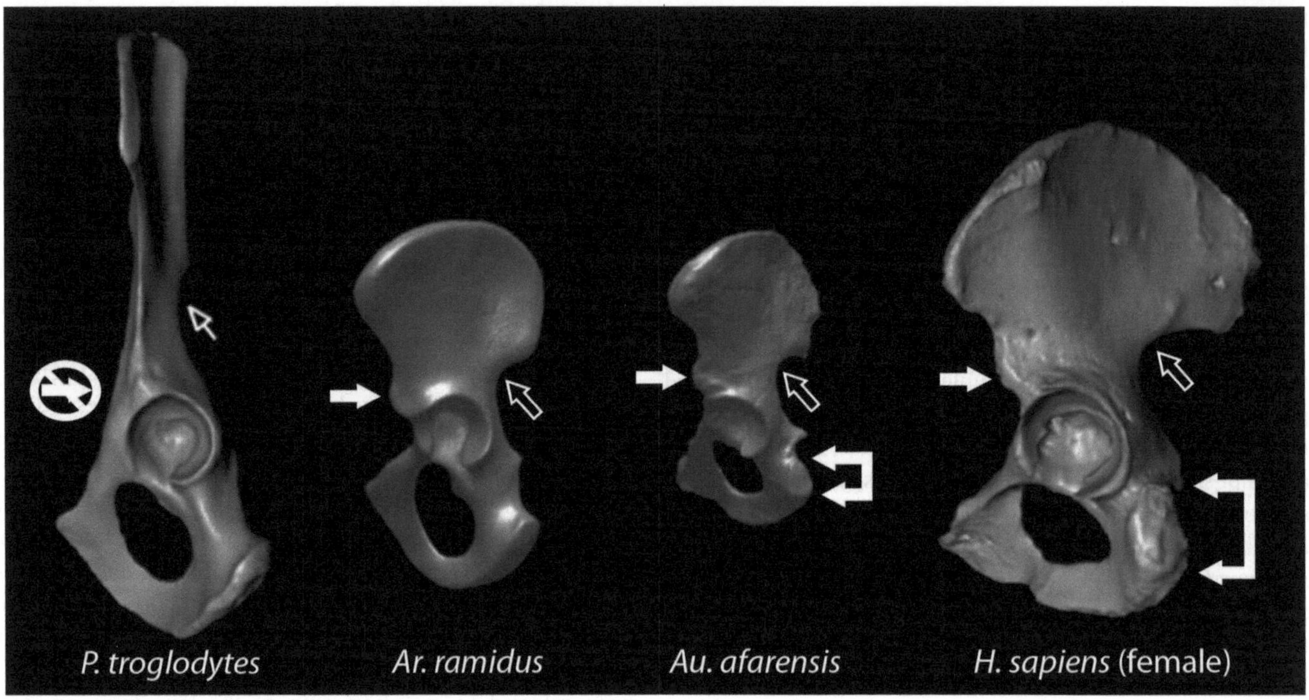

Fig. 8.21 A general presentation of the most critical changes in pelvic structure between *Ar. ramidus*, *Australopithecus*, and *Homo*, compared to the typical pelvis of a chimpanzee. The double arrow points to the angulated ischial surface in both *Homo* and *Australopithecus*, which is absent and planar in *Ar. ramidus* and *P. troglodytes*. In bipedal running, the force exerted by the hamstrings to brake the limb at the end of swing phase can be very substantial and as a result cause subluxation of the ischial apophysis prior to its fusion to the ischial ramus. This is a fairly common pathology in modern human adolescents, even with the more favorable position of the ischial apophysis in humans (and *Australopithecus*) than in *Ardipithecus*. This is the most likely cause of the reduced ischial length in later bipedal hominins. Also shown in this figure are the secondary center for the Anterior Inferior Iliac Spine (closed single arrows) and a distinct sciatic notch (open arrows). Both of these structures are prominently present in KSD-VP-1/1. For further discussion, see text. Figure modified from Lovejoy et al. (2009c)

arm" although such abbreviation has rarely received any satisfactory explanation, especially since it obviously reduces the potential torque that can be exerted by the hamstrings at the end of swing phase.

However, an obvious reason why selection has produced this fairly dramatic change can be found in a review of the clinical literature relating to avulsion injuries of the pelvis in childhood and adolescence (Gidwani et al. 2004; Ferlic et al. 2014). In fact, avulsion of the apophysis for the ischial tuberosity is the singular most frequent avulsion injury to the pelvis during this period of rapid growth, and these injuries clearly occur *despite* modifications that have occurred with respect to the positioning of the ischial ramus apophysis as just mentioned. Were the ischial body to have remained in its more ancient position as found in both pongids and *Ardipithecus* (Lovejoy 2009b), the frequency of avulsion injury would seem to have been much greater than it actually is today, and as it most likely was in *Ardipithecus*, if its pelvis was positioned in its typical hominin orientation during upright walking and running which is likely to have been the case (Lovejoy et al. 2009b), even though simple lordosis alone does encourage some favorable repositioning of the hamstring origin.

From simple geometry alone, it seems almost certain that the division of the ischial apophyseal surface into two portions and the angulation of the upper portion tend to somewhat reduce tension on its apophysis on the ischial body during hamstring contraction. This contrasts with what must have been almost pure tension and shear in the less favorably disposed ischial ramus apophysis of *Ardipithecus*. The fact that avulsion injury still occurs with moderate frequency in living humans seems ample evidence that such injuries must have been much more frequent prior to reorientation of the semimembranosus origin as seen in KSD-VP-1/1 and all later hominins (see Fig. 8.21).

Some readers might object that ischial repositioning *also* tends to reduce hamstring moment arm, and this is almost certainly the case (Le Gros Clark 1978). In fact, it is quite likely that the alterations of the growth process by which the apophyseal surface has reached its angulated state (in contrast to what must have been its previous structure as seen in *Ardipithecus*) are quite likely the primary reason why the ischial tuber has indeed shortened in post-*Ardipithecus* hominins. We have argued that a second factor in its further (relative) shortening may have been an enlargement of the acetabulum so as to bring its inferior edge closer to the hamstring surface. It is quite likely that both processes "cooperated" to achieve the novel position *and length* of the ischial tuber in hominins. True, this has resulted in some minor loss of torque generated about the hip by the hamstrings as the end of swing phase approaches. However, given that not only the hamstring position has changed but also that there has been an obviously dramatic reduction in thigh mass, simple compensation for loss of moment arm has apparently been achieved by only slightly increasing the muscle force. It should be remembered that selection, relying on relative individual fitness as its only arbiter, is not an engineering process but rather one which must simply "choose" among variations which are themselves generated by slight shifts in the positional information ultimately underlying any musculoskeletal structure. And it must do so by also failing to alter other closely related successfully generated structures. There must be many cases where what *can* be lost is actually sacrificed in order to achieve a more important "target," such as the loss of the thumb in atelines in deference to elongation of its posterior manual digits (Reno et al. 2008).

One last aspect that needs to be stressed with respect to the shift in the position of the hamstring origin in KSD-VP-1/1 is that it has sometimes been claimed that members of the species *Au. afarensis* were "incompletely adapted" to upright walking, including the argument that they must have relied on a "bent hip bent knee" (BHBK) gait (see e.g., Stern and Susman 1983; Stern 2000). It is too often ignored as to what such a gait pattern would require in terms of *running*. Not only would it have been exceptionally energy costly and likely physically exhausting, it would have also likely been comparatively slow relative to the particularly ample gaits of carnivore predators. It therefore seems unlikely that a species as successful as *Au. afarensis* would have been able to successfully occupy grasslands and savannas without at least near full development of a running gait that involved complete extension of the hip and knee. The shift in pelvic structure now known to have occurred between *Ar. ramidus* and *Au. afarensis* would seem to further establish the likely absence of any such disability in the latter taxon. This is the same conclusion that can be drawn from studies of both the pelvis and vertebral column (McCollum et al. 2009; Lovejoy and McCollum 2010). Moreover, it seems more than a bit illogical to suppose that any mammal could adopt and evolve a locomotor strategy that did not permit it to escape predators at least in some less than rudimentary fashion; the concept that *Australopithecus* would have to have "waddled" with a BHBK gait away from a threatening predator seems nothing less than abjectly silly.

In any case, the issue is now fully resolved by KSD-VP-1/1's ischial tuber. The force exerted by the hamstrings during walking is almost certainly below the threshold of what would be required to avulse its apophysis. Only during active and likely high-energy running is such a mishap likely to occur. The new positioning of the apophysis in KSD-VP-1/1 thereby establishes a threshold for highly forceful bipedal running to a minimum age of at least the fossil itself at 3.6 Ma.

The Natural History of Hominin Limb Proportions

The partial skeleton of *Ar. ramidus* from Aramis, ARA-VP-6/500, preserves both a virtually complete tibia (missing only a small portion of its distal end) and a complete radius (requiring only slight correction for crushing but otherwise intact) (White et al. 2014). We previously used these two bones to directly compare relative fore and hind limb lengths in living apes, middle Miocene hominoid specimens (such as *Proconsul*), and generalized above branch quadrupeds such as cercopithecoids and hominins. These comparisons demonstrate that the limb proportions of the Chimpanzee/Human last common ancestor (CLCA) were very likely primitive (i.e., hindlimbs and forelimbs of approximately equal length) and that they almost certainly did not conform to ratios that would reflect excessive elongation of the forelimb as now seen in apes specialized for suspension including *Pan*, *Gorilla*, and *Pongo*. These data received substantial support from a plethora of other primitive morphological characters present in the hand, foot, and cranial skeletons (Lovejoy et al. 2009a, b, c; White et al. 2009).

Such a conclusion runs counter to traditional assumptions about the nature of the fundamental *bauplan* of the CLCA that historically stretch back to the earliest formal speculation by such early twentieth century figures as Sir Arthur Keith (Lovejoy 2014). However, that *Ar. ramidus* provides the most direct evidence of early hominin limb proportions and *bauplan* now available is almost incontrovertible given that it is the earliest specimen known (4.4 Ma) that preserves element length for both forelimb and hindlimb (White et al. 2014).

KSD-VP-1/1 now provides a further direct test of the hypothesis that the CLCA's limb proportions were primitive and did not reflect specialization for suspension. Even though the specimen does not preserve the entire length of any forelimb element for comparison with its tibia (of known length – see earlier), it does nevertheless include intact portions of measurable joint size in three forelimb elements (humerus, ulna, and scapula) that can be metrically assessed for body size (Table 8.6).

We therefore calculated three size parameters of the distal humerus and a total of seven different size parameters for the humerus, ulna, and scapula, which we then used to calculate a geometric mean for humeral size and one for upper limb size (see Figs. 8.22, 8.23, and Table 8.6). These two size

Table 8.6 Metrics used for the construction of graphs showing relative upper limb size and tibia length

TAXON	ULNA: Depth of trochlear notch	ULNA: AP distance from Coronoid tip to shaft	SCAPULA: Breadth of Glenoid	SCAPULA: Height of Glenoid	HUMERUS: Biepicondylar breadth	HUMERUS: Trochlear breadth	HUMERUS: Distal articular breadth	TIBIA: Maximum length
P. troglodytes	11.2 (1.3)	37.9 (3.3)	23.2 (1.8)	32.5 (2.5)	63.0 (5.1)	27.2 (2.3)	45.7 (3.0)	254.3 (12.4)
G. gorilla	13.3 (2.5)	48.0 (7.4)	32.9 (4.9)	46.7 (7.2)	89.5 (13.7)	39.2 (6.4)	60.9 (8.8)	288.5 (32.0)
H. sapiens	12.0 (1.8)	36.6 (3.5)	25.9 (2.7)	37.3 (3.3)	60.8 (5.4)	27 (2.9)	44.9 (4.6)	376.2 (27.3)
A.L. 288-1	7.9	22.4	18	26	41	16	30	232
KSD-VP-1/1[a]	14.5 *(13.2)*	33.7 *(32.7)*	28 *(25.5)*	37 *(35.6)*	62 *(58.8)*	27	48	355

[a]Numbers for KSD-VP-1/1 in parentheses and in italics are from the CT scans. Differences probably represent cast enlargement or method of caliper orientation or both. Only the larger numbers are used in Figs. 8.22 and 8.23

variables could then be compared to tibia length. It should be noted here that although the tibia for A.L. 288-1 was not intact, its length can be estimated from its almost complete femur using the stable crural index of hominoids (Table 8.4) (Lovejoy et al. 2009a), so this specimen can also be included in such comparisons (i.e., A.L. 288-1 also preserves the seven dimensions we have chosen; Table 8.6). Metrics used included the biepicondylar breadth of the distal humerus (medial to lateral epicondyles), the mediolateral breadth of the humeral trochlea, the articular breadth of the distal humerus (breadth of trochlea and capitulum combined), the superoinferior length of the scapular glenoid as well as its maximum mediolateral breadth, and two metrics from the proximal ulna used by Drapeau and Ward (2007) for similar calculations [Trochlear Notch Depth (their UDT) and Maximum AP distance from the tip of the coronoid to the posterior margin of the shaft (their UCP)]. Data for *Pan*, *Gorilla*, *Homo*, and the two hominin specimens are shown in Figs. 8.22, 8.23 and Table 8.6. The data taken by us using calipers differ slightly from those reported by Ryan and Sukhdeo (2016). Ours, if different, are slightly greater, resulting from either differences in caliper versus linear distance metrics obtained from the CT, or possibly some enlargement of the specimen by the cast. Since our data were consistently greater and because our hypothesis is that the KSD-VP-1/1 specimen will have a more gracile forelimb (in comparison to tibia length), we choose the more conservative route of using the larger dimensions when the two methods differed.

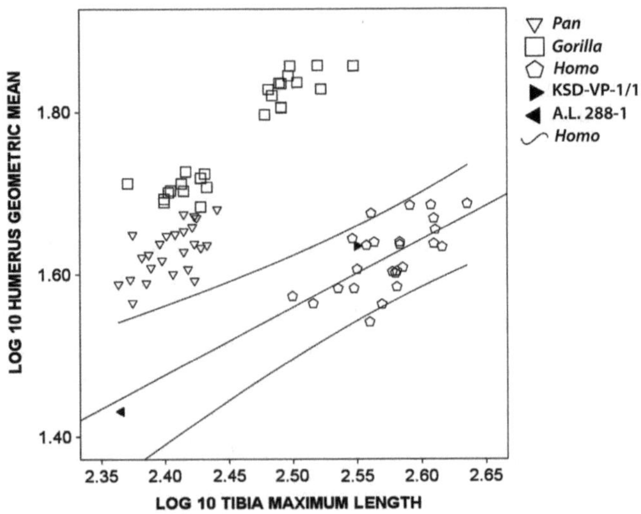

Fig. 8.22 Comparison of geometric means of four dimensions of the distal humerus (see text) compared to tibial length in KSD-VP-1/1. Tibial length for A.L. 288-1 was obtained using crural index and its known femur length (see Lovejoy, et al. 2009a for details of method and range of crural index in hominoids). These data suggest elongation of the hominin lower limb and either a primitive degree of forelimb mass or some forelimb reduction following separation from the LCA were both present and expressed in KSD-VP-1/1. A least-squares-regression line has been fit to the human cloud with 95% confidence intervals inscribed. While those confidence limits substantially exceed the original human distribution, A.L. 288-1 nevertheless lies very close to the regression line, suggesting it bore the same morphological complex as did KSD-VP-1/1. Note that no African ape specimen occurs within the 95% confidence intervals

The results are quite striking. KSD-VP-1/1 falls well within the human distribution and well outside the distributions of the two African apes. The data for A.L. 288-1 are almost equally striking. They show that despite having a tibia of approximately equal length to those of many chimpanzees, the upper limb geometric size data for A.L. 288-1 are substantially less robust. A virtually identical conclusion is implied by the tibial length and forelimb robusticity (as defined by the seven metrics listed earlier) in KSD-VP-1/1. In the case of the latter, however, because its body size is also well within the human distribution, the only alternative interpretation would be that its tibia is unusually long for the species, and had already experienced the elongation known to have occurred later in human evolution by 2.5 Ma based on the limb proportions of BOU-VP-1/1 (Asfaw et al. 1999).

In Figs. 8.22 and 8.23, a regression line has been entered for only the cloud of human specimens. The 95% confidence limits for individuals are shown in each and both specimens fall comfortably within the human distributions. The A.L. 288-1 point is well away from the actual *Homo*

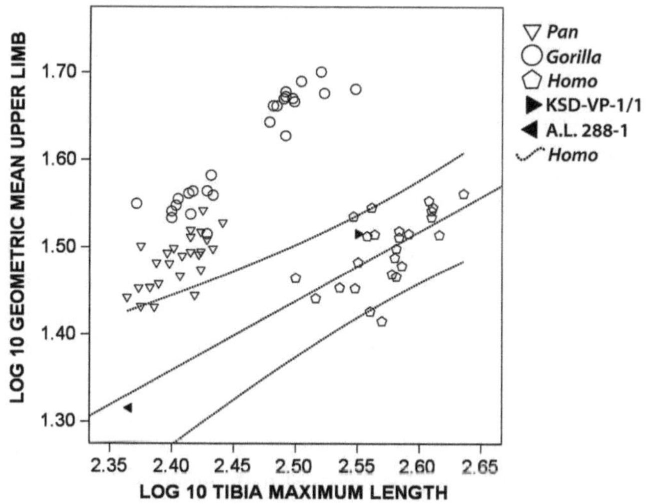

Fig. 8.23 Comparison of geometric means of forelimb elements (eight linear dimensions of the distal humerus, proximal ulna, and scapular glenoid) compared to tibial length in KSD-VP-1/1. Conclusions identical to those drawn in the text and legend of Fig. 8.22 are the same, although the 95% confidence intervals include several *Pan* specimens. However, no *Pan* specimen lies as near the regression line as does A.L. 288-1

distribution of specimens, which might call into question its actual fit to the population, but conversely, the specimen is nearly exactly on the *Homo* regression line, confirming that its proportions (hindlimb length to forelimb robusticity as judged by the cited series of dimensional metrics) are fully within the modern human range.

Assuming that each of the two *Au. afarensis* specimens represents means from their local populations would imply that either the *Au. afarensis* tibia is elongated as in *Homo* or that the upper limb size variable is considerably more diminutive than in either of the extant African apes, or both. Note that at least with respect to the humerus, no living ape falls within the human distribution, whereas KSD-VP-1/1 does and A.L. 288-1 falls on the regression line itself.

While limb proportions certainly may have fluctuated between the CLCA and ARA-VP-6/500 and as well during the time interval separating the latter at 4.4 Ma and KSD-VP-1/1 at 3.6 Ma, the most parsimonious interpretation of the data from these three partially articulated skeletons (i.e., Aramis, Hadar, and Woranso-Mille) is that hominins have never exhibited limb proportions specialized for suspension as seen in living apes. Moreover, the combination of more gracile upper limbs than are found in the African apes, elongated lower limbs, loss of the abducent great toe, dramatic restructuring of the ilium and shortening of its isthmus, when added to the equally dramatic reduction in the length of the ischium and modification of its morphology to improve resistance to subluxation during the latter part of swing phase together suggests a species that is *entirely committed to terrestrial locomotion and one which entered or used the arboreal canopy for anything other than simple feeding bouts so infrequent that they had no impact on selection for arboreal competency*. This conclusion is fully consistent with the novel interpretations about *bauplan* favored by KSD-VP-1/1's possible ancestor ARA-VP-6/500 (Lovejoy et al. 2009b; see Haile-Selassie et al. 2016).

"Models" of Skeletal Change in Human Evolution

During the past several decades, a number of important trends have continued in various scientific disciplines that bear some relationship to issues of morphological evolution. Imaging has become dramatically advanced, as has the technological capacity to examine external morphology in terms of size and detail (e.g., by means of a three-dimensional digitizer). Engineering "style" analysis has likewise matured and anatomical structures can increasingly be represented by such procedures as finite element modeling, which has, indeed, become quite prevalent in studies of human evolution.

It is therefore perhaps more than a little ironic that during this same period even greater strides have been achieved in studies of genomic expression and their bearing on the embryogenesis of morphological structures. Even more important is the fact that the technological advances just cited above are exactly that – technological – they do not, a priori, bear any necessary relationship to biological structures, and in fact in most cases were developed to study non-biological configurations (e.g., finite element scaling). To the contrary, studies of genomic expression have been dedicated exactly to fundamental biological processes.

Much of what has been learned in this regard bears a direct relationship to our understanding of skeletal evolution simply because the vertebrate limb skeleton itself has been one of the chief experimental modeling agencies. During the same period in which it has become possible to model a long bone that has been presumed to be an engineered structure, it has also become possible to examine the possibility that such a structure can be, in fact, evolved in terms of the embryogenetic process. We used the term "ironic" at the beginning of this paragraph because, as evolutionary discovery would have it, these two modes of analysis have diverged in an almost spectacular fashion.

The musculoskeletal system is, almost by definition, a mechanical construct because it must necessarily conform to Newton's Laws. As such it can therefore be subjected to virtually any form of mechanical analysis, or more importantly, mechanical theory. This is not, however, evidence that structural modifications of the skeleton are necessarily responsive to the predictions of such theories and/or analyses (Gould and Lewontin 1975).

Perhaps no better example of this phenomenon can be found in the recent description and analysis of the two partial pelves recovered from the site of Malapa, South Africa, to which we earlier referred. Its authors present a "new theory" with respect to the selective agencies primarily responsible for shaping the hominin pelvis over the past 4 million years or so. Termed the "kinematic-shift model," they posit that changes in the hominin pelvis over this long evolutionary period do not primarily reflect adjustments to its relatively precarious relationship to successful parturition imposed by bipedality, but instead to reflect the "emergence of a more effective and efficient form of full-striding bipedalism."

Their evidence for this new model is the observation that some traits, which they attribute to more "modern" pelves, are found in the newly named species from Malapa, which they argue does not show evidence of significant brain enlargement. An examination of the features that are listed as being shared synapomorphically or homoplastically with *Homo*, however, lists none that is likely to have had any potential effect on gait pattern (Kibii 2011). Instead, each is, perhaps, simply some slightly varied degree of expression of morphological characters which often figure prominently in post hoc analyses of specimens, but which were of no consequence to an individual's reproductive success during his/her lifetime. The mammalian skeleton is famously variable and we can always expect to find odd expressions of any variable that most probably exists only in the mind of the investigator. Given any single modern skeleton, any number of morphological perturbations can be found (upon sufficiently intensive search) that rank as less "advanced" or more "primitive" than the mean of modern human expression for that character.

More importantly, many of the claimed increases in "efficiency" are largely beyond the capacity of the vertebrate anlagen "transforming genomic mechanisms" of development to even mold – even if they were under selective control. For example, one claim is that the Malapa specimen is more *Homo*-like than other *Australopithecus* pelves because it has "reduced relative distance from the sacroiliac to hip joints" and that this region also has *Homo*-like robusticity. First, the distance could have little effect on gait pattern, or its efficiency, even if it were not largely negated by the fact that the region being defined is largely vertical when the pelvis assumes its lordotic rotation during gait (i.e., the anterior superior iliac spines and pubic tubercle define a nearly vertical plane). Second, of what consequence is its supposed robusticity unless this region is subject to failure? Such failure, short of violent trauma, has never been seen in any primate.

It has now been decades since Gould and Lewontin (1979) attempted to warn against excessive adaptationism, but there can hardly be any more salient examples than such lists of particularized "definitions." Moreover, among the characters described as being plesiomorphic in the Malapa pelvis is that the specimen has a large biacetabular diameter – a feature that *does* have significant implications with respect to successful parturition. In the hundreds of thousands of years during which the platypelloid pelvis of *Australopithecus* was being transformed into one with a more birth-friendly conformation (i.e., one nearer the more ideal, largely circular, form; Rosenberg and Trevathan 2002), many pleiotropic changes can be expected to have occurred in the pelvis – some of which can easily be interpreted as "advances" but some equally likely to be regarded as "regressions" – simply because none by itself was a target of selection – positive or negative – only the *true* target of selection which was, for the most part, an ample birth canal.

While there are likely a few changes in the pelvis over the last 4 million years that *were* related to possible failure (size of the sacroiliac joint) or to a possible increase in gait efficiency (e.g., moment arm of the quadriceps), most are as likely to have been negative effects (such as shortening of the hamstring moment arm – see earlier) as positive – the latter occurring only in the mind of the analyst in these cases. In short, with the exception of the few minor modifications just noted, the gait patterns of *Australopithecus* and those of *Homo sapiens* are unlikely to have been differentiable except by sophisticated modern equipment, and apparently largely silent with respect to the effects of natural selection.

Conclusion

Almost every aspect of the pelvis of KSD-VP-1/1 exhibits the same fundamental adaptations to upright walking as are found in *Homo*. As would be expected, however, there are several minor elements that still appear more primitive and which have likely been modified by selection over the course of evolution in *Australopithecus* to *Homo*. These include a small increment in the size of the acetabulum, a marginal increase in the length of the patellar moment arm at the knee (Lovejoy 2005), a possible enlargement in the surface area of the auricular surface, and potentially increased ligamentous rugosity (and therefore competency) in the postauricular region of the sacroiliac articulation. It should be noted that none of these are likely to have significantly affected the *Au. afarensis* gait pattern, but only its potential to resist musculoskeletal degradation and joint damage over time, i.e., they represent characters of only minimal selective valence which must have therefore taken very long periods of time for selection to successfully execute.

Importantly, KSD-VP-1/1 establishes the minimum age of a nearly complete adaptation to what has been termed full bipedality and a "striding gait," at least until more complete elements are found for *Au. anamensis*, its more ancient and likely precursor. The difference in age between KSD-VP-1/1 and *Ar. ramidus* (800 Kyr) suggests that either this transition to a fully committed terrestrial bipedality took place at a relatively rapid pace or that a more ancient and possibly more direct ancestor of the genus *Australopithecus* has yet to be found.

References

Abel, R., & Macho, G. A. (2011). Ontogenetic changes in the internal and external morphology of the ilium in modern humans. *Journal of Anatomy, 218*, 324–335.

Asfaw, B., White, T. D., Lovejoy, C. O., Latimer, B., Simpson, S. W., & Suwa, G. (1999). *Australopithecus garhi*: A new species of early hominid from Ethiopia. *Science, 284*, 629–635.

Begun, D. R. (1992). Miocene fossil hominids and the chimp-human clade. *Science, 257*, 1929–1933.

Benefit, B. R., & McCrossin, M. L. (1995). Miocene hominoids and hominid origins. *Annual Review of Anthropology, 24*, 237–256.

Cunningham, C. A., & Black, S. M. (2009). Anticipating bipedalism: trabecular organization in the newborn ilium. *Journal of Anatomy, 214*, 817–829.

DeSilva, J. M. (2009). Functional morphology of the ankle and the likelihood of climbing in early hominins. *Proceedings of the National Academy of Sciences USA, 106*, 6567–6572.

DeSilva, J. M., & Throckmorton, Z. J. (2010). Lucy's flat feet: The relationship between the ankle and rearfoot arching in early hominins. *PLoS ONE, 5*, 1–8.

Drapeau, M. S. M. (2004). Functional anatomy of the olecranon process in hominoids and plio-pleistocene hominins. *American Journal of Physical Anthropology, 124*, 297–314.

Drapeau, M. S. M. (2008). Articular morphology of the proximal ulna in extant and fossil hominoids and hominins. *Journal of Human Evolution, 55*, 86–102.

Drapeau, M. S. M., Ward, C. V., Kimbel, W. H., Johanson, D. C., & Rak, Y. (2005). Associated cranial and forelimb remains attributed to *Australopithecus afarensis* from Hadar, Ethiopia. *Journal of Human Evolution, 48*, 593–642.

Drapeau, M. S. M., & Ward, C. V. (2007). Forelimb segment length proportions in extant hominoids and *Australopithecus afarensis*. *American Journal of Physical Anthropology, 132*, 327–343.

Ferlic, P. W., Sadoghi, P., Singer, G., Kraus, T., & Eberl, R. (2014). Treatment for ischial tuberosity avulsion fractures in adolescent athletes. *Knee Surgery, Sports Traumatology, Arthroscopy: Official Journal of the ESSKA, 22*, 893–897.

Gibson, V., & Prieskorn, D. (2007). The valgus ankle. Foot and *Ankle Clinics of North America, 12*, 15–27.

Gidwani, S., Jagiello, J., & Bircher, M. (2004). Avulsion fracture of the ischial tuberosity in adolescents – an easily missed diagnosis. *British Medical Journal, 329*, 99–100.

Gould, S. J., & Lewontin, R. C. (1979). The spandrels of San Marco and the Panglossian paradigm: a critique of the adaptationist programme. *Proceedings of the Royal Society of London (Biology), 205*, 147–164.

Haile-Selssie, Y., Latimer, B. M., Lovejoy, C. O., Meyer, M. R. (2016). Conclusion: Implications of KSD-VP-1/1 for early hominin paleobiology and insights into the last common ancestor. In Y. Haile-Selassie & D.F. Su (Eds.), *The postcranial anatomy of* Australopithecus afarensis: *New insights from KSD-VP-1/1* (pp. 179–188). Dordrecht: Springer.

Haile-Selassie, Y., Latimer, B. M., Alene, M., Deino, A. L., Gibert, L., Melillo, S. M., et al. (2010a). An early *Australopithecus afarensis* postcranium from Woranso-Mille, Ethiopia. *Proceedings of the National Academy of Sciences USA, 107*, 12121–12126.

Haile-Selassie, Y., Saylor, B. Z., Deino, A., Alene, M., & Latimer, B. M. (2010b). New hominid fossils from Woranso-Mille (Central Afar, Ethiopia) and taxonomy of early *Australopithecus*. *American Journal of Physical Anthropology, 141*, 406–417.

Hamrick, M. W. (1999). A chondral modeling theory revisited. *Journal of Theoretical Biology, 201*, 201–298.

Harrison, T., & Rook, L. (1997). Enigmatic anthropoid or misunderstood ape? The phylogenetic status of *Oreopithecus bambolii* reconsidered. In D. R. Begun, C. V. Ward, & M. D. Rose (Eds.), *Function, phylogeny, and fossils: Miocene hominid evolution and adaptations* (pp. 327–362). New York: Plenum.

Johanson, D. C., Lovejoy, C. O., Kimbel, W. H., White, T. D., Ward, S. C., Bush, M. E., et al. (1982). Morphology of the Pliocene partial hominid skeleton (A.L. 288-1) from the Hadar formation, Ethiopia. *American Journal of Physical Anthropology, 57*, 403–452.

Kibii, J. M., Churchill, S. E., Schmid, P., Carlson, K. J., Reed, N. D., de Ruiter, D. J., & Berger, L. R. (2011). A partial pelvis of *Australopithecus sediba*. *Science, 333*, 1407–1411.

Konin, G. P., & Walz, D. M. (2010). Lumbosacral transitional vertebrae: Classification, imaging findings, and clinical relevance. *American Journal of Neuroradiology, 31*, 1778–1786.

Lague, M. R. (2014). The pattern of hominin postcranial evolution reconsidered in light of size-related shape variation of the distal humerus. *Journal of Human Evolution, 75*, 90–109.

Le Gros Clark, W. E. (1978). *The fossil evidence for human evolution* (3rd ed.). Chicago: University of Chicago Press.

Latimer, B. M., Lovejoy, C. O., Spurlock, L., & Haile-Selassie, Y. (2016). The thoracic cage of KSD-VP-1/1. In Y. Haile-Selassie & D.F. Su (Eds.), *The postcranial anatomy of* Australopithecus afarensis: *New insights from KSD-VP-1/1* (pp. 143–153). Dordrecht: Springer.

Lovejoy, C. O. (1974). The gait of australopithecines. *Yearbook of Physical Anthropology, 16*, 18–30.

Lovejoy, C. O. (1975). Biomechanical perspectives on the lower limb of early hominids. In R. H. Tuttle (Ed.), *Primate morphology and evolution* (pp. 291–326). The Hague: Mouton.

Lovejoy, C. O. (1979). A reconstruction of the pelvis of A.L. 288-1 (Hadar formation, Ethiopia). *American Journal of Physical Anthropology, 50*, 460.

Lovejoy, C. O. (1988). Evolution of human walking. *Scientific American, 259*, 118–125.

Lovejoy, C. O. (2005). The natural history of human gait and posture II: Hip and thigh. *Gait and Posture, 21*, 129–151.

Lovejoy, C. O. (2007a). Evolution of the human lumbopelvic region and its relationship to some clinical deficits of the spine and pelvis. In A. Vleeming, V. Mooney, & R. Stoeckart (Eds.), *Movement, stability, and lombopelvic pain* (pp. 141–158). Paris: Churchill Livingston.

Lovejoy, C. O. (2007b). The natural history of human gait and posture III: Knee. *Gait and Posture, 25*, 325–341.

Lovejoy, C. O. (2014). *Ardipithecus* and earliest human evolution in light of twenty-first century developmental biology. *Journal of Anthropological Research, 70*, 337–363.

Lovejoy, C. O., Cohn, M. J., & White, T. D. (1999). Morphological analysis of the mammalian postrcranium: A developmental perspective. *Proceedings of the National Academy of Sciences USA, 96*, 13247–13252.

Lovejoy, C. O., Johanson, D. C., & Coppens, Y. (1982). Hominid upper limb bones recovered from the Hadar formation: 1974–1977 collections. *American Journal of Physical Anthropology, 57*, 637–650.

Lovejoy, C. O., & McCollum, M. (2010). Spinopelvic pathways to bipedality: Why no hominids ever relied on a bent-hip–bent-Knee gait. *Philosopical Transactions of the Royal Society of London, 365*, 3289–3299.

Lovejoy, C. O., Meindl, R. S., Ohman, J. C., Heiple, K. G., & White, T. D. (2002). The Maka femur and its bearing on the antiquity of human walking: Applying contemporary concepts of morphogenesis to the human fossil record. *American Journal of Physical Anthropology, 119*, 97–133.

Lovejoy, C. O., Simpson, S. W., White, T. D., Asfaw, B., & Suwa, G. (2009a). Careful climbing in the Miocene: The forelimbs of *Ardipithecus ramidus* and humans are primitive. *Science, 326,* 70e71–70e78.

Lovejoy, C. O., Suwa, G., Simpson, S. W., Matternes, J. H., & White, T. D. (2009b). The great divides: *Ardipithecus ramidus* reveals the postcrania of our last common ancestors with African apes. *Science, 326*, 100–106.

Lovejoy, C. O., Suwa, G., Spurlock, L., Asfaw, B., & White, T. D. (2009c). The pelvis and femur of *Ardipithecus ramidus*: The emergence of upright walking. *Science, 326,* 71e71–71e76.

McCollum, M. A., Rosenman, B. A., Suwa, G., Meindl, R. S., & Lovejoy, C. O. (2009). The vertebral formula of the last common ancestor of African apes and humans. *Journal of Experimental Zoology (B) Molecular and Developmental Evolution, 314*, 123–134.

McHenry, H., & Corrucini, R. (1975). Distal humerus in hominoid evolution. *Folia Primatologica, 23*, 227–244.

Reno, P. L., McCollum, M. A., Cohn, M. J., Meindl, R. S., Hamrick, M., & Lovejoy, C. O. (2008). Patterns of correlation and covariation of anthropoid distal forelimb segments correspond to Hoxd expression territories. *Journal of Experimental Zoology (B) Molecular and Developmental Evolution, 310*, 240–258.

Rosenberg, K., & Trevathan, W. (2002). Birth, obstetrics, and human evolution. *BJOG: An International Journal of Obstetrics and Gynaecology, 109*, 1199–1206.

Rose, M. D. (1988). Another look at the anthropoid elbow. *Journal of Human Evolution, 17*, 193–224.

Rose, M. D. (1993). Functional anatomy of the elbow and forearmin primates. In D. L. Gebo (Ed.), *Postcranial adaptation innonhuman primates* (pp. 70–95). DeKalb: Northern Illinois UniversityPress.

Ryan, T. M., & Sukhdeo, S. (2016). KSD-VP 1/1: Analysis of the postcranial skeleton using high-resolution computed tomography. In Y. Haile-Selassie & D. F. Su (Eds.), *The Postcranial Anatomy of* Australopithecus afarensis: *New Insights from KSD-VP-1/1* (pp. 39–62). Dordrecht: Springer.

Selby, M. S., & Lovejoy, C. O. (2014). Hominoid humeral trochlear morphology is unrelated to suspensory locomotion.

Simpson, S. W., Quade, J., Levin, N. E., Butler, R., Dupont-Nivet, G., Everett, M., et al. (2008). A female *Homo erectus* pelvis from Gona, Ethiopia. *Science, 322*, 1089–1092.

Stern, J. T. (2000). Climbing to the top: A personal memoir of *Australopithecus afarensis*. *Evolutionary Anthropology, 9*, 113–133.

Stern, J. T., & Susman, R. L. (1983). The locomotor anatomy of *Australopithecus afarensis*. *American Journal of Physical Anthropology, 60*, 279–317.

Uçar, B. Y., Uçar, D. E., Bulut, M., Azboy, I., & Demitraş, A. (2013). Lumbosacral transitional vertebrae in low back pain population. *Spine, 2*, 125–128.

Ward, C. V. (1991). Functional anatomy of the lower back and pelvis of the Miocene hominoid *Proconsul nyanze* from Mfangano Island, Kenya. Ph. D. Dissertation, Johns Hopkins University.

White, T. D., Asfaw, B., Beyene, Y., Haile-Selassie, Y., Lovejoy, C. O., Suwa, G., et al. (2009). *Ardipithecus ramidus* and the paleobiology of early hominids. *Science, 326*, 75–86.

White, T. D., Lovejoy, C. O., Asfaw, B., Carlson, J., & Suwa, G. (2014). Neither Chimpanzee nor human: *Ardipithecus* reveals the surprising ancestry of both. *Proceedings of the National Academy of Sciences USA, 112*, 4877–4887.

White, T. D., Suwa, G., Hart, W. K., Walter, R. C., WoldeGabriel, G., de Heinzelin, J., et al. (1993). New discoveries of *Australopithecus* at Maka in Ethiopia. *Nature, 366*, 261–265.

Wiltse, L. L. (1972). Valgus deformity of the ankle: A sequel to acquired or congenital abnormalities of the fibula. *Journal of Bone and Joint Surgery (Am), 54*, 595–606.

Wolpert, L., Beddington, R., Brockes, J., Jessell, T., Lawrence, P., & Meyerowitz, E. (1998). *Principles of development*. London: Current Biology Ltd.

Chapter 9
Conclusion: Implications of KSD-VP-1/1 for Early Hominin Paleobiology and Insights into the Chimpanzee/Human Last Common Ancestor

Yohannes Haile-Selassie, Bruce Latimer, C. Owen Lovejoy, Stephanie M. Melillo, and Marc R. Meyer

Abstract KSD-VP-1/1 is a 3.6 million years old (Ma) partial skeleton of *Australopithecus afarensis* recently discovered from the Woranso-Mille study area in the Afar region of Ethiopia. The recovered elements of this specimen, which include cervical vertebrae, a complete scapula, clavicle, numerous ribs, pelvis, and elements of the fore- and hindlimbs, greatly enhance our understanding of the paleobiology of early *Australopithecus afarensis*. Detailed analyses of the cervical vertebrae indicate that *Australopithecus afarensis* had a highly mobile neck, signaling human-like kinematics consistent with habitual upright posture and bipedalism. Elements of the shoulder girdle exhibit some primitive morphology but are overall more similar to humans than has been previously understood. This similarity is inconsistent with the notion that the *Australopithecus afarensis* shoulder retained primitive morphology from an African ape-like chimpanzee/human last common ancestor. Morphology of the thorax also indicates that while some individual traits may appear to superficially suggest arboreality, *Australopithecus afarensis* did not have an abundance of functionally significant morphological traits that would suggest high canopy arboreality as found today in large-bodied apes. Most of the inconsistencies in interpretations of early hominin paleobiology appear to stem from methodological differences, incorrect *a priori* assumptions, or incomplete information derived from fragmentary specimens.

Keywords KSD-VP-1/1 • *Australopithecus afarensis* • Cervical vertebrae • Shoulder girdle • Thorax shape • Woranso-Mille

Introduction

Partial skeletons of early hominins are extremely rare in the fossil record, particularly for the time predating 3 Ma. When they are found, however, they are significant not only for testing existing hypotheses in human origins and evolution, but also in generating new data that are otherwise unavailable from fragmentary fossils. The paleobiology of *Australopithecus afarensis* has been the subject of intensive research and great debate since its naming in the late 1970s. Many researchers have studied its fossil remains and traces from Hadar (Ethiopia) and Laetoli (Tanzania) in an attempt to understand its overall body shape, locomotor adaptation, and phylogenetic position. Despite the relative abundance of fossils representing this species, there are only a handful of adult partial skeletons, most notably A.L. 288-1 (Lucy), that have been extensively used in studying the paleobiology of the species.

Prior to the discovery of KSD-VP-1/1, A.L. 288-1 was the only adult partial skeleton recovered with associated forelimb and hindlimb elements. The distinctive anatomy of this small individual may have been influenced by the allometric effects of its diminutive size (Jungers 1982;

Y. Haile-Selassie (✉) · B. Latimer · C.O. Lovejoy
Department of Physical Anthropology, Cleveland Museum of Natural History, 1 Wade Oval Drive, Cleveland, OH 44106, USA
e-mail: yhailese@cmnh.org

B. Latimer
Department of Orthodontics, School of Dental Medicine, Case Western Reserve University, Cleveland, OH 44106, USA
e-mail: bxlatimer@aol.com

C.O. Lovejoy
Department of Anthropology and Division of Biomedical Sciences, Kent State University, Kent, OH 44242, USA
e-mail: olovejoy@aol.com

S.M. Melillo
Department of Human Evolution, Max Planck Institute for Evolutionary Anthropology, Deutscher Platz 6, 04103 Leipzig, Germany
e-mail: stephanie_melillo@eva.mpg.de

M.R. Meyer
Department of Anthropology, Chaffey College, 5885 Haven Ave, Rancho Cucamonga, CA 91739, USA
e-mail: marc.meyer@chaffey.edu

Yohannes Haile-Selassie and Denise F. Su (eds.), *The Postcranial Anatomy of Australopithecus afarensis: New Insights from KSD-VP-1/1*, Vertebrate Paleobiology and Paleoanthropology, DOI 10.1007/978-94-017-7429-1_9

Wolpoff 1983; Jungers and Stern 1983; Aiello 1992). For example, some of the specimen's fragmentary elements such as the scapula, which preserves the bar-glenoid angle and glenoid, were used to infer arboreal locomotor behavior in the species (Stern and Susman 1983; Susman et al. 1984; see Stern 2000 for detailed review), suggesting an African ape-like shoulder girdle anatomy (Larson 2007, 2009) and thoracic form (Schmid 1991; Schmid et al. 2013). Others, such as Mensforth et al. (1990) and Inouye and Shea (1997), argued that the orientation of the glenoid fossa in A.L. 288-1 was more likely a result of the specimen's small size and not necessarily indicative of arboreal adaptation. The recovery of more complete specimens of *Au. afarensis* from eastern Africa (Alemseged et al. 2006; Haile-Selassie et al. 2010; Green and Alemseged 2012) and other hominin species from South Africa (Churchill et al. 2013; Schmid et al. 2013), led researchers to some degree of consensus that the *Australopithecus* scapular glenoid was probably more cranially oriented than is typical of modern humans – regardless of size. However, the debate still continues today, and no agreement has surfaced in terms of the overall shoulder girdle anatomy and thorax shape of pre-*Homo* ancestors. Detailed accounts of these debates are presented in Melillo (2016; see also Melillo 2011) and Latimer et al. (2016).

KSD-VP-1/1 is a partial skeleton of a large-bodied individual of *Au. afarensis* which preserves both forelimb and hindlimb elements and more complete elements of the axial skeleton including scapula, clavicle, ribs, and pelvis. These are significant as a means of testing existing hypotheses in relation to the paleobiology and locomotor adaptation of the species, and the skeletal *bauplan* of the chimpanzee/human last common ancestor (CLCA). This chapter presents the broader implications of KSD-VP-1/1 for the paleobiology of *Au. afarensis* and reviews some of the current interpretations of the shoulder girdle anatomy, thoracic form, and locomotor behavior of early hominins before and after *Au. afarensis*.

Anatomy of the Early Hominin Neck

Early hominin cervical vertebrae are rare elements in the fossil record largely due to their fragile and delicate nature. As a result, only a few isolated cervical vertebrae are known for *Australopithecus* (Lovejoy et al. 1982; Cook et al. 1983; Williams et al. 2013). KSD-VP-1/1 includes the oldest series of cervical vertebrae of this genus. Although recent discoveries of a juvenile *Au. afarensis* specimen from Dikika (Alemseged et al. 2006) and two fragmentary skeletons from Malapa (Berger et al. 2010) include serial cervical vertebrae, their detailed descriptions and interpretation have not yet been published. Therefore, the KSD-VP-1/1 cervical vertebrae provide the first opportunity to examine serial cervical functional anatomy in an adult *Au. afarensis* (Meyer 2016).

Analyses of these vertebrae demonstrate that the most cranial cervical levels are smaller and more ape-like in shape than the larger, more derived caudal levels. One potential explanation for this gradation would be a range of intra-individual variation in *Au. afarensis* not expressed in extant apes and modern humans, but the absence of other hominin fossil cervical columns currently available for comparison renders this hypothesis difficult to test with precision. However, like KSD-VP-1/1's upper vertebrae, isolated upper cervical vertebrae of other *Australopithecus* and *Paranthropus* species and *Homo erectus* are relatively primitive in many aspects of their morphology. This might allow proposing an alternative hypothesis that the relative size and shape of the upper cervical vertebrae in KSD-VP-1/1 can be explained by an ontogenetic link between the cranium and cervical vertebrae in this small-brained, bipedal hominin.

In mammals, this ontogenetic link is rooted in the cellular structure of the upper cervical levels, which develop under somite influences of cranial origin, while subaxial vertebrae each derive from single somites unrelated to the cranium (Muller and O'Rahilly 2003; Ogden and Ganey 2003). The result is that upper cervical vertebral size and shape correlate with cranial dimensions, while lower cervical elements develop independently of cranial influences (Solow et al. 1982; Huggare and Houghton 1996; Grave et al. 1999; Grave and Townsend 2003). Thus, the basicranium and upper cervical vertebrae are developmentally connected, and many features of the occipital basicranium (e.g., foramen magnum, cranial condyles) are actually modified vertebral elements derived from somites shared with the most cranial cervical elements (Ogden and Ganey 2003; Lieberman 2011).

As in other early hominins, the size and shape of the KSD-VP-1/1 upper cervical vertebrae may be somewhat *Pan*-like in many respects due to a developmental link with a somewhat *Pan*-like cranial anatomy (for example, basicranial flexion) in *Au. afarensis*, as the most cranial centra in KSD-VP-1/1 are primitive (relatively narrower than humans) and reciprocate the primitive, relatively narrow shape of *Au. afarensis* basicrania. But the ontogenetic independence of the subaxial vertebrae in the KSD-VP-1/1 cervical column likely explains why they are larger, transversely expanded, and overall more human-like. Free from the constraints of cranial morphogenesis, these more caudal cervical elements reflect adaptation to the human-like pattern of orthograde axial loading with incrementally higher magnitudes in the caudal direction.

Large subaxial cervical bodies in both humans and primates typically signal a heavy prognathic head (Katz et al.

1975; Nakatsukasa et al. 2003) and for this reason KSD-VP-1/1 would be more likely to have had subaxial vertebrae that are relatively larger than the human mean. Yet, despite their large size, the KSD-VP-1/1 subaxial bodies are superoinferiorly short and transversely wide as in humans, indicating a similar loading regime, as relatively short bodies are best suited to resist vertical loading, buckling, and bending moments, all of which intensify with orthograde posture and are further exacerbated during sustained bipedal locomotion (Rose 1975; Shi et al. 1995, 1999). Because the KSD-VP-1/1 vertebrae fall closer to the human distribution for centrum shape than the more diminutive fossils of *Au. afarensis*, *H. erectus*, and *Paranthropus* (which fall closer to, or within, the distribution for *Pan*), KSD-VP-1/1 provides limited support to the suggestion that aspects of skeletal anatomy in early *Homo* augmented the existing *Australopithecus* pattern largely through size increases (Holliday 2012). This hypothesis is consistent with the observation that the KSD-VP-1/1 vertebrae are more derived than those of smaller and more recent hominins, such as *H. erectus*. In short, the more caudal vertebrae in the KSD-VP-1/1 cervical spine mirror the functional anatomy of the genus *Homo,* whereas the more plesiomorphic cranial elements are likely developmental reciprocates of *Australopithecus* cranial morphogenesis.

Holliday's (2012) suggestion that body and brain size increases explain the derived anatomy of *Homo* is further supported by Kimbel and Rak's (2010) analysis of the *Au. afarensis* basicranium where the large male A.L. 444-2 exhibits derived characters while the smaller female A.L. 822-1 cranium presents a more primitive morphology. The hominin basicranium is generally broader, shorter, and more flexed than those of apes (hylobatids, *Pan*, *Gorilla*, and *Pongo*) with low nuchal lines on more horizontal nuchal planes (e.g., Dean and Wood 1981, 1982; Nevell and Wood 2008; Bastir et al. 2010). Unlike humans, A.L. 822-1 exhibits very high nuchal lines and a very steep nuchal plane, closer to the condition seen in *Pan* and *Gorilla* (although smaller) indicating enlarged nuchal musculature relative to A.L. 444-2. The cranial base in A.L. 822-1 is also more primitive in that it is narrower than that of A.L. 444-2, whose cranial base is transversely wider and more derived toward humans. These observations match those seen in the cervical vertebrae of *Au. afarensis* associated with KSD-VP-1/1, a large male, exhibiting human-like derived characters in the centrum, articular processes, neural arch and lamina, while the plesiomorphic A.L. 333-106 female does not. Because the cranial base and vertebrae share common developmental pathways and functional dependence, it is not surprising that cervical vertebrae generally accord with the pattern of cranial sexual dimorphism noted for the *Au. afarensis* hypodigm.

Anatomy of the Early Hominin Shoulder Girdle

In recent reviews of hominin shoulder evolution, Larson (2007, 2009, 2013) articulates the traditionally accepted hypothesis that the *Australopithecus* shoulder retained primitive morphology from an African ape-like chimpanzee/human last common ancestor (CLCA). Historically, this idea was supported by comparative observations on two very fragmentary scapulae attributed to *Au. afarensis* (A.L. 288-1 l) and *Au. africanus* (Sts 7) (Broom and Robinson 1950; Campbell 1966; Oxnard 1968; Ciochon and Corruccini 1976; Vallois 1977; Vrba 1979; Johanson et al. 1982; Stern 2000), with particular emphasis on the cranial orientation of the glenoid (Vrba 1979; Stern and Susman 1983; Stern 2000). However, much more complete shoulder girdle material has been recovered during the past decade. The resulting studies differ in the degree that they support the African ape-like conception of the *Australopithecus* shoulder. The juvenile *Au. afarensis* scapulae of the DIK-1-1 partial skeleton are consistently described as morphologically most similar to *Gorilla* (Alemseged et al. 2006; Green and Alemseged 2012), whereas the adult *Au. sediba* scapula (MH2) is described as most similar to *Pongo* or *Gorilla* (Churchill et al. 2013). Churchill and colleagues (2013) further suggest that the morphology of the MH2 clavicle demonstrates that the *Au. sediba* shoulder girdle was elevated, as in non-human apes. In contrast, the initial description of the KSD-VP-1/1 partial skeleton showed that the scapula was more similar to *Homo* than to either of the African apes (Haile-Selassie et al. 2010). Unfortunately, whenever comparisons among the new fossil specimens have been made, they were complicated by methodological differences.

Melillo (2011, 2016) shows that the KSD-VP-1/1 scapula is intermediate between modern humans and orangutans with respect to the orientation of the scapular spine and glenoid. Importantly, this specimen lacks the pronounced cranial orientation of these structures that is shared in common among the African extant apes. Further, the infraspinous fossa is craniocaudally expanded in KSD-VP-1/1, which is shared uniquely with extant and fossil *Homo*. These results are consistent, regardless of whether traditional or geometric morphometric (GM) methods are used to quantify these features.

Until recently, it was widely accepted that the CLCA shoulder girdle was morphologically similar to *Pan* – or perhaps more conscientiously, possessed those features that are shared in common between *Pan* and *Gorilla*. In this framework, the morphology of KSD-VP-1/1 scapula signals that the *Au. afarensis* scapula is derived toward the human condition – with the cranial orientation of the spine and glenoid considerably reduced and the infraspinous fossa craniocaudally expanded. In humans, these features are associated with use of the upper limb in a somewhat pendant position and with an emphasis on manipulation (Inman et al. 1944, Larson 2013). A similar evolutionary scenario is discussed in the description of the MH2 scapula, in which evolution from an African ape-like CLCA toward human morphology produces *Australopithecus* morphology that is "homoplastically similar to *Pongo* (Churchill et al. 2013, p. 1233477-4)." Churchill and colleagues (2013) suggest the morphology of MH2 indicates "the retention of substantial climbing and suspensory ability (p. 1233477-1)" in *Australopithecus,* whereas Melillo (2016) suggests that departure from the primitive condition, in the direction of human morphology, signals selection for human-like function in *Australopithecus* or its ancestors.

The KSD-VP-1/1 clavicle is also very similar to *Homo* in terms of relative length, but differs from living *Homo* with regard to curvature. The shaft exhibits pronounced internal and external curvatures in the transverse plane, similar to *Homo* and *Pan*, but it also has a superior curvature in the coronal plane that is most similar to KNM-WT 15000 C (left side) (Walker and Leakey 1993), *Pongo* or *Pan*. Superior coronal curvature is also reported for some Neanderthal clavicles (Voisin 2006, 2008). Like other *Australopithecus* clavicles, the origin of the deltoid on KSD-VP-1/1 clavicle is located within the external curve, at the junction of the superior surface and anterior border.

Observations on clavicle coronal curvature have been used to reconstruct shoulder girdle height in extinct hominins (Ohman 1986; Voisin 2006; Churchill et al. 2013). However, a predictive relationship between clavicle coronal curvature and shoulder girdle height has yet to be demonstrated and reconstructions based on qualitative assessments of curvature do not sufficiently consider within-taxon variability. Melillo (2016) uses GM methods to quantify coronal curvature, and shows that there is considerable overlap between humans and non-human great apes. This suggests that coronal curvature may not be a very precise predictor of girdle height. Individual-level observations on girdle height are needed to better address this question. Accordingly, the implications of the superior coronal curvature in *Australopithecus* and other fossil hominins are currently unclear.

The KSD-VP-1/1 clavicle and scapula show more similarity to living or fossil *Homo* than was previously appreciated from more fragmentary specimens. The fossil evidence, both from this specimen and from MH2, is not compatible with a scenario of human shoulder evolution that both suggests a CLCA with African ape-like morphology and depicts *Australopithecus* morphology as primitive. Similarity in shoulder girdle morphology between early hominins and *Pongo* is either homoplastic or represents the primitive condition for early hominins.

Finally, the postcranial morphology of *Ardipithecus ramidus* challenges the African ape model of the CLCA. Lovejoy et al. (2009a, b, c) suggest that the CLCA was adapted to a generalized form of arboreality without advanced suspensory adaptation. Given the lack of adequate shoulder girdle fossils from the late Miocene, however, it is difficult to hypothesize the shoulder girdle morphology of such an ancestor.

Early Hominin Thoracic Shape

As the oldest relevant fossil currently available, the KSD-VP 1/1 partial skeleton provides the most reliable information regarding a variety of issues relating to early hominin thoracic shape. Schultz (1961, 1969a, b) illustrated the differences in thoracic shape among hominoids and cercopithecids and noted that the great apes have "funnel-shaped" thoraces whereas humans and hylobatids have "barrel-shaped" thoraces. Schultz (1969a, b), Preuschoft (2004), and Barker and Ward (2008) stress that the shape of the thorax is influenced by differences in the length and curvature of the ribs (see also Kagaya et al. 2008). Likewise, the abbreviation and reduced mobility of the lumbar spine present in the large-bodied clambering and vertical-climbing apes never occurred in early hominins. In addition, the stiff torso observed in the great apes probably arose independently. Thus, all extant apes show, to some degree, a reduction of the lumbar column relative to cercopithecids and their early Miocene ancestors (Ward 1993) along with other mechanisms of restricting motion. The features preserved in both the KSD-VP-1/1 and the 4.4 Ma ARA-VP-6/500 (*Ar. ramidus*) individuals imply that the last common ancestor of *Gorilla*, *Pan*, and *Homo* (GLCA) probably did not have a constricted upper thorax or a short, immobile lumbar spine and that early hominins never possessed these features.

These issues necessarily raise the question as to why only the great apes have markedly narrowed their upper thoracic cages. Traditionally, this thoracic shape has been attributed to climbing behaviors (see Hunt 1991; Schmid 1991; Schmid et al. 2013) with the implication that *Australopithecus* (and perhaps other early hominins) maintained this so-called primitive "funnel-shaped" thorax. Furthermore, this was considered as evidence for adaptively significant amount of arboreality in this genus. However, the KSD-VP-1/1 ribs, shoulder, and pelvis fail to support this scenario. Because features like rib declination and torsion are evidence of a long and flexible vertebral column in KSD-VP-1/1 and because these features differ dramatically from the highly derived inflexible spine of extant great apes, it seems very unlikely that given its skeletal anatomy, *Australopithecus* ever possessed the variety of arboreal adaptations practiced by any of the extant great apes. In fact, such arboreal adaptation would have been exceedingly awkward for these early hominins, and they show no morphological adaptations to any form of ape-like arboreality (e.g., elongation of the brachium, antebrachium, and metacarpus, stiffened torso, shortened hindlimb with abducent hallux).

A re-examination of the mechanical reasons for the "funnel-shaped" thorax of the great apes can now be addressed. Instead of being viewed as a climbing, suspensory adaptation, the narrowed operculum can instead be seen as an adaptation to the unique locomotor form of knuckle-walking. While the transversely broad thorax in hylobatids and great apes realigns scapular orientation to permit enhanced shoulder mobility and suspensory climbing (Erikson 1963; Ankel 1967; Benton 1967, 1974; Sarmiento 1985; Harrison 1986; Ward 1993), it also places the frequently unstable glenohumeral joint in an exceptionally vulnerable position during terrestrial locomotion.

Owing to the extremely low coefficient of friction in any healthy synovial joint, it is necessary that the major transarticular forces remain perpendicular across the adjoining chondral surfaces (Burstein and Wright 1994). This requirement has profound consequences for large-bodied animals seeking to combine both arboreal and terrestrial locomotor modalities. In chimpanzees and gorillas, the scapulae are positioned on the dorsum of the thorax such that the glenohumeral joints are directed superolaterally, an orientation suitable for heightened shoulder mobility and suspensory climbing. However, this same orientation becomes highly deleterious during knuckle-walking because, unless modified, it subjects the glenohumeral joint to considerable shear stresses when loaded by the ground reaction impulse forces that occur during the knuckle-strike phase. This is problematic in an extremely mobile articulation like the glenohumeral joint as the muscles and ligaments surrounding the joint that maintain its integrity provide little resistance to shear. The stiff, extended elbow used during knuckle-walking and the abbreviated olecranon process in apes (Drapeau 2004; Simpson et al. submitted) both greatly reduce the capacity of the *M. triceps brachii* to eccentrically and/or isometrically contract, further exacerbating this deficiency.

The tapering of the upper thorax in the African apes is therefore likely a parallelism (as is knuckle-walking; Dainton and Macho 1999; Kivell and Schmitt 2009; see also Williams 2010 for further discussions) that ameliorates this fundamental problem by allowing the scapulae to more freely rotate about the top of the inverted "funnel-shaped" thorax (reduced radius of curvature). This permits the joint to align and face both ventrally and inferiorly, thereby opposing the ground reaction forces and thus reducing potentially damaging shear stress across the shoulder. This altered position of the scapulae would also enhance the ability of *M. serratus anterior* to participate in impact attenuation through isometric and eccentric contraction.

The hands and wrists of African apes also demonstrate adaptations for knuckle-walking (Tuttle 1969; Inouye 1994; Dainton and Macho 1999; Simpson et al. submitted). As these apes are relatively large bodied and spend greater than 80% of their total locomotor repertoire engaging in bouts of knuckle-walking (Doran 1997), it is not surprising that numerous skeletal and soft tissue adaptations to this activity are evident (see Tuttle 1969 for details). Importantly, the fact that these highly specialized adaptations are solely associated with and independently acquired in *Pan* and *Gorilla* indicate that the GLCA probably never knuckle-walked, nor did hominins evolve from a *Pan*-like knuckle-walking ancestor (contra Richmond and Strait 2000; Richmond et al. 2001; Begun 2004). Furthermore, available morphological evidence suggests that this unusual locomotor mode (knuckle-walking) was independently derived in *Pan* and *Gorilla* (Dainton and Macho 1999; White et al. 2009; Kivell and Schmitt 2009).

Recent analyses of the *Au. sediba* postcrania have suggested that this 1.9 million-year-old hominin possessed an African ape-like, conically shaped upper rib cage (Schmid et al. 2013) and, furthermore, that this represents the primitive condition in hominins. In contrast, the inferior thoracic cage is described as "not like that of an ape" implying a peculiar chimera of an upper-half, ape-like thorax juxtaposed to a lower half, human-like rib cage. They also note that the lower ribs of *Au. sediba* demonstrate twisting (torsion), an observation in agreement with the associated hominin pelvis (Kibii et al. 2011) and a long, lordosed lower back (Williams et al. 2013). Importantly, none of these features, save perhaps the proposed upper thoracic shape, are similar to the condition in either genus of the African apes. They further suggest that *Au. sediba* "had habitually elevated, 'shrugged' shoulders like that of a chimpanzee," (Schmid et al. 2013, p. 1234598-2) creating an odd amalgam of ape and human characters.

Regarding the suggestion of "shrugged" shoulders, it is noteworthy that the first rib of *Au. sediba* is described as having a single facet (Schmid et al. 2013), a character usually considered to be indicative of a hominin-like shoulder carriage (Ohman 1986). In view of basic rib cage mechanics and the necessity for the rib bodies to remain roughly parallel, it is difficult to reconstruct a functional thoracic cage (shrugged shoulders over declining ribs) or even to imagine how this animal would have been capable of breathing. This conundrum was appreciated by those authors in noting that their described combination of features would result in "limiting the ability of *Au. sediba* to engage in heavy breathing during fast walking or running" (Schmid et al. 2013, p. 1234598-4). This latter issue, that of a reduced respiratory vital capacity in *Au. sediba* (Schmid 1991, Schmid et al. 2013) is troublesome as no primate has increased its reproductive fitness by decreasing its ability to breathe. Moreover, the ability to walk quickly and to run can be reasonably seen as exceedingly important adaptations for any primate that is a terrestrial biped. Suggestions to the contrary simply strain credulity and cast more than serious doubt on the proposed reconstruction of the *Au. sediba* thorax. Future research on these important fossils will clarify the unlikely combination of ape and human features. At the very least, the peculiar amalgamation of characters described for *Au. sediba*, even if real, can in no way be presumed to represent the primitive condition among hominins.

The Special Significance of the Axial *Bauplan* of KSD-VP-1/1

KSD-VP-1/1 is one of the few early hominin specimens to preserve elements of both the thorax and the pelvis and greatly increases our understanding of the *bauplan* of the GLCA, which the earliest emergent hominins must have shared with ancestors of *Pan* (bonobos and chimpanzees) (the CLCA) or those of both *Pan* and *Gorilla* (the GLCA) (Lovejoy et al. 2009a, b, c; White et al. 2009). As demonstrated by Latimer et al. (2016), analysis of the thoracic skeleton of KSD-VP-1/1 shows it to have been demonstrably unlike those of *Pan* and *Gorilla*, which both appear to be derived for each taxon's specialization to vertical climbing, suspension, and knuckle-walking. Each of the three successively more caudal elements; thorax, lower abdomen/lumbar spine, and pelvis, has been uniquely and divergently specialized in African apes and hominins in the following ways:

1. The hominin pelvis has been completely transformed for upright walking, while those of the African apes also underwent important modifications, but these differ in *Pan* and *Gorilla* (see below).
2. Conversely, the lumbar column of hominins is more similar to the GLCA's primitive state simply because it is more similar in number (5–6) to those of known Middle and possibly Late Miocene precursors from which the GLCA must have evolved (Ward 1993; McCollum et al. 2009; Lovejoy and McCollum 2010).
3. Until the recovery of KSD-VP-1/1, little was known about the CLCA's thoracic skeleton or even that of *Australopithecus*. Reconstruction was largely based on the presumption that the chimpanzee axial *bauplan* was primitive (e.g., Washburn 1967; Wrangham and Pilbeam 2002). This error underlies the reconstruction of the Malapa specimen's axial *bauplan*, which has now been shown to be incorrect (Latimer et al. 2016). Some resolution of these conflicting views on the degree of specialization in the GLCA thorax can be resolved by parsimony with respect to the likely evolutionary sequence of morphological evolution in hominins and the African apes.

The sacrum might be regarded as the most pivotal element in distinguishing the directionality of *bauplan* modification in hominins and great apes. Because of their large body mass (as compared to most other potentially ancestral Late Miocene apes) and their obvious specialization for vertical climbing as their primary means of arboreal locomotion (Thorpe and Crompton 2006), both *Pan* and *Gorilla* have greatly modified their lumbar spine in several ways. This modification includes a reduction in the craniocaudal height of each lumbar body (Ward 1991) and a dramatic reduction in the number of lumbar vertebrae (from a likely 6–7 in the GLCA) to an average of 3.5–4 in each taxon (Ward 1993; McCollum et al. 2009). Furthermore, their sacrum became mediolaterally narrow to such an extent that the most cranial portions of their tall ilia "entrap" the lowest and sometimes the two lowest lumbar vertebrae so that they are syndesmotically immobilized (Lovejoy et al. 2009b; Lovejoy and McCollum 2010).

In contrast, hominins have broadened their sacrum in order to achieve the opposite result – freeing of the lower lumbar vertebrae allowing lordosis in the lumbar region – a change necessary to allow bipedal walking with an extended hip and knee and unlike the bipedality of the African apes. Although a small sacral fragment was recovered from ARA-VP-6/500, it was not complete enough to provide a direct indication of the probable sacral breadth in the specimen (Lovejoy et al. 2009c). Thus, the presence of a broad, rather than a narrow sacrum in *Ar. ramidus* had to be inferred by the form of its reconstructed os coxae, where a sacrum with a narrow breadth would not have allowed for sufficient parturition space in females (Lovejoy et al. 2009c). Fortunately, KSD-VP-1/1 includes a nearly complete first, or more likely, second sacral segment whose breadth is demonstrable and is similar to other hominins when normalized by body size (see Fig. 8.20 in Lovejoy et al. 2016).

Given the current fossil evidence, two evolutionary scenarios are possible for the evolution of the hominin axial *bauplan*. First, the number of lumbar vertebrae in the GLCA was already reduced from their Late Miocene precursor. If this were the case, then an equally likely part of this adaptation would have been that the sacrum of the GLCA was already narrowed similar to its state in extant great apes. This implies that (1) hominins evolved bipedality as a habitual terrestrial locomotor pattern despite the disadvantage of doing so while relying on a bent-hip-bent-knee (BHBK) gait; (2) sacral breadth had been narrowed to a condition like that seen in extant great apes before reversing direction and subsequently broadening as observed in ARA-VP-6/500, KSD-VP-1/1, and A.L. 288-1. The alternative scenario is that the lumbar column of the GLCA was still long at the time of the separation of the GLCA and its sacrum was still sufficiently broad to allow at least partial lordosis during bipedal locomotion such as not to require a pronounced BHBK gait pattern. This implies that the adaptation of a narrowed sacrum and reduced lumbar column in the African apes was not accompanied by all of the other major changes in the African ape skeleton that directly reflect suspensory locomotion, including those of the arm, elbow, wrist, carpus, and metacarpus (Lovejoy et al. 2009b).

In order to resolve which of these scenarios is likely, rigorous parsimony analysis that includes outgroups such as hylobatids, orangutans, and other Early Miocene hominoids would be necessary. While this is outside the scope of this study, the newly available data from KSD-VP-1/1, combined with those from ARA-VP-6/500, suggest that the second scenario is more likely than the first, as the former implies multiple anatomical reversals and is inconsistent with known fossil evidence. The combination of a primitive thorax and derived pelvis in KSD-VP-1/1 and a pelvis that is derived cranially and primitive caudally in ARA-VP-6/500 suggest that a moderately broad sacrum, relatively long lumbar column (5–6 lumbars), and generally primitive thorax and pelvis constituted the axial *bauplan* of the GLCA. From this shared state, *Pan* and *Gorilla* both shortened their lumbar columns and narrowed their sacrum. In addition, each taxon's thorax was modified in response to shortening of the lumbar column for spinal rigidity and mediolateral expansion of its ilia for increased body mass in the face of a shortened abdomen, the latter being much more dramatic in *Gorilla* than in *Pan*. Conversely, the iliac isthmus of *Pan* appears to have been elongated in possible compensation for the shortening of its lumbar column and appears somewhat less likely in the case of *Gorilla*. In hominins, an initially long iliac isthmus was subsequently shortened to its state in ARA-VP-6/500 (Lovejoy et al. 2009c).

Conclusion

The historically significant works of the great evolutionary theorists of the late nineteenth and early twentieth centuries such as Darwin, Huxley, and Keith, were created without the benefit of even the barest fossil record. Even Dart and Broom were limited in their understanding of the forebears of *Australopithecus* by the absence of the now abundant Miocene fossil hominoids of Africa and Asia, nor did they have the plethora of individual skeletal elements that can now be attributed to this genus. In the absence of associated skeletons of sufficiently great geological age and in light of our close genetic relationship with *Pan*, an anatomically *Pan*-like ancestor was presumed (e.g., Pilbeam and Young 2004). However, new data derived from KSD-VP-1/1 challenge these long-held assumptions. KSD-VP-1/1 is unique among early hominin fossils in terms of the completeness of its preserved skeletal elements that include a nearly intact scapula, clavicle, cervical vertebrae, tibia of known length, and the pelvic elements, all of which contribute greatly toward improving our understanding of the paleobiology of *Au. afarensis*. The cervical vertebrae of KSD-VP-1/1 shed, for the first time, some light on the neck anatomy of early hominins in general, and of *Au. afarensis* in particular. The complete scapula and preserved elements of the thorax allow detailed metric and morphological analyses of the shoulder girdle anatomy of *Au. afarensis* and show that despite, unsurprisingly, some extant ape-like traits, the overall morphology of the *Au. afarensis* shoulder girdle and thorax was more human-like, contrary to previous inferences of African ape-like morphology.

Although full knowledge of the GLCA will remain unknown until the discovery of even older fossils, we can surmise, based on the evidence presented here and in Ryan and Sukhdeo (2016), Meyer (2016), Melillo (2016), Latimer et al. (2016), and Lovejoy et al. (2016), along with data from the recently published *Ar. ramidus* partial skeleton (ARA-VP-6/500; Lovejoy et al. 2009b, White et al. 2009), that early hominins were decidedly unlike *Pan* in their skeletal *bauplan*, locomotion, and behavior. KSD-VP-1/1 therefore provides a new glimpse into our origins and a deeper understanding of early *Australopithecus* paleobiology.

Acknowledgments We thank the Authority for Research and Conservation of Cultural Heritage (ARCCH) of the Ministry of Culture and Tourism of Ethiopia and administrative offices of the Afar Regional State of Ethiopia for field and laboratory research permits. We thank E. Delson, T. Harrison, and S. Simpson for constructive comments. The Woranso-Mille project was financially supported by grants from The Leakey Foundation, The Wenner-Gren Foundation, The National Geographic Society, The Cleveland Museum of Natural History, and The National Science Foundation (BCS-0234320, BCS-0321893, BCS-0542037, BCS-1124705, BCS-1124713, BCS-1124716, BCS-1125157, and BCS-1125345).

References

Aiello, L. C. (1992). Allometry and the analysis of size and shape in human evolution. *Journal of Human Evolution, 22*, 127–147.

Alemseged, Z., Spoor, F., Kimbel, W. H., Bobe, R., Geraads, D., Reed, D., & Wynn, J. G. (2006). A juvenile early hominin skeleton from Dikika, Ethiopia. *Nature, 443*, 296–301.

Ankel, F. (1967). *Morphologie von Wirbelsaule und Brustkorb. Primatologia, 4*, 1–120.

Barker, K. B., & Ward, C. V. (2008). Patterns of upper rib morphology in hominoids. *American Journal of Physical Anthropology, Supplement, 46*, 64.

Bastir, M., Antonio Rosas, A., Stringer, C., Cuétara, J.M., Kruszynski, R., & Weber, G. W., et al. (2010). Effects of brain and facial size on basicranial form in human and primate evolution. *Journal of Human Evolution, 58*, 424–431.

Begun, D. R. (2004). Knuckle-walking and the origin of human bipedalism. In D. J. Meldrum & C. E. Hilton (Eds.), *Biped to strider: The emergence of modern human walking, running, and resource transport* (pp. 9–33). New York: Academic/Plenum Publishers.

Benton, R. (1967). Morphological evidence for adaptations within the epaxial region of the primates. In H. Vagt-borg (Ed.), *The baboon in medical research* (pp. 10–20). Houston: University of Texas Press.

Benton, R. S. (1974). Structural patterns in the Pongidae and Cercopithecidae. *Yearbook of Physical Anthropology, 18*, 65–88.

Berger, L. R., de Ruiter, D. J., Churchill, S. E., Schmid, P., Carlson, K. J., Dirks, P. H., & Kibii, J. M. (2010). *Australopithecus sediba*: A new species of *Homo*-like australopith from South Africa. *Science, 328*, 195–204.

Broom, R., & Robinson, J. T. (1950). Further evidence of the structure of the Sterkfontein ape-man *Plesianthropus*. (Vol. 4, Transvaal Museum Memoir). Pretoria: Transvaal Musuem.

Burstein, A. H., & Wright, T. M. (1994). *Fundamentals of orthopaedic biomechanics*. New York: Williams & Wilkins.

Campbell, B. G. (1966). *Human evolution: An introduction to man's adaptations*. Chicago: Aldine Pub. Co.

Churchill, S. E., Holliday, T. W., Carlson, K. J., Jashashvili, T., Macias, M. E., & Mathews, S., et al. (2013). The upper limb of *Australopithecus sediba*. *Science, 340*, 1233477.

Ciochon, R. L., & Corruccini, R. S. (1976). Shoulder joint of sterkfontein *Australopithecus*. *South African Journal of Science, 72*, 80–82.

Cook, D. C., Buikstra, J. E., DeRousseau, C. J., & Johanson, D. C. (1983). Vertebral pathology in the Afar australopithecines. *American Journal of Physical Anthropology, 60*, 83–101.

Dainton, M., & Macho, G. A. (1999). Did knuckle walking evolve twice? *Journal of Human Evolution, 36*, 171–194.

Dean, M. C., & Wood, B. A. (1981). Metrical analysis of the basicranium of extant hominoids and *Australopithecus*. *American Journal of Physical Anthropology, 54*, 63–71.

Dean, M. C., & Wood, B. A. (1982). Basicranial anatomy of Plio-Pleistocene hominids from East and South Africa. *American Journal of Physical Anthropology, 59*, 157–174.

Doran, D. M. (1997). Ontogeny of locomotion in mountain gorillas and chimpanzees. *Journal of Human Evolution, 32*, 323–344.

Drapeau, M. S. (2004). Functional anatomy of the olecranon process in hominoids and plio-pleistocene hominins. *American Journal of Physical Anthropology, 124*, 297–314.

Erikson, G. E. (1963). Brachiation in new world monkeys and in anthropoid apes. *Symp. Zoological Society of London, 10*, 135–164.

Grave, K., & Townsend, G. (2003). Cervical vertebral maturation as a predictor of the adolescent growth spurt. *Australian Orthodontic Journal, 19*, 25–32.

Grave, B., Brown, T., & Townsend, G. (1999). Comparison of cervicovertebral dimensions in Australian Aborigines and Caucasians. *European Journal of Orthodontics, 21*, 127–135.

Green, D. J., & Alemseged, Z. (2012). *Australopithecus afarensis* scapular ontogeny, function, and the role of climbing in human evolution. *Science, 338*, 514–517.

Haile-Selassie, Y., Latimer, B. M., Alene, M., Deino, A. L., Gibert, L., & Melillo, S. M., et al. (2010). An early *Australopithecus afarensis* postcranium from Woranso-Mille, Ethiopia. *Proceedings of the National Academy of Sciences USA, 107*, 12121–12126.

Harrison, T. (1986). A reassessment of the phylogenetic relationships of *Oreopithecus bambolii* Gervais. *Journal of Human Evoution, 15*, 541–583.

Holliday, T. W. (2012). Body size, body shape, and the circumscription of the genus *Homo*. *Current Anthropology, 53*, S330–S345.

Huggare, J., & Houghton, P. (1996). Associations between atlantoaxial and craniomandibular anatomy. *Growth Development Aging, 60*, 21–30.

Hunt, K. D. (1991). Mechanical implications of chimpanzee positional behavior. *American Journal of Physical Anthropology, 86*, 521–536.

Inman, V. T., Saunders, M., & Abbott, L. C. (1944). Observations on the function of the shoulder joint. *The Journal of Bone and Joint Surgery, 26*(1), 1–30.

Inouye, S. E. (1994). Ontogeny of knuckle-walking hand postures in African apes. *Journal of Human Evolution, 26*, 459–485.

Inouye, S. E., & Shea, B. T. (1997). What's your angle? Size correction and bar-glenoid orientation in "Lucy"(AL 288-1). *International Journal of Primatology, 18*, 629–650.

Johanson, D. C., Lovejoy, C. O., Kimbel, W. H., White, T. D., Ward, S. C., & Bush, M. E., et al. (1982). Morphology of the Pliocene partial hominid skeleton (A.L. 288-1) from the Hadar Formation, Ethiopia. *American Journal of Physical Anthropology, 57*, 403–452.

Jungers, W. L. (1982). Lucy's limbs: Skeletal allometry and locomotion in *Australopithecus afarensis*. *Nature, 297*, 676–678.

Jungers, W. L., & Stern, J. T. (1983). Body proportions, skeletal allometry and locomotion in the Hadar hominids: A reply to Wolpoff. *Journal of Human Evolution, 12*, 673–684.

Kagaya, M., Ogihara, N., & Nakatsukasa, M. (2008). Morphological study of the anthropoid thoracic cage: Scaling of thoracic width and an analysis of rib curvature. *Primates, 49*, 89–99.

Katz, P. R., Reynolds, H. M., Foust, D. R., & Baum, J. K. (1975). Mid-sagittal dimensions of cervical vertebral bodies. *American Journal of Physical Anthropology, 43*, 319–326.

Kibii, J. M., Churchill, S. E., Schmid, P., Carlson, K. J., Reed, N. D., & de Ruiter, D. J., et al. (2011). A partial pelvis of *Australopithecus sediba*. *Science, 333*, 1407–1411.

Kimbel, W. H., & Rak, Y. (2010). The cranial base of *Australopithecus afarensis*: New insights from the female skull. *Philosophical Transactions of the Royal Society of London B, Biological Sciences, 365*, 3365–3376.

Kivell, T. L., & Schimtt, D. (2009). Independent evolution of knuckle-walking in African apes shows than humans did not evolve from a knuckle-walking ancestor. *Proceedings of the National Academy of Sciences USA, 106*, 14241–14246.

Larson, S. G. (2007). Evolutionary transformation of the hominin shoulder. *Evolutionary Anthropology, 16*, 172–187.

Larson, S. G. (2009). Evolution of the hominin shoulder: Early *Homo*. In *The First Humans-origin and early evolution of the genus Homo* (pp. 65–75). Netherlands: Springer.

Larson, S. G. (2013). Shoulder morphology in early hominin evolution. In K. E. Reed, J. G. Fleagle, & R. E. Leakey (Eds.), *The paleobiology of Australopithecus* (pp. 247–261). Netherlands: Springer.

Latimer, B. M., Lovejoy, C. O., Spurlock, L., & Haile-Selassie, Y. (2016). The thoracic cage of KSD-VP-1/1. In Y. Haile-Selassie & D. F. Su (Eds.), *The postcranial anatomy of* Australopithecus afarensis: *New insights from KSD-VP-1/1* (pp. 143–153). Dordrecht: Springer.

Lieberman, D. (2011). *The evolution of the human head*. Cambridge, Mass.: Belknap Press of Harvard University Press.

Lovejoy, C. O., & McCollum, M. (2010). Spinopelvic pathways to bipedality: Why no hominids ever relied on a bent-hip-bent-knee gait. *Philosophical Transactions of the Royal Society of London, 365*, 3289–3299.

Lovejoy, C. O., Johanson, D. C., & Coppens, Y. (1982). Elements of the axial skeleton recovered from the Hadar formation: 1974–1977 collections. *American Journal of Physical Anthropology, 57*, 631–635.

Lovejoy, C. O., Simpson, S. W., White, T. D., Asfaw, B., & Suwa, G. (2009a). Careful climbing in the Miocene: The forelimbs of *Ardipithecus ramidus* and humans are primitive. *Science, 326*, 70e71–70e78.

Lovejoy, C. O., Suwa, G., Simpson, S. W., Matternes, J. H., & White, T. D. (2009b). The Great Divides: *Ardipithecus ramidus* Reveals the Postcrania of Our Last Common Ancestors with African Apes. *Science, 326*, 100–106.

Lovejoy, C. O., Suwa, G., Spurlock, L., Asfaw, B., & White, T. D. (2009c). The pelvis and femur of *Ardipithecus ramidus*: The emergence of upright walking. *Science, 326*, 71e71–71e76.

Lovejoy, C. O., Latimer, B. M., Spurlock, L., & Haile-Selassie, Y. (2016). The pelvic girdle and limb bones of KSD-VP-1/1. In Y.

Haile-Selassie & D. F. Su (Eds.), *The postcranial anatomy of* Australopithecus afarensis*: New insights from KSD-VP-1/1* (pp. 155–178). Dordrecht: Springer.

McCollum, M. A., Rosenman, B. A., Suwa, G., Meindl, R. S., & Lovejoy, C. O. (2009). The vertebral formula of the last common ancestor of African apes and humans. *Journal of Experimental Zoology Part B Molecular and Developmental Evolution, 314B*, 123–134.

Melillo, S. M. (2011). *Anatomy of the Australopithecus scapula and evolution of the human shoulder: Insights from a new fossil specimen, KSD-VP-1/1G*. Thesis: Stanford University, Stanford (CA).

Melillo, S. M. (2016). The shoulder girdle of KSD-VP-1/1. In Y. Haile-Selassie & D. F. Su (Eds.), *The postcranial anatomy of* Australopithecus afarensis: *New insights from KSD-VP-1/1* (pp. 113–141). Dordrecht: Springer.

Mensforth, R. P., Latimer, B., & Senturia, S. (1990). A review of the functional significance of the AL-288 axilloglenoid angle. *American Journal of Physical Anthropology, 81*, 267–268.

Meyer, M. R. (2016). The cervical vertebrae of KSD-VP-1/1. In Y. Haile-Selassie & D. F. Su (Eds.), *The postcranial anatomy of* Australopithecus afarensis: *New insights from KSD-VP-1/1* (pp. 63–111). Dordrecht: Springer.

Muller, F., & O'Rahilly, R. (2003). Segmentation in staged human embryos: The occipitocervical region revisited. *Journal of Anatomy, 203*, 297–315.

Nakatsukasa, M., Kunimatsu, Y., Nakano, Y., & Ishida, H. (2003). *Morphology of the axial skeleton of* Nacholapithecus *from the Middle Miocene of Kenya*. Paper presented at the AAPA 73rd Annual Meeting, Tempe, AZ.

Nevell, L., & Wood, B. (2008). Cranial base evolution within the hominin clade. *Journal of Anatomy, 212*, 455–468.

Ogden, J. A., & Ganey, T. M. (2003). Development and maturation of the spine and spinal cord. In V. Arlet, A. L. Carl, & R. L. DeWald (Eds.), *Spinal deformities: The comprehensive text* (pp. 54–70). New York: Thieme.

Ohman, J. C. (1986). The first rib of hominoids. *American Journal of Physical Anthropology, 70*, 209–229.

Oxnard, C. E. (1968). A note on the fragmentary sterkfontein scapula. *American Journal of Physical Anthropology, 28*, 213–217.

Pilbeam, D., & Young, N. (2004). Hominoid evolution: Synthesizing disparate data. *Comptes Rendus Palevol, 3*, 305–321.

Preuschoft, H. (2004). Mechanisms for the acquisition of habitual bipedality: Are there biomechanical reasons for the acquisition of upright bipedal posture. *Journal of Anatomy, 204*, 363–384.

Richmond, B. G., & Strait, D. S. (2000). Evidence that humans evolved from a knuckle-walking ancestor. *Nature, 404*, 382–385.

Richmond, B. G., Begun, D., & Strait, D. S. (2001). Origin of human bipedalism: The knuckle-walking hypothesis revisited. *Yearbook of Physical Anthropology, 44*, 70–105.

Rose, M. D. (1975). Functional proportions of primate lumbar vertebral bodies. *Journal of Human Evolution, 4*, 21–38.

Ryan, T. M., & Sukhdeo, S. (2016). Chapter 4: KSD-VP-1/1: Analysis of the postcranial skeleton using high-resolution computed tomography. In: Y. Haile-Selassie & D. F. Su (Eds.), *The postcranial anatomy of* Australopithecus afarensis: *New insights from KSD-VP-1/1* (pp. 39–63). Dordrecht: Springer.

Sarmiento, E. E. (1985). Functional differences in the skeleton of wild and captive orangutans and their adaptive significance. Ph.D. thesis, New York University.

Schmid, P. (1991). The trunk of the *Australopithecus*. In Y. Coppens & B. Senut (Eds.), *Origine(s) de la bipedie chez les hominides* (pp. 225–234). Cahiers de Paleoanthropologie, Editions du CNRS. Paris.

Schmid. P., Churchill, S. E., Nalla, S., Weissen, E., Carlson, K. J., & de Ruiter, D. J., et al. (2013). Mosaic morphology in the thorax of *Australopithecus sediba*. *Science, 340*, 1234598.

Schultz, A. H. (1961). *Vertebral column and thorax. Primatologia, 4*, 1–66.

Schultz, A. H. (1969a). *The life of primates*. New York: Universe Books.

Schultz, A. H. (1969b). The Skeleton of the Chimpanzee. In G. H. Bourne (Ed.), *The chimpanzee: Anatomy, behaviour, and diseases of chimpanzees* (Vol. 1, pp. 50–103). Basel: Karger.

Shi, C., Nishizawa, S., Adachi, K., & Tashuku, K. (1995). A comparative morphological analysis of vertebral bodies in humans and some other mammals. *Anthropological Science, 103*, 467–484.

Shi, C., Nishizawa, S., Adachi, K., & Tashuku, K. (1999). Morphological comparison of the vertebrae in humans and some other mammals. *Anthropological Science, 107*, 3–19.

Simpson, S. W., Lovejoy, C. O., & Latimer, B. M. (submitted). Why do knuckle-walking African apes knuckle-walk? *The Anatomical Record*.

Solow, B., Barrett, M. J., & Brown, T. (1982). Craniocervical morphology and posture in Australian aboriginals. *American Journal of Physical Anthropology, 59*, 33–45.

Stern, J. T. (2000). Climbing to the top: A personal memoir of *Australopithecus afarensis*. *Evolutionary Anthropology, 9*, 113–133.

Stern, J. T., & Susman, R. L. (1983). The locomotor anatomy of *Australopithecus afarensis*. *American Journal of Physical Anthropology, 60*, 279–317.

Susman, R. L., Stern, J. T., & Jungers, W. L. (1984). Arboreality and bipedality in the Hadar hominids. *Folia Primatologica, 43*, 113–156.

Thorpe, S. K. S., & Crompton, R. H. (2006). Orangutan positional behavior and the nature of arboreal locomotion in Hominoidea. *American Journal of Physical Anthropology, 131*, 384.

Tuttle, R. H. (1969). Knuckle-walking and the problem of human origins. *Science, 166*, 953–961.

Vallois, H. V. (1977). Interpretation of the Scapula of *Plesianthropus transvaalensis*. *Journal of Human Evolution, 6*, 675–679.

Voisin, J.-L. (2006). Krapina and other Neanderthal clavicles: A peculiar morphology? *Periodicum Biologorum, 108*(3), 331–339.

Voisin, J.-L. (2008). The Omo 1 hominin clavicle: Archaic or modern? *Journal of Human Evolution, 55*(3), 438–443.

Vrba, E. S. (1979). A new study of the scapula of *Australopithecus africanus* from Sterkfontein. *American Journal of Physical Anthropology, 51*, 117–129.

Walker, A., & Leakey, R. E. (Eds.). (1993). *The Nariokotome Homo erectus skeleton*. Cambridge: Harvard University Press.

Ward, C. V. (1991). The functional anatomy of the lower back and pelvis of the Miocene hominoid *Proconsul nyanzae* from the Miocene of Mfangano Island, Kenya. Thesis: The John Hopkins University, Baltimore (MD).

Ward, C. V. (1993). Torso morphology and locomotion in *Proconsul nyanzae*. *American Journal of Physical Anthropology, 92*, 291–328.

Washburn, S. L. (1967). Behavior and the origin of man. *Proceedings of the Royal Anthropological Institute of Great Britain and Ireland, 3*, 21–27.

White, T. D., Asfaw, B., Beyene, Y., Haile-Selassie, Y., Lovejoy, C. O., & Suwa, G. (2009). *Ardipithecus ramidus* and the paleobiology of early hominids. *Science, 326*, 75–86.

Williams, S. A. (2010). Morphological integration and the evolution of knuckle-walking. *Journal of Human Evolution, 58*, 432–440.

Williams, S. C., Ostrofsky, K. R., Frater, N., Churchill, S. E., Schmid, P., & Berger, L. R. (2013). The vertebral column of *Australopithecus sediba*. *Science, 340*, 1232996.

Wolpoff, M. H. (1983). Lucy's little legs. *Journal of Human Evolution, 12*, 443–453.

Wrangham, R., & Pilbeam, D. (2002). African apes as time machines. In B. M. F. Galdikas, N. E. Briggs, L. K. Sheeran, G. L. Shapiro, & J. Goodall (Eds.), *All apes great and small* (pp. 5–17). New York: Kluwer Acad./Plenum.

Index

Note: Page numbers followed by *f* and *t* indicate figures and tables, respectively

Yohannes Haile-Selassie and Denise F. Su (eds.), *The Postcranial Anatomy of Australopithecus afarensis: New Insights from KSD-VP-1/1*, Vertebrate Paleobiology and Paleoanthropology, DOI 10.1007/978-94-017-7429-1